Classical Mechanics

Lecture notes

Classical Mechanics

Lecture notes

Konstantin K Likharev

IOP Publishing, Bristol, UK

ISBN 978-0-7503-1398-8 (ebook)
ISBN 978-0-7503-1399-5 (print)
ISBN 978-0-7503-1400-8 (mobi)

DOI 10.1088/978-0-7503-1398-8

Version: 20171201

IOP Expanding Physics
ISSN 2053-2563 (online)
ISSN 2054-7315 (print)

British Library Cataloguing-in-Publication Data: A catalogue record for this book is available from the British Library.

Published by IOP Publishing, wholly owned by The Institute of Physics, London

IOP Publishing, Temple Circus, Temple Way, Bristol, BS1 6HG, UK

US Office: IOP Publishing, Inc., 190 North Independence Mall West, Suite 601, Philadelphia, PA 19106, USA

Contents

Preface to the EAP Series

Essential Advanced Physics

Essential Advanced Physics (EAP) is a series of lecture notes and problems with solutions, consisting of the following four parts[1]:

- *Part CM*: *Classical Mechanics* (a one-semester course),
- *Part EM*: *Classical Electrodynamics* (two semesters),
- *Part QM*: *Quantum Mechanics* (two semesters), and
- *Part SM*: *Statistical Mechanics* (one semester).

Each part includes two volumes: *Lecture Notes* and *Problems with Solutions*, and an additional file *Test Problems with Solutions*.

Distinguishing features of this series—in brief

- condensed lecture notes (~250 pp per semester)—much shorter than most textbooks
- emphasis on simple explanations of the main notions and phenomena of physics
- a focus on problem solution; extensive sets of problems with detailed model solutions
- additional files with test problems, freely available to qualified university instructors
- extensive cross-referencing between all parts of the series, which share style and notation

Level and precursors

The goal of this series is to bring the reader to a general physics knowledge level necessary for professional work in the field, regardless on whether the work is theoretical or experimental, fundamental or applied. From the formal point of view, this level (augmented by a few special topic courses in a particular field of concentration, and of course by an extensive thesis research experience) satisfies the typical PhD degree requirements. Selected parts of the series may be also valuable for graduate students and researchers of other disciplines, including astronomy, chemistry, mechanical engineering, electrical, computer and electronic engineering, and material science.

The entry level is a notch lower than that expected from a physics graduate from an average US college. In addition to physics, the series assumes the reader's familiarity with basic calculus and vector algebra, to such an extent that the meaning of the formulas listed in appendix A, 'Selected mathematical formulas' (reproduced at the end of each volume), is absolutely clear.

[1] Note that the (very ambiguous) term *mechanics* is used in these titles in its broadest sense. The acronym *EM* stems from another popular name for classical electrodynamics courses: *Electricity and Magnetism*.

Origins and motivation

The series is a by-product of the so-called 'core physics courses' I taught at Stony Brook University from 1991 to 2013. My main effort was to assist the development of students' problem-solving skills, rather than their idle memorization of formulas. (With a certain exaggeration, my lectures were not much more than introductions to problem solution.) The focus on this main objective, under the rigid time restrictions imposed by the SBU curriculum, had some negatives. First, the list of covered theoretical methods had to be limited to those necessary for the solution of the problems I had time to discuss. Second, I had no time to cover some core fields of physics—most painfully general relativity[2] and quantum field theory, beyond a few quantum electrodynamics elements at the end of *Part QM*.

The main motivation for putting my lecture notes and problems on paper, and their distribution to students, was my desperation to find textbooks and problem collections I could use, with a clear conscience, for my purposes. The available graduate textbooks, including the famous *Theoretical Physics* series by Landau and Lifshitz, did not match the minimalistic goal of my courses, mostly because they are far too long, and using them would mean hopping from one topic to another, picking up a chapter here and a section there, at a high risk of losing the necessary background material and logical connections between the course components—and the students' interest with them. In addition, many textbooks lack even brief discussions of several traditional and modern topics that I believe are necessary parts of every professional physicist's education[3].

On the problem side, most available collections are not based on particular textbooks, and the problem solutions in them either do not refer to any background material at all, or refer to the included short sets of formulas, which can hardly be used for systematic learning. Also, the solutions are frequently too short to be useful, and lack discussions of the results' physics.

Style

In an effort to comply with the Occam's Razor principle[4], and beat Malek's law[5], I have made every effort to make the discussion of each topic as clear as the time/space (and my ability :-) permitted, and as simple as the subject allowed. This effort has resulted in rather succinct lecture notes, which may be thoroughly read by a student during the semester. Despite this briefness, the introduction of every new

[2] For an introduction to this subject, I can recommend either a brief review by S Carroll, *Spacetime and Geometry* (2003, New York: Addison-Wesley) or a longer text by A Zee, *Einstein Gravity in a Nutshell* (2013, Princeton University Press).

[3] To list just a few: the statics and dynamics of elastic and fluid continua, the basics of physical kinetics, turbulence and deterministic chaos, the physics of computation, the energy relaxation and dephasing in open quantum systems, the reduced/RWA equations in classical and quantum mechanics, the physics of electrons and holes in semiconductors, optical fiber electrodynamics, macroscopic quantum effects in Bose–Einstein condensates, Bloch oscillations and Landau–Zener tunneling, cavity quantum electrodynamics, and density functional theory (DFT). All these topics are discussed, if only briefly, in my lecture notes.

[4] *Entia non sunt multiplicanda praeter necessitate*—Latin for 'Do not use more entities than necessary'.

[5] 'Any simple idea will be worded in the most complicated way'.

physical notion/effect and of every novel theoretical approach is always accompanied by an application example or two.

The additional exercises/problems listed at the end of each chapter were carefully selected[6], so that their solutions could better illustrate and enhance the lecture material. In formal classes, these problems may be used for homework, while individual learners are strongly encouraged to solve as many of them as practically possible. The few problems that require either longer calculations, or more creative approaches (or both), are marked by asterisks.

In contrast with the lecture notes, the model solutions of the problems (published in a separate volume for each part of the series) are more detailed than in most collections. In some instances they describe several alternative approaches to the problem, and frequently include discussions of the results' physics, thus augmenting the lecture notes. Additional files with sets of shorter problems (also with model solutions) more suitable for tests/exams, are available for qualified university instructors from the publisher, free of charge.

Disclaimer and encouragement

The prospective reader/instructor has to recognize the limited scope of this series (hence the qualifier *Essential* in its title), and in particular the lack of discussion of several techniques used in current theoretical physics research. On the other hand, I believe that the series gives a reasonable introduction to the *hard core* of physics—which many other sciences lack. With this hard core knowledge, today's student will always feel at home in physics, even in the often-unavoidable situations when research topics have to be changed at a career midpoint (when learning from scratch is terribly difficult—believe me :-). In addition, I have made every attempt to reveal the remarkable logic with which the basic notions and ideas of physics subfields merge into a wonderful single construct.

Most students I taught liked using my materials, so I fancy they may be useful to others as well—hence this publication, for which all texts have been carefully reviewed.

[6] Many of the problems are original, but it would be silly to avoid some old good problem ideas, with long-lost authorship, which wander from one textbook/collection to another one without references. The assignments and model solutions of all such problems have been re-worked carefully to fit my lecture material and style.

Preface to *Classical Mechanics: Lecture Notes*

A fairer title for this part of the series could be *Classical Mechanics and Dynamics*, because the notions of non-relativistic[7] mechanics and dynamics, though much intertwined, are still somewhat different. The term *mechanics*, in its narrow sense, means the derivation of the equations of motion of particles and their systems (including continua such as solids and fluids), followed by the solution of these equations, and interpretation of the results. *Dynamics* is a more ambiguous term; it may mean, in particular:

(i) the part of mechanics that deals with motion (in contrast to *statics*);
(ii) the part of mechanics that deals with reasons for motion (in contrast to *kinematics*); or
(iii) the part of mechanics that focuses on its two last tasks, i.e. the solution of the equations of motion and discussion of the results.

The last definition invites a question. It may look like mechanics and dynamics are just two sequential steps of a single analytical process; why should they be considered separate disciplines[8]? The main reason is that the many differential equations of motion, obtained in classical mechanics, also describe processes in different systems, so their analysis may reveal important features of these systems as well.

To summarize, the term *dynamics* is so ambiguous that, after some hesitation, I have opted to use for this course the traditional name *Classical Mechanics*, implying a broader meaning of the term *mechanics* (also very ambiguous) which includes many results important to non-mechanical systems as well. Note that in other parts of the series (*Quantum Mechanics* and *Statistical Mechanics*) this term is understood in a similarly broad sense.

---

[7] Following tradition, an introduction to special relativity is included in *Part EM* of these notes. As a preliminary, the relativistic effects are small if the speed of every particle of the system is much lower than the speed of light, $c \approx 3.00 \times 10^8$ m s^{-1}, and the system size is much larger than the system's *Schwarzschild radius*, $r_s \equiv 2Gm/c^2$, where $G \approx 6.67 \times 10^{-11}$ SI units (m^3 kg s) is the Newtonian gravity constant, and m is the system's mass. (More exact values of c, G, and some other physical constants may be found in appendix B.)

[8] Another important issue concerning definition (iii) of dynamics is that it is suspiciously close to the part of mathematics devoted to differential equation analysis; what is the difference? To answer, we have to dip, for just a second, into the philosophy of physics. This discipline may be described as the art (and a bit of science :-) of the description of Mother Nature by mathematical means; hence in many cases the approaches of a mathematician and a physicist to a problem are very similar. The main difference is that physicists try to express the results of their analysis in less formal terms of a *system's motion* rather than *function properties*. As a result they develop some sort of intuition ('gut feeling') about how other, apparently similar, systems may behave, even if their exact equations of motion are somewhat different—or not known at all. The intuition thus developed has an enormous heuristic power, and most discoveries in physics have been made through gut-feeling-based insights rather than just by plugging one formula into another one.

Perhaps the most distinguishing feature of this course is the substantial attention on the mechanics of continua, including discussions of 1D waves in chapter 6, deformations and elasticity (including 3D waves) in chapter 7, and fluid dynamics in chapter 8. In most classical mechanics courses, these topics are either neglected completely or covered only as an introduction to field theory, without discussion of even basic real-world problems. However, I believe that familiarity with these topics is a necessary part of any professional physicist's education. The discussions of deformation and elasticity in the current chapter 7, and the fluid dynamics in chapter 8, are close to the introductory parts of the corresponding volumes (6 and 7) of the Landau and Lifshitz *Theoretical Physics* series.

Another not-quite-standard feature of this course is the introduction to analytical mechanics, starting with the Lagrangian formalism in chapter 2, from the experiment-based Newton's laws rather from some general ideas, such as the Hamilton principle, with this principle discussed only at the end of the course (section 10.3). I feel that this route emphasizes better the experimental roots of physics, and the secondary nature of any general principles—regardless of their important aesthetic and heuristic values.

Acknowledgments

I am extremely grateful to my faculty colleagues and other readers of the preliminary (circa 2013) version of this series, who provided feedback on certain sections; here they are listed in alphabetical order[9]: A Abanov, P Allen, D Averin, S Berkovich, P-T de Boer, M Fernandez-Serra, R F Hernandez, A Korotkov, V Semenov, F Sheldon, and X Wang. (Obviously, these kind people are not responsible for any remaining deficiencies.)

A large part of my scientific background and experience, reflected in these materials, came from my education, and then research, in the Department of Physics of Moscow State University from 1960 to 1990. The Department of Physics and Astronomy of Stony Brook University provided a comfortable and friendly environment for my work during the following 25+ years.

Last but not least, I would like to thank my wife Lioudmila for all her love, care, and patience—without these, this writing project would have been impossible.

I know very well that my materials are still far from perfection. In particular, my choice of covered topics (always very subjective) may certainly be questioned. Also, it is almost certain that despite all my efforts, not all typos have been weeded out. This is why all remarks (however candid) and suggestions from readers will be greatly appreciated. All significant contributions will be gratefully acknowledged in future editions.

Konstantin K Likharev
Stony Brook, NY

[9] I am very sorry for not keeping proper records from the beginning of my lectures at Stony Brook, so I cannot list all the numerous students and TAs who have kindly attracted my attention to typos in earlier versions of these notes. Needless to say, I am very grateful to all of them as well.

Notation

Abbreviations	Fonts	Symbols
c.c. complex conjugate	F, $\mathcal{F}$ scalar variables[10]	$\dot{}$ time differentiation operator (*d*/*dt*)
h.c. Hermitian conjugate	$\mathbf{F}$, $\boldsymbol{\mathcal{F}}$ vector variables	∇ spatial differentiation vector (*del*)
	$\hat{F}$, $\hat{\mathcal{F}}$ scalar operators	$\approx$ approximately equal to
	$\hat{\mathbf{F}}$, $\hat{\boldsymbol{\mathcal{F}}}$ vector operators	$\sim$ of the same order as
	F matrix	$\propto$ proportional to
	$F_{jj'}$ matrix element	$\equiv$ equal to by definition (or evidently)
		$\cdot$ scalar ('dot-') product
		$\times$ vector ('cross-') product
		$\overline{}$ time averaging
		$\langle\ \rangle$ statistical averaging
		[,] commutator
		{ , } anticommutator

Prime signs

The prime signs (′, ″, etc) are used to distinguish similar variables or indices (such as j and j' in the matrix element above), rather than to denote derivatives.

Parts of the series

Part CM: Classical Mechanics *Part EM: Classical Electrodynamics*
Part QM: Quantum Mechanics *Part SM: Statistical Mechanics*

Appendices

Appendix A: Selected mathematical formulas
Appendix B: Selected physical constants

Formulas

The abbreviation Eq. may mean any displayed formula: either the equality, or inequality, or equation, etc.

[10] The same letter, typeset in different fonts, typically denotes different variables.

IOP Publishing

Classical Mechanics
Lecture notes
Konstantin K Likharev

Chapter 1

Review of fundamentals

This short introductory chapter reviews the basic notions and facts of non-relativistic classical mechanics that should be known to the readers from their undergraduate studies[1]*. For this reason, the discussion is very brief.*

1.1 Kinematics: basic notions

The basic notions of kinematics may be defined in various ways, and some mathematicians pay a lot of attention to analyzing such systems of axioms and the relations between them. In physics, we typically stick to less rigorous methods (in order to proceed faster to particular problems), and end up debating any definition as soon as 'everybody in the room' agrees that we are all speaking about the same thing—at least in the context in which they are discussed. Let me hope that the following notions used in classical mechanics do satisfy this criterion in our 'room':

(i) All the *Euclidean geometry* notions, including the *point*, the straight line, the plane, etc.

(ii) *Reference frames*: platforms for observation and mathematical description of physical phenomena. A reference frame includes a *coordinate system* used for measuring the point's position (namely, its *radius-vector*, $\mathbf{r}$, which connects the coordinate origin to the point, see figure 1.1) and a clock measuring *time*, t. A coordinate system may be understood as a certain method of expressing the radius-vector $\mathbf{r}$ of a point as a set of its *scalar*

[1] The reader is advised to perform (perhaps after reading this chapter as a reminder) a self-check by solving a few problems from the dozen listed in section 1.6. If the results are not satisfactory, it may make sense to start from some remedial reading. For that, I can recommend, for example [1–3].

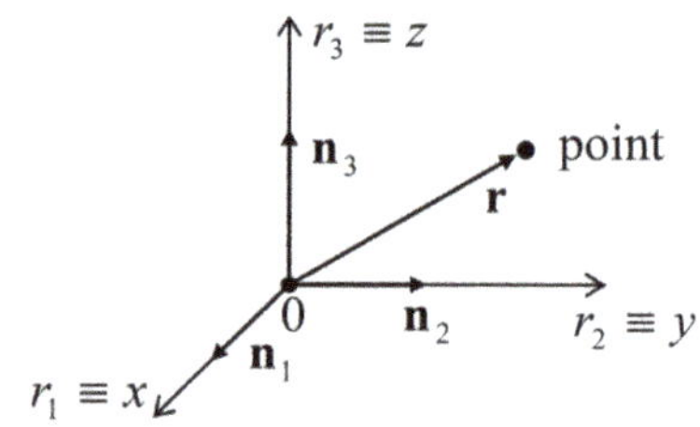

Figure 1.1. Cartesian coordinates of a point.

coordinates. The most important of such systems (but by no means the only one[2]) is the *Cartesian* (orthogonal, linear) *coordinates*[3], r_j, of a point, in which its radius-vector may be represented as the following sum:

$$\mathbf{r} = \sum_{j=1}^{3} \mathbf{n}_j r_j, \tag{1.1}$$

where $\mathbf{n}_1$, $\mathbf{n}_2$, $\mathbf{n}_3$ are unit vectors directed along the coordinate axis—see figure 1.1.

(iii) The *absolute* ('Newtonian') *space/time*, which does not depend on the matter distribution. The space is assumed to have the *Euclidean metric*, which may be expressed as the following relation between the length r of any radius-vector $\mathbf{r}$ and its Cartesian coordinates:

$$r \equiv |\mathbf{r}| = \left(\sum_{j=1}^{3} r_j^2\right)^{1/2}, \tag{1.2}$$

while time t is assumed to run similarly in all reference frames.

(iv) The (instant) *velocity* of the point,

$$\mathbf{v}(t) \equiv \frac{d\mathbf{r}}{dt} \equiv \dot{\mathbf{r}}, \tag{1.3}$$

and its *acceleration*:

$$\mathbf{a}(t) \equiv \frac{d\mathbf{v}}{dt} \equiv \dot{\mathbf{v}} = \ddot{\mathbf{r}}. \tag{1.4}$$

(v) *Transfer between reference frames*. The above definitions of vectors $\mathbf{r}$, $\mathbf{v}$, and $\mathbf{a}$ depend on the chosen reference frame (they are 'reference-frame-specific'), but we frequently need to relate these vectors as observed in different frames. Within Euclidean geometry, the relation between the radius-vectors in

[2] For example, the cylindrical/polar and spherical coordinates (see, e.g. appendix A, section A.10), which will be extensively used in this series, are not Cartesian; they are locally orthogonal, but not linear.

[3] In these notes, the Cartesian coordinates are denoted as either $\{r_1, r_2, r_3\}$ or $\{x, y, z\}$, depending on convenience in the particular case. Note that axis numbering is important for operations such as the vector ('cross') product; the 'correct' (meaning generally accepted) numbering order is such that the rotation $\mathbf{n}_1 \to \mathbf{n}_2 \to \mathbf{n}_3 \to \mathbf{n}_1 \ldots$ looks counter-clockwise if watched from a point with all $r_i > 0$—see figure 1.1.

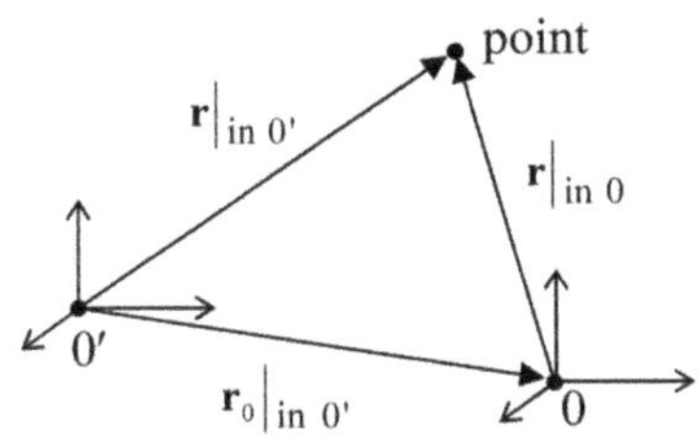

Figure 1.2. Transfer between two reference frames.

two frames with the corresponding axes parallel in the moment of interest (figure 1.2) is very simple:

$$\mathbf{r}|_{\text{in }0'} = \mathbf{r}|_{\text{in }0} + \mathbf{r}_0|_{\text{in }0'}. \tag{1.5}$$

If the frames move versus each other by *translation* only (no mutual rotation!), similar relations are valid for the velocity and the acceleration as well:

$$\mathbf{v}|_{\text{in }0'} = \mathbf{v}|_{\text{in }0} + \mathbf{v}_0|_{\text{in }0'}, \tag{1.6}$$

$$\mathbf{a}|_{\text{in }0'} = \mathbf{a}|_{\text{in }0} + \mathbf{a}_0|_{\text{in }0'}. \tag{1.7}$$

Note that in the case of mutual rotation of the reference frames, notions such as $\mathbf{v}_0|_{\text{in }0'}$ are not well defined, because different points of a rigid body connected to frame 0 may have different velocities when observed in frame 0′. As a result, generally the transfer laws for velocities and accelerations are more complex than those given by Eqs. (1.6) and (1.7). It will be more natural for me to discuss them at the end of chapter 4, which is devoted to rigid-body motion.

(vi) A *particle* (or 'point particle'): a localized physical object whose size is negligible, and whose shape is irrelevant *for the given problem.* Note that the last qualification is extremely important. For example, the size and shape of the Space Shuttle are not too important for the discussion of its orbital motion, but are paramount when its landing procedures are being developed. Since classical mechanics neglects quantum-mechanical uncertainties[4], in classical mechanics the position of a particle, at any particular instant t, may be identified with a single point, i.e. with a single radius-vector $\mathbf{r}(t)$. The formal final goal of classical mechanics is finding the *laws of motion* $\mathbf{r}(t)$ of all particles participating in a given problem.

1.2 Dynamics: Newton's laws

Generally, classical dynamics is fully described (in addition to the kinematic relations discussed above) by the three *Newton laws*[5]. In contrast to the impression

[4] This approximation is legitimate, crudely, when the product of the coordinate and momentum scales of the particle motion is much larger than the Planck's constant $\hbar \approx 1.054 \times 10^{-34}$ J s. More detailed conditions of the applicability of classical mechanics depend on a particular system—see, e.g. *Part QM* of this lecture note series.

[5] Due to the genius of Sir Isaac Newton, these laws were formulated as early as in the 1680s, far ahead of the science of that time.

that some textbooks on theoretical physics try to create, these laws are experimental in nature, and cannot be derived from *purely* theoretical arguments.

I am confident that the reader of these notes is already familiar with Newton's laws, in one or another formulation. Let me note only that in some formulations *Newton's first law* looks just like a particular case of the second law—for the case of zero net force acting on a particle. In order to avoid this duplication, the first law may be formulated as the following postulate:

> There exists at least one reference frame, called *inertial*, in which any *free particle* (i.e. a particle isolated from the rest of the Universe) moves with $\mathbf{v}$ = const, i.e. with $\mathbf{a} = 0$.

Note that according to Eq. (1.7), this postulate immediately means that there is also an infinite number of inertial frames, because all frames 0′ moving without rotation or acceleration relative to the postulated inertial frame 0 (i.e. having $\mathbf{a}_0|_{\text{in } 0'} = 0$) are also inertial.

On the other hand, the Newton's second and third laws may be postulated *together* in the following elegant way. Each particle, say number k, may be characterized by a scalar constant (called *mass*, m_k), such that at any interaction of N particles (isolated from the rest of the Universe), in any inertial system,

$$\mathbf{P} \equiv \sum_{k=1}^{N} \mathbf{p}_k \equiv \sum_{k=1}^{N} m_k \mathbf{v}_k = \text{const.} \tag{1.8}$$

(Each component of this sum,

$$\mathbf{p}_k \equiv m_k \mathbf{v}_k, \tag{1.9}$$

is called the *mechanical momentum*[6] of the corresponding particle, and the whole sum $\mathbf{P}$, the *total momentum* of the system.)

Let us apply this postulate to just two interacting particles. Differentiating Eq. (1.8), written for this case over time, we obtain

$$\dot{\mathbf{p}}_1 = -\dot{\mathbf{p}}_2. \tag{1.10}$$

Let us give the derivative $\dot{\mathbf{p}}_1$ (which is a vector) the name of *force* exerted on particle 1. In our current case, when the only possible source of force is particle 2, the force may be denoted as $\mathbf{F}_{12}$: $\dot{\mathbf{p}}_1 = \mathbf{F}_{12}$. Similarly, $\mathbf{F}_{21} \equiv \dot{\mathbf{p}}_2$, so that Eq. (1.10) becomes the *Newton's third law*:

$$\mathbf{F}_{12} = -\mathbf{F}_{21}. \tag{1.11}$$

Plugging Eq. (1.9) into these force definitions, and differentiating the products $m_k \mathbf{v}_k$, taking into account that the particles' masses are constants[7], we obtain that for k and k' taking any of values 1, 2,

$$m_k \dot{\mathbf{v}}_k \equiv m_k \mathbf{a}_k = \mathbf{F}_{kk'}, \quad \text{where } k' \neq k. \tag{1.12}$$

---

[6] The more extended term *linear momentum* is typically used in cases when there is a chance of its confusion with the *angular momentum* of the same particle/system—see below.

[7] Note that this may not be true for composite bodies of varying total mass M (e.g. rockets emitting jets, see problem 1.11); in these cases the momentum's derivative may differ from $M\mathbf{a}$.

Now, returning to the general case of several interacting particles, and making an additional (but very natural) assumption that all partial forces $\mathbf{F}_{kk'}$ acting on particle k add up as vectors, we may generalize Eq. (1.12) into *Newton's second law*:

$$m_k \mathbf{a}_k \equiv \dot{\mathbf{p}}_k = \sum_{k' \neq k} \mathbf{F}_{kk'} \equiv \mathbf{F}_k, \tag{1.13}$$

which allows a clear interpretation of the mass as a measure of the particle's *inertia*.

As a matter of principle, if the dependence of all pair forces $\mathbf{F}_{kk'}$ of particle positions (and generally of time as well) is known, Eq. (1.13), augmented with the kinematic relations (1.2) and (1.3), allows the calculation of the laws of motion $\mathbf{r}_k(t)$ of all particles of the system. For example, for one particle the second law (1.13) gives an ordinary differential equation of the second order,

$$m\ddot{\mathbf{r}} = \mathbf{F}(\mathbf{r}, t), \tag{1.14}$$

which may be integrated—either analytically or numerically.

For certain cases, this is very simple. As an elementary example, Newton's gravitational force

$$\mathbf{F} = -G\frac{mm'}{R^3}\mathbf{R} \tag{1.15}$$

(where $\mathbf{R} \equiv \mathbf{r} - \mathbf{r}'$ is the distance between particles of masses m and m')[8], is virtually uniform and may be approximated as

$$\mathbf{F} = m\mathbf{g}, \tag{1.16}$$

with the vector $\mathbf{g} \equiv (Gm'/r'^3)\mathbf{r}'$ being constant, for local motions with $r \ll r'$[9]. As a result, m in Eq. (1.13) cancels, it is reduced to just $\ddot{\mathbf{r}} = \mathbf{g} = \text{const}$, and may be easily integrated twice:

$$\begin{aligned} \dot{\mathbf{r}}(t) \equiv \mathbf{v}(t) &= \int_0^t \mathbf{g}\, dt' + \mathbf{v}(0) = \mathbf{g}t + \mathbf{v}(0), \\ \mathbf{r}(t) &= \int_0^t \mathbf{v}(t')dt' + \mathbf{r}(0) = \mathbf{g}\frac{t^2}{2} + \mathbf{v}(0)t + \mathbf{r}(0), \end{aligned} \tag{1.17}$$

[8] The fact that the masses participating in Eqs. (1.14) and Eq. (1.16) are equal, the so-called *weak equivalence principle*, is actually highly non-trivial, but has been verified experimentally to the relative accuracy of at least 10^{-13}. Due to the conceptual significance of the principle, new experiments, such as MISCROSCOPE (http://smsc.cnes.fr/MICROSCOPE/), are being planned for a substantial, nearly 100-fold accuracy improvement.

[9] Of course, the most important particular case of Eq. (1.16) is the motion of objects near the Earth's surface. In this case, using the fact that Eq. (1.15) remains valid for the gravity field created by a heavy sphere, we obtain $g = GM_E/R_E^2$, where M_E and R_E are the Earth's mass and radius. Plugging in their values, $M_E \approx 5.92 \times 10^{24}$ kg, $R_E \approx 6.37 \times 10^6$ m, we obtain $g \approx 9.74$ m s^{-2}. The experimental value of g varies from 9.78 to 9.83 m s^{-2} at various locations on the Earth's surface (due to the deviations of Earth's shape from a sphere, and the location-dependent effect of the centrifugal 'inertial force'—see section 4.6 below), with an average value of $g \approx 9.807$ m s^{-2}.

thus giving the generic solution for all those undergraduate problems on projectile motion, which should be so familiar to the reader.

All this looks (and indeed is) very simple, but in most other cases Eq. (1.13) leads to more complex calculations. As an example, let us think about how we would solve another simple problem: a bead of mass m sliding, without friction, along a round ring of radius R in a gravity field obeying Eq. (1.16)—see figure 1.3. (This system is of course equivalent to the usual *point pendulum*, i.e. a point mass suspended from point 0 on a light rod or string, and in addition constrained to move in one vertical plane.)

Suppose we are only interested in bead's velocity, v, in the lowest point, after it has been dropped from rest at the rightmost position. If we want to solve this problem using only Newton's laws, we have to take the following steps:

(i) consider the bead in an arbitrary intermediate position on a ring, described, for example, by the angle θ shown in figure 1.3;

(ii) draw all the forces acting on the particle—in our current case, the gravity force $m\mathbf{g}$ and the reaction force $\mathbf{N}$ exerted by the ring—see figure 1.3;

(iii) write the Cartesian components of Newton's second law (1.14) for the bead acceleration: $ma_x = N_x$, $ma_y = N_y - mg$,

(iv) recognize that in the absence of friction, the force $\mathbf{N}$ should be normal to the ring, so that we can use two additional equations, $N_x = -N\sin\theta$ and $N_y = N\cos\theta$;

(v) eliminate unknown variables N, N_x, and N_y from the resulting system of four equations, thus obtaining a single second-order differential equation for one variable, for example θ:

$$mR\ddot{\theta} = -mg \sin \theta; \tag{1.18}$$

(vi) use the mathematical identity $\ddot{\theta} \equiv (d\dot{\theta}/d\theta)\,\dot{\theta}$ to integrate this equation over θ once to obtain an expression relating the velocity $\dot{\theta}$ and the angle θ; and finally

(vii) using our specific initial condition ($\dot{\theta} = 0$ at $\theta = \pi/2$), find the final velocity as $v = R\dot{\theta}$ at $\theta = 0$.

All this is very much doable, but please agree that the procedure it too cumbersome for such a simple problem. Moreover, in many other cases even writing equations of motion along relevant coordinates is very complex, and any help the general theory may provide is highly valuable. In many cases, such help is given by *conservation laws*; let us review the most general of them.

1.3 Conservation laws

Energy conservation is arguably the most general law of physics, but in mechanics it takes the humbler form of *mechanical energy conservation*, which has limited applicability. To derive it, we first have to define the *kinetic energy* of a particle as

$$T \equiv \frac{m}{2}v^2, \tag{1.19}$$

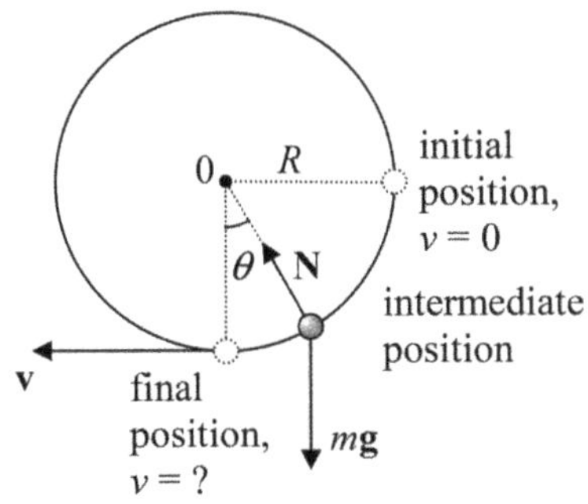

Figure 1.3. A bead sliding along a vertical ring.

and then recast its differential as[10]

$$dT = d\left(\frac{m}{2}v^2\right) \equiv d\left(\frac{m}{2}\mathbf{v}\cdot\mathbf{v}\right) = m\mathbf{v}\cdot d\mathbf{v} = m\frac{d\mathbf{v}\cdot d\mathbf{r}}{dt} = \frac{d\mathbf{p}}{dt}\cdot d\mathbf{r}. \quad (1.20)$$

Now plugging in the momentum's derivative from Newton's second law, $d\mathbf{p}/dt = \mathbf{F}$, where $\mathbf{F}$ is the full force acting on the particle, we obtain relation $dT = \mathbf{F}\cdot d\mathbf{r}$. Its integration along the particle's trajectory between some points A and B gives the formula that is sometimes called the *work–energy principle*:

$$\Delta T \equiv T(\mathbf{r}_B) - T(\mathbf{r}_A) = \int_A^B \mathbf{F}\cdot d\mathbf{r}, \quad (1.21)$$

where the integral on the right-hand is called the *work* of the force $\mathbf{F}$ on the path from A to B.

A further step may be taken only for *potential* (also called 'conservative') force that may be represented as a (minus) gradient of some scalar function $U(\mathbf{r})$, called the *potential energy*[11]. The vector operator ∇ (called either 'del' or 'nabla') of spatial differentiation[12] allows a very compact expression of this fact:

$$\mathbf{F} = -\nabla U. \quad (1.22)$$

For example, for the uniform gravity field (1.16),

$$U = mgh + \text{const}, \quad (1.23)$$

where h is the vertical coordinate directed 'up'—opposite to the direction of the vector $\mathbf{g}$.

Integrating the tangential component F_τ of the vector $\mathbf{F}$, given by Eq. (1.22), along an arbitrary path connecting points A and B, we obtain

$$\int_A^B F_\tau dr \equiv \int_A^B \mathbf{F}\cdot d\mathbf{r} = U(\mathbf{r}_A) - U(\mathbf{r}_B), \quad (1.24)$$

[10] Symbol $\mathbf{a}\cdot\mathbf{b}$ denotes the scalar (or 'dot-') product of vectors $\mathbf{a}$ and $\mathbf{b}$—see, e.g. Eq. (A.43).

[11] Note that because of its definition via the gradient, the potential energy is only defined to an arbitrary additive constant.

[12] Its basic properties are listed in appendix A, section A.8.

i.e. the work of potential forces may be represented as the difference of values of the function $U(\mathbf{r})$ in the initial and final point of the path. (Note that according to Eq. (1.24), work of a potential force on any closed path, with $\mathbf{r}_A = \mathbf{r}_B$, is zero.)

Now returning to Eq. (1.21) and comparing it to Eq. (1.24), we see that

$$T(\mathbf{r}_B) - T(\mathbf{r}_A) = U(\mathbf{r}_A) - U(\mathbf{r}_B), \tag{1.25}$$

so that the *total mechanical energy* E, defined as

$$E \equiv T + U, \tag{1.26}$$

is indeed conserved,

$$E(\mathbf{r}_A) \equiv E(\mathbf{r}_B), \tag{1.27}$$

but for conservative forces only. (Non-conservative forces may change E by either transferring energy from its mechanical form into another form, e.g. heat in the case of friction, or by pumping the energy into the system under consideration from another, 'external' system.)

The mechanical energy conservation allows us to return for a second to the problem shown in figure 1.3 and solve it in one shot by writing Eq. (1.27) for the initial and final points[13]:

$$0 + mgR = \frac{m}{2}v^2 + 0. \tag{1.28}$$

Solving Eq. (1.28) for v immediately gives us the desired answer. Let me hope that the reader agrees that this method of problem solving is much simpler, and I have secured his or her attention to discuss other conservation laws—which may be equally effective.

Momentum. Actually, the conservation of the full linear momentum of any system of particles isolated from the rest of the world has already been discussed and may serve as the basic postulate of classical dynamics—see Eq. (1.8). In the case of one free particle the law is reduced to a trivial result $\mathbf{p} = \text{const}$, i.e. $\mathbf{v} = \text{const}$. If a system of N particles *is* affected by external forces $\mathbf{F}^{(\text{ext})}$, we may write

$$\mathbf{F}_k = \mathbf{F}_k^{(\text{ext})} + \sum_{k=1}^{N} \mathbf{F}_{kk'}. \tag{1.29}$$

If we sum up the resulting Eqs. (1.13) for all particles of the system then, due to Newton's third law (1.11), valid for any indices $k \neq k'$, the contributions of all internal forces $\mathbf{F}_{kk'}$ to the resulting double sum in the right-hand part cancel, and we obtain the equation

$$\dot{\mathbf{P}} = \mathbf{F}^{(\text{ext})}, \quad \text{where } \mathbf{F}^{(\text{ext})} \equiv \sum_{k=1}^{N} \mathbf{F}_k^{(\text{ext})}, \tag{1.30}$$

[13] Here the arbitrary constant in Eq. (1.23) is chosen so that the potential energy is zero in the finite point.

which tells us that the translational motion of the system as the whole is similar to that of a single particle, under the effect of the *net external force* $\mathbf{F}^{(\text{ext})}$. As a simple sanity check, if the external forces have a zero sum, we return to postulate Eq. (1.8). Just one reminder: Eq. (1.30), just as its precursor Eq. (1.13), is only valid in an inertial reference frame.

I hope that the reader knows numerous examples of the application of the linear momentum conservation law, including the problems on car collisions, where the large collision forces are typically not known, so that the direct application of Eq. (1.13) to each car is impracticable.

The *angular momentum* of a particle[14] is defined as the following vector:

$$\mathbf{L} \equiv \mathbf{r} \times \mathbf{p}, \tag{1.31}$$

where $\mathbf{a} \times \mathbf{b}$ means the vector (or 'cross-') product of the vector operands[15]. Differentiating Eq. (1.31) over time, we obtain

$$\dot{\mathbf{L}} = \dot{\mathbf{r}} \times \mathbf{p} + \mathbf{r} \times \dot{\mathbf{p}}. \tag{1.32}$$

In the first product, $\dot{\mathbf{r}}$ is just the velocity vector $\mathbf{v}$, parallel to the particle momentum $\mathbf{p} = m\mathbf{v}$, so that this product vanishes, since the vector product of any two parallel vectors equals zero. In the second product, $\dot{\mathbf{p}}$ is equal to the full force $\mathbf{F}$ acting on the particle, so that Eq. (1.32) is reduced to

$$\dot{\mathbf{L}} = \boldsymbol{\tau}, \tag{1.33}$$

where vector

$$\boldsymbol{\tau} \equiv \mathbf{r} \times \mathbf{F}, \tag{1.34}$$

is called the *torque* exerted by force $\mathbf{F}$[16]. (Note that the torque is reference-frame-specific—and again, the frame has to be inertial for Eq. (1.33) to be valid, because we have used Eq. (1.13) for its derivation.) For an important particular case of a *central* force $\mathbf{F}$ that is parallel to the radius-vector $\mathbf{r}$ of a particle, the torque vanishes, so that (in that particular reference frame only!) the angular momentum is conserved:

$$\mathbf{L} = \text{const}. \tag{1.35}$$

For a system of N particles, the total angular momentum is naturally defined as

$$\mathbf{L} \equiv \sum_{k=1}^{N} \mathbf{L}_k. \tag{1.36}$$

---

[14] Here we imply that the internal motions of the particle, including its rotation about its own axis, are negligible. (Otherwise it could not be represented by a point, as was postulated in section 1.1.)

[15] See, e.g. Eq. (A.45).

[16] Alternatively, especially in mechanical engineering, the torque is called the *force moment*.

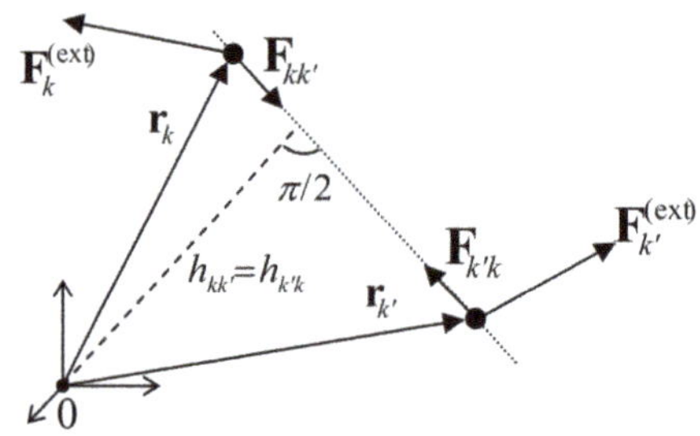

Figure 1.4. Internal and external forces, and the internal torque cancellation in a system of two particles.

Differentiating this equation over time, using Eq. (1.33) for each $\dot{\mathbf{L}}_k$, and again partitioning each force in a accordance with Eq. (1.29), we obtain

$$\dot{\mathbf{L}} = \sum_{\substack{k,k'=1 \\ k'\neq k}}^{N} \mathbf{r}_k \times \mathbf{F}_{kk'} + \boldsymbol{\tau}^{(\text{ext})}, \quad \text{where } \boldsymbol{\tau}^{(\text{ext})} \equiv \sum_{k=1}^{N} \mathbf{r}_k \times \mathbf{F}_k^{(\text{ext})}. \tag{1.37}$$

The first (double) sum may always be divided into pairs of the type $(\mathbf{r}_k \times \mathbf{F}_{kk'} + \mathbf{r}_{k'} \times \mathbf{F}_{k'k})$. With a natural assumption of the central forces, $\mathbf{F}_{kk'} \parallel (\mathbf{r}_k - \mathbf{r}_{k'})$, each of these pairs equals zero. Indeed, in this case both components of the pair are vectors perpendicular to the plane passing through the positions of both particles and the reference-frame origin, i.e. to the plane of the drawing in figure 1.4.

Also, due to Newton's third law (1.11) the two forces are equal and opposite, and the magnitude of each term in the sum may be represented as $|F_{kk'}|h_{kk'}$, equal to the 'lever arms' $h_{kk'} = h_{k'k}$. As a result, each sum $(\mathbf{r}_k \times \mathbf{F}_{kk'} + \mathbf{r}_{k'} \times \mathbf{F}_{k'k})$, and hence the whole double sum in Eq. (1.37) vanishes, and it is reduced to a very simple result,

$$\dot{\mathbf{L}} = \boldsymbol{\tau}^{(\text{ext})}, \tag{1.38}$$

which is similar to Eq. (1.33) for a single particle, and is the angular analog of Eq. (1.30). In particular, Eq. (1.38) shows that if the full external torque $\boldsymbol{\tau}^{(\text{ext})}$ vanishes for some reason (e.g. if the system of particles is isolated from the rest of the Universe), the conservation law (1.35) is valid for the full angular momentum $\mathbf{L}$, even if its individual components $\mathbf{L}_k$ are not conserved due to inter-particle interactions.

Note again that since the conservation laws may be derived from Newton's laws, they do not introduce anything new to the dynamics of any system. Indeed, from the mathematical point of view, the conservation laws discussed above are just *the first integrals of motion*, which may liberate us from the necessity to integrate the second-order differential equations of motion, following from Newton's laws, twice.

1.4 Potential energy and equilibrium

Another important role of the potential energy U, in particular for dissipative systems whose total mechanical energy E is *not* conserved because it may be drained to the environment, is finding the positions of equilibrium (sometimes called the *fixed points* of the system under analysis) and analyzing their stability with respect to small perturbations. For a single particle, this is very simple: the force (1.22) vanishes at each extremum (minimum or maximum) of the potential energy[17]. Of those fixed points, only the minimums of $U(\mathbf{r})$ are stable—see section 3.2 below for a discussion of this point.

A slightly subtler case is a particle with potential energy $U(\mathbf{r})$, subjected to an *additional* external force $\mathbf{F}^{(\text{ext})}(\mathbf{r})$. In this case, the stable equilibrium is reached at the minimum of not the function $U(\mathbf{r})$, but of what is sometimes called the *Gibbs potential energy*[18]

$$U_{\text{G}}(\mathbf{r}) \equiv U(\mathbf{r}) - \int^{\mathbf{r}} \mathbf{F}^{(\text{ext})}(\mathbf{r}') \cdot d\mathbf{r}', \tag{1.39}$$

which is defined, just as $U(\mathbf{r})$ is, to an arbitrary constant. The proof of Eq. (1.39) is very simple: in an extremum of this function, the total force acting on the particle,

$$\mathbf{F}^{(\text{tot})} = \mathbf{F} + \mathbf{F}^{(\text{ext})} \equiv -\nabla U + \nabla \int^{\mathbf{r}} \mathbf{F}^{(\text{ext})}(\mathbf{r}') \cdot d\mathbf{r}' = -\nabla U_{\text{G}}, \tag{1.40}$$

vanishes, as it should.

Physically, the difference $U_{\text{G}} - U$ specified by Eq. (1.39) is the $\mathbf{r}$-dependent part of the potential energy $U^{(\text{ext})}$ of the external system responsible for the force $\mathbf{F}^{(\text{ext})}$, so that U_{G} is just the total potential energy $U + U^{(\text{ext})}$, excluding the part that does not depend on $\mathbf{r}$ and hence is irrelevant for the analysis. According to Newton's third law, the force exerted by the particle on the external system equals $(-\mathbf{F}^{(\text{ext})})$, so that its work (and hence the change of $U^{(\text{ext})}$ due to the change of $\mathbf{r}$) is given by the second term on the right-hand side of Eq. (1.39). Thus the condition of equilibrium, $-\nabla U_{\text{G}} = 0$, is just the condition of an extremum of the total potential energy, $U + U^{(\text{ext})}$, of the two interacting systems.

For the simplest (and very frequent) case when the applied force is independent of the particle's position, the Gibbs potential energy is just[19]

$$U_{\text{G}}(\mathbf{r}) \equiv U(\mathbf{r}) - \mathbf{F}^{(\text{ext})} \cdot \mathbf{r} + \text{const}. \tag{1.41}$$

As the simplest example, consider a 1D deformation of the usual elastic spring providing the returning force $(-\kappa x)$, where x is the deviation from its equilibrium.

---

[17] Assuming that the additional, non-conservative forces (such as viscosity), responsible for the mechanical energy drain vanish at equilibrium—as they typically do. (The static friction is one counter-example.)

[18] Unfortunately, in most textbooks the association of the (unavoidably used) notion of U_{G} with the glorious name of Josiah Willard Gibbs is postponed until a course on statistical mechanics and/or thermodynamics, where U_{G} is a part of the *Gibbs free energy*, in contrast to U, which is a part of the *Helmholtz free energy*—see, e.g. *Part SM* section 1.4. In contrast, I use this name throughout my series, because the difference between U_{G} and U, and hence that between the Gibbs and Helmholtz free energies, has nothing to do with statistics or thermal motion, and belongs to the whole of physics, including not only mechanics, but also electrodynamics and quantum mechanics.

[19] Note that Eq. (1.41) is a particular case of what mathematicians call the *Legendre transformations*.

As follows from Eq. (1.22), its potential energy is $U = \kappa x^2/2$ + const, so that its minimum corresponds to $x = 0$. Now let us apply an additional external force F, say independent of x. Then the equilibrium deformation of the spring, $x_0 = F/\kappa$, corresponds to the minimum not of U, but rather of the Gibbs potential energy (1.41), taking the form

$$U_G \equiv U - Fx = \frac{\kappa x^2}{2} - Fx. \tag{1.42}$$

1.5 OK, we've got it—can we go home now?

Not yet. In many cases the conservation laws discussed above provide little help, even in systems without dissipation. Consider, for example, a generalization of the bead-on-the-ring problem shown in figure 1.3, in which the ring is rotated by external forces, with a constant angular velocity ω, about its vertical diameter[20]. In this problem (to which I will repeatedly return below, using it as an analytical mechanics 'testbed'), none of the three conservation laws listed in the last section holds. In particular, the bead's energy,

$$E = \frac{m}{2}v^2 + mgh, \tag{1.43}$$

is *not* constant, because the external forces rotating the ring may change it. Of course, we can still solve the problem using Newton's laws, but this is even more complex than for the above case of the ring at rest, in particular because the force **N** exerted on the bead by the ring now may have three rather than two Cartesian components, which are not simply related. On the other hand, it is clear that the bead still has just one degree of freedom (angle θ), so that its dynamics should not be too complicated.

This fact gives a clue as to how situations such as this one could be simplified: if we could only exclude the so-called *reaction forces* such as **N** (that take into account the *external constraints* imposed on the particle motion) in advance, it would help a lot. Such a constraint exclusion may be provided by analytical mechanics, in particular its Lagrangian formulation, to we will now proceed.

Of course, the value of the Lagrangian approach goes far beyond simple systems such as the bead on a ring. Indeed, this system has just two externally imposed constrains: the fixed distance of the bead from the center of the ring, and the instant angle of rotation of the ring about its vertical diameter. Now let us consider the motion of a rigid body. It is essentially a system of a very large number, $N \gg 1$, of particles ($\sim 10^{23}$ of them if we think about atoms in a 1 cm scale body). If the only way to analyze its motion would be to write Newton's laws for each of the particles, the situation would be completely hopeless. Fortunately, the number of constraints imposed

[20] This is essentially a simplified model of the famous mechanical control device called the *centrifugal* (or 'flyball' or 'centrifugal flyball') *governor*—see, e.g. http://en.wikipedia.org/wiki/Centrifugal_governor. Sometimes the device is called the 'Watt's governor', after the famous engineer J Watts who used it in 1788 in one of his first steam engines, but it had been used in European windmills since at least the early 1600s.

on its motion is almost similarly huge. (At negligible deformations of the body, the distances between each pair of its particles should be constant.) As a result, the number of actual degrees of freedom of such a body (at negligible deformations, just 6—see section 4.1), so that with kind help from analytical mechanics, the motion of the body may be, in many important cases, analyzed even without numerical calculations.

One more important motivation for analytical mechanics is given by the dynamics of 'non-mechanical' systems, for example, of the electromagnetic field—possibly interacting with charged particles, conducting bodies, etc. In many such systems, the easiest (and sometimes the only practicable) way to find the equations of motion is to derive then from either the Lagrangian or Hamiltonian function of the system. Moreover, the Hamiltonian formulation of the analytical mechanics (to be discussed in chapter 10) offers a direct pathway to deriving quantum-mechanical Hamiltonian operators of various systems, which are necessary for the analysis of their quantum properties.

1.6 Problems

Problem 1.1. A bicycle, ridden with velocity v on a wet pavement, has no mudguards on its wheels. How far behind should the following biker ride to avoid being splashed over? Neglect the effects of air resistance.

Problem 1.2. Two round disks of radius R are firmly connected with a coaxial cylinder of a smaller radius r, and a thread is wound on the resulting spool. The spool is placed on a horizontal surface, and the thread's end is pooled out at angle ϕ—see the figure below. Assuming that the spool does not slip on the surface, what direction would it roll?

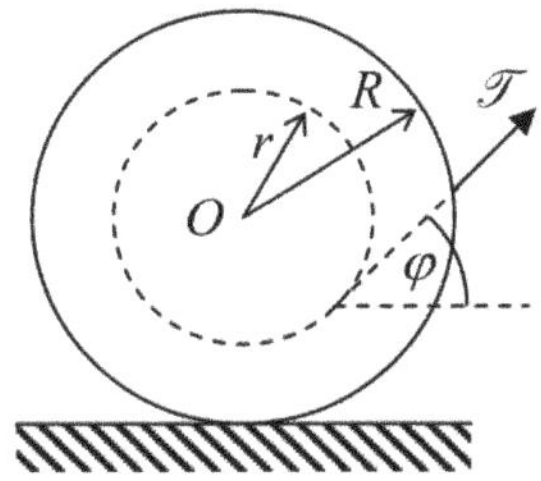

Problem 1.3.* Calculate the equilibrium shape of a flexible, heavy rope of length l, with a constant mass μ per unit length, if it is hung in a uniform gravity field between two points separated by a horizontal distance d—see the figure below.

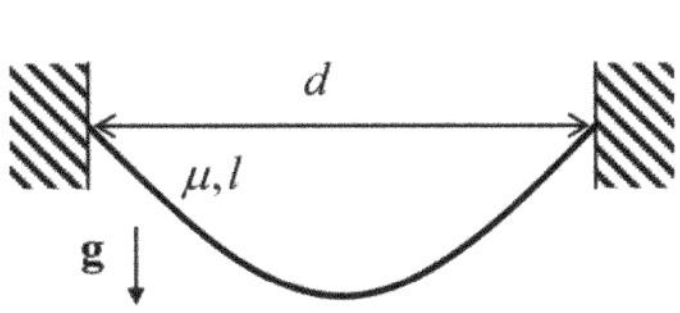

Problem 1.4. A uniform, long, thin bar is placed horizontally on two similar round cylinders rotating toward each other with the same angular velocity ω and displaced by distance d—see the figure below. Calculate the laws of relatively slow horizontal motions of the bar within the plane of the drawing for both possible directions of cylinder rotation, assuming that the friction force between the slipping surfaces of the bar and each cylinder obeys the simple *Coulomb approximation*[21] $|F| = \mu N$, where N is the normal pressure force between them, and μ is a constant (velocity-independent) coefficient. Formulate the condition of validity of your result.

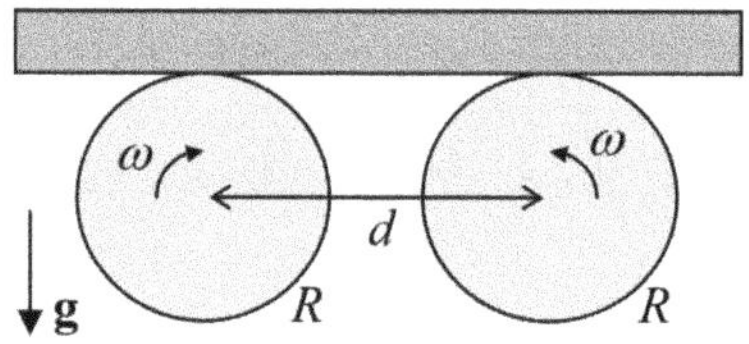

Problem 1.5. A small block slides, without friction, down a smooth slide that ends with a round loop of radius R—see the figure below. What smallest initial height h allows the block to make its way around the loop without dropping from the slide, if it is launched with negligible initial velocity?

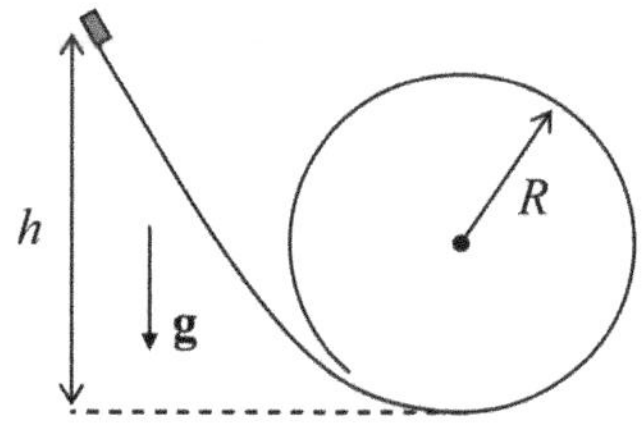

Problem 1.6. A satellite of mass m is being launched from height H over the surface of a spherical planet with radius R and mass $M \gg m$—see the figure below. Find the range of initial velocities $\mathbf{v}_0$ (normal to the radius) providing closed orbits above the planet's surface.

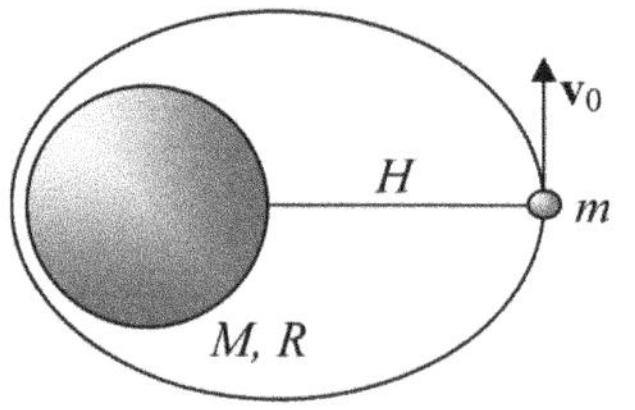

Problem 1.7. Prove that the thin-uniform-disk model of a galaxy describes small harmonic oscillations of stars inside it along the direction normal to the disk,

[21] This approximation was suggested in 1785 by the same C-A de Coulomb who discovered the famous *Coulomb law* of electrostatics, and hence pioneered the whole qualitative science of electricity.

and calculate the frequency of these oscillations in terms of Newton's gravitational constant G and the average density ρ of the star/dust matter of the galaxy.

Problem 1.8. Derive the differential equations of motion for small oscillations of two similar pendula coupled with a spring (see the figure below), within the vertical plane. Assume that at the vertical position of both pendula, the spring is not stretched ($\Delta L = 0$).

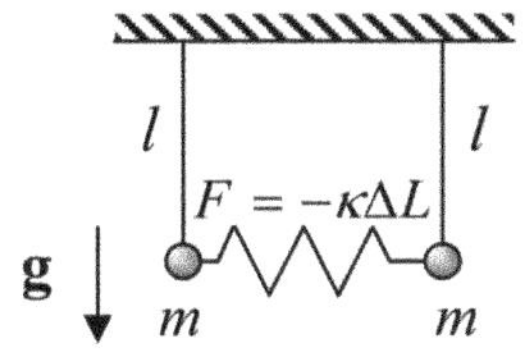

Problem 1.9. One popular futuristic concept of travel is digging a straight railway tunnel through the Earth and letting a train go through it without initial velocity—driven only by gravity. Calculate the train's travel time through such a tunnel, assuming that the Earth's density ρ is constant, and neglecting the effects of friction and planetary rotation.

Problem 1.10. A small bead of mass m may slide, without friction, along a light string, stretched with a force $\mathscr{T} \gg mg$ between two points separated by a horizontal distance $2d$—see the figure below. Calculate the frequency of horizontal oscillations of the bead about its equilibrium position.

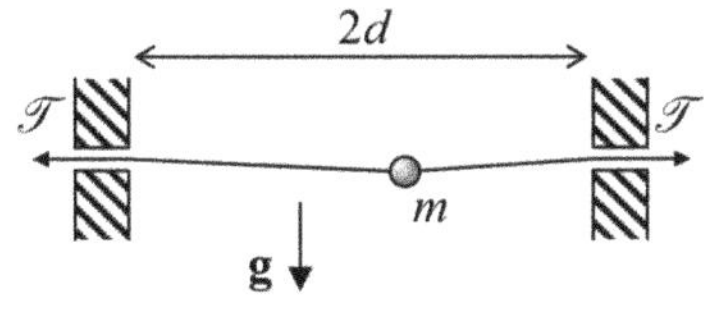

Problem 1.11. For a rocket accelerating due to a working jet motor (and hence spending its fuel), calculate the relation between its velocity and the remaining mass. *Hint*: For the sake of simplicity, consider 1D motion.

Problem 1.12. Prove the following *virial theorem*[22]. For a set of N particles performing a periodic motion,

$$\overline{T} = -\frac{1}{2}\sum_{k=1}^{N}\overline{\mathbf{F}_k \cdot \mathbf{r}_k},$$

where the top bar means time averaging, in this case over the motion period. What does the virial theorem say about:

[22] It was first stated by R Clausius in 1870.

(i) the 1D motion of a particle in a confining potential $U(x) = ax^{2s}$, with $a > 0$ and $s > 0$, and
(ii) the orbital motion of a particle moving in a central potential $U(r) = -C/r$?

Hint: Explore the time derivative of the following scalar function of time:

$$G(t) \equiv \sum_{k=1}^{N} \mathbf{p}_k \cdot \mathbf{r}_k.$$

References

[1] Fowles G and Cassiday G 2004 *Analytical Mechanics* 7th edn (Brooks Cole)
[2] Marion J and Thornton S 2003 *Classical Dynamics of Particles and Systems* 5th edn (Brooks Cole)
[3] Symon K 1971 *Mechanics* 3rd edn (Reading, MA: Addison-Wesley)

IOP Publishing

Classical Mechanics
Letcture notes
Konstantin K Likharev

Chapter 2

Lagrangian analytical mechanics

The goal of this chapter is to describe the Lagrangian formalism of analytical mechanics, which is extremely useful for obtaining the differential equations of motion (and sometimes their first integrals) not only for mechanical systems with holonomic constraints, but also other dynamic systems.

2.1 Lagrange equations

In many cases, the constraints imposed on the 3D motion of a system of N particles may be described by N vector (i.e. $3N$ scalar) algebraic equations

$$\mathbf{r}_k = \mathbf{r}_k(q_1, q_2, \ldots, q_j, \ldots, q_J, t), \qquad \text{with } 1 \leqslant k \leqslant N, \tag{2.1}$$

where q_j are certain *generalized coordinates* which (together with constraints) completely define the system position. Their number $J \leqslant 3N$ is called the number of the actual *degrees of freedom* of the system. The constraints that allow such a description are called *holonomic*[1].

For example, for the problem discussed briefly in section 1.5, namely the bead sliding along a rotating ring (figure 2.1), $J = 1$, because with the constraints imposed by the ring, the bead's position may be uniquely determined by just one generalized coordinate—for example, its polar angle θ.

Indeed, selecting the reference frame as shown in figure 2.1 and using the well-known formulas for the spherical coordinates[2], we see that in this case, Eq. (2.1) has the form

$$\begin{aligned} &\mathbf{r} \equiv \{x, y, z\} = \{R \sin\theta \cos\varphi,\ R \sin\theta \sin\varphi,\ R\cos\theta\}, \\ &\quad \text{where } \varphi = \omega t + \text{const}, \end{aligned} \tag{2.2}$$

[1] Possibly, the simplest example of a *non-holonomic* constraint is a set of inequalities describing the hard walls confining the motion of particles in a closed volume. Non-holonomic constraints are better dealt with using other methods, e.g. by imposing proper boundary conditions on the (otherwise unconstrained) motion.
[2] See, e.g. Eq. (A.65).

doi:10.1088/978-0-7503-1398-8ch2

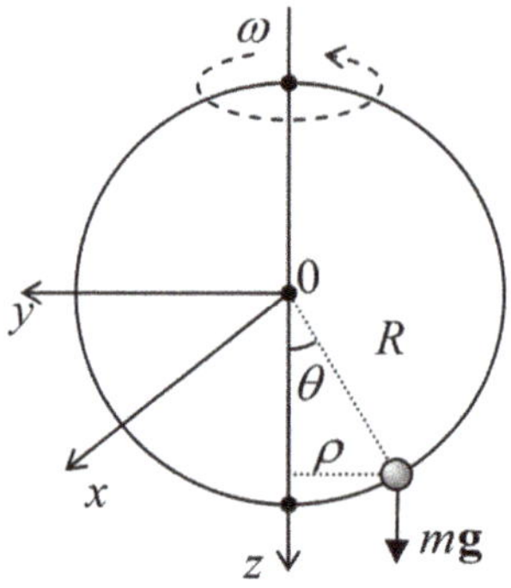

Figure 2.1. A bead on a rotating ring as an example of a system with just one degree of freedom ($J = 1$).

with the last constant depending on the exact selection of axes x and y and the time origin. Since $\varphi(t)$ is a fixed function of time, and R is a fixed constant, the position of a particle in space at any instant t is completely determined by the value of its only generalized coordinate θ. Note that the dimensionality of the generalized coordinate may be different from that of Cartesian coordinates (meters)!

Now, returning to the general case of J degrees of freedom, let us consider a set of small *variations* (alternatively called 'virtual displacements') δq_j allowed by the constraints. Virtual displacements differ from the actual small displacements (described by *differentials* dq_j proportional to time variation dt) in that δq_j describes not the system's motion as such, but rather its possible variation—see figure 2.2.

Generally, operations with variations are the subject of a special field of mathematics, the calculus of variations[3]. However, the only math background necessary for our current purposes is the understanding that operations with variations are similar to those with the usual differentials, although we need to pay careful attention to what each variable is a function of. For example, if we consider the variation of the radius-vectors (2.1), at a fixed time t, as a function of independent variations δq_j, we may use the usual formula for the differentiation of a function of several arguments[4]:

$$\delta \mathbf{r}_k = \sum_j \frac{\partial \mathbf{r}_k}{\partial q_j} \delta q_j. \tag{2.3}$$

Now let us break the force acting upon the kth particle into two parts: the frictionless, constraining part $\mathbf{N}_k$ of the reaction force and the remaining part $\mathbf{F}_k$—including the force components from other sources and possibly the friction part of

[3] For a concise introduction to the field see, e.g. [1] or [2]. An even shorter review may be found in chapter 17 of Arfken *et al* [3]. For a more detailed discussion, using many examples from physics, see [4].

[4] See, e.g. Eq. (A.23). In all the formulas of this section, all summations over index j are from 1 to J, while those over the particle number k are from 1 to N, so for the sake of brevity, these limits are not explicitly specified.

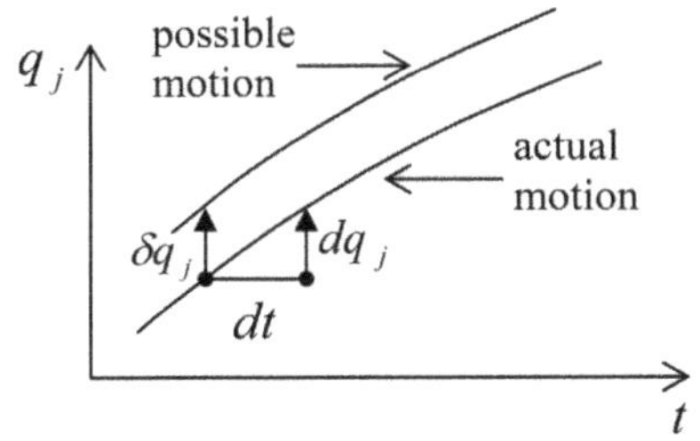

Figure 2.2. Actual displacement dq_j versus the virtual one (i.e. variation) δq_j.

the reaction force. Then Newton's second law for the kth particle of the system may be written as

$$m_k \dot{\mathbf{v}}_k - \mathbf{F}_k = \mathbf{N}_k. \tag{2.4}$$

Since any variation of the motion has to be allowed by the constraints, its $3N$-dimensional vector with N 3D-vector components $\delta\mathbf{r}_k$ has to be perpendicular to the $3N$-dimensional vector of the constraining forces, also with N 3D-vector components $\mathbf{N}_k$. (For example, for the problem shown in figure 2.1, the virtual displacement vector $\delta\mathbf{r}_k$ may be directed only along the ring, while the constraining force $\mathbf{N}$, exerted by the ring, has to be perpendicular to that direction.) This condition may be expressed as

$$\sum_k \mathbf{N}_k \cdot \delta\mathbf{r}_k = 0, \tag{2.5}$$

where the scalar product of $3N$-dimensional vectors is defined exactly as that of 3D vectors, i.e. as the sum of the products of the corresponding components of the operands. The substitution of Eq. (2.4) into Eq. (2.5) results in the so-called *D'Alembert principle*[5]:

$$\sum_k (m_k \dot{\mathbf{v}}_k - \mathbf{F}_k) \cdot \delta\mathbf{r}_k = 0. \tag{2.6}$$

Now we may plug Eq. (2.3) into Eq. (2.6) to obtain

$$\sum_j \left\{ \sum_k m_k \dot{\mathbf{v}}_k \cdot \frac{\partial \mathbf{r}_k}{\partial q_j} - \mathscr{F}_j \right\} \delta q_j = 0 \tag{2.7}$$

where the scalars $\mathscr{F}_j$, called the *generalized forces*, are defined as follows[6]:

$$\mathscr{F}_j \equiv \sum_k \mathbf{F}_k \cdot \frac{\partial \mathbf{r}_k}{\partial q_j}. \tag{2.8}$$

Now we may use the standard argument of the calculus of variations: in order for the left-hand side of Eq. (2.7) to be zero for an arbitrary selection of independent

[5] It was spelled out in a 1743 work by J-B le Rond d'Alembert, although the core of this result has been traced to an earlier work by J Bernoulli (1667–1748)—not to be confused with his son, D Bernoulli, who is credited, in particular, with the *Bernoulli equation* for ideal fluids, which will be discussed in section 8.4 below.

[6] Note that since the dimensionality of generalized coordinates may be arbitrary, that of generalized forces may also differ from the newton.

variations δq_j, the expressions in the curly brackets, for every j, should equal zero. This gives us the desired set of $J \leqslant 3N$ equations

$$\sum_k m_k \dot{\mathbf{v}}_k \cdot \frac{\partial \mathbf{r}_k}{\partial q_j} - \mathcal{F}_j = 0; \tag{2.9}$$

what remains is just to recast them in a more convenient form. First, using the differentiation by parts to calculate the following time derivative:

$$\frac{d}{dt}\left(\mathbf{v}_k \cdot \frac{\partial \mathbf{r}_k}{\partial q_j}\right) = \dot{\mathbf{v}}_k \cdot \frac{\partial \mathbf{r}_k}{\partial q_j} + \mathbf{v}_k \frac{d}{dt}\left(\frac{\partial \mathbf{r}_k}{\partial q_j}\right), \tag{2.10}$$

we may notice that the first term on the right-hand side is exactly the scalar product in the first term of Eq. (2.9). Second, let us use another key fact of the calculus of variations (which is, essentially, evident from figure 2.3): the differentiation of a variable over time and over the generalized coordinate variation (at fixed time) are interchangeable operations.

As a result, in the second term on the right-hand side of Eq. (2.10) we may write

$$\frac{d}{dt}\left(\frac{\partial \mathbf{r}_k}{\partial q_j}\right) = \frac{\partial}{\partial q_j}\left(\frac{d\mathbf{r}_k}{dt}\right) \equiv \frac{\partial \mathbf{v}_k}{\partial q_j}. \tag{2.11}$$

Finally, let us differentiate Eq. (2.1) over time:

$$\mathbf{v}_k \equiv \frac{d\mathbf{r}_k}{dt} = \sum_j \frac{\partial \mathbf{r}_k}{\partial q_j}\dot{q}_j + \frac{\partial \mathbf{r}_k}{\partial t}. \tag{2.12}$$

This equation shows that particle velocities $\mathbf{v}_k$ may be considered as linear functions of the generalized velocities $\dot{q}_j$ *considered as independent variables*, with proportionality coefficients

$$\frac{\partial \mathbf{v}_k}{\partial \dot{q}_j} = \frac{\partial \mathbf{r}_k}{\partial q_j}. \tag{2.13}$$

With the account of Eqs. (2.10), (2.11), and (2.13), Eq. (2.9) turns into

$$\frac{d}{dt}\sum_k m_k \mathbf{v}_k \cdot \frac{\partial \mathbf{v}_k}{\partial \dot{q}_j} - \sum_k m_k \mathbf{v}_k \frac{\partial \mathbf{v}_k}{\partial q_j} - \mathcal{F}_j = 0. \tag{2.14}$$

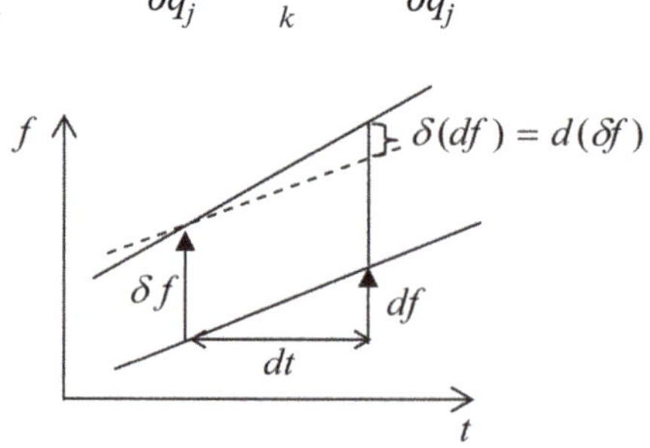

Figure 2.3. The variation of the differential (of any function f) is equal to the differential of its variation.

This result may be further simplified by making, for the total kinetic energy of the system,

$$T \equiv \sum_k \frac{m_k}{2} v_k^2 = \frac{1}{2} \sum_k m_k \mathbf{v}_k \cdot \mathbf{v}_k, \tag{2.15}$$

the same commitment as for $\mathbf{v}_k$, i.e. considering T a function of not only the generalized coordinates q_j and time t, but also of the generalized velocities $\dot{q}_i$—as variables independent of q_j and t. Then we may calculate the partial derivatives of T as

$$\frac{\partial T}{\partial q_j} = \sum_k m_k \mathbf{v}_k \cdot \frac{\partial \mathbf{v}_k}{\partial q_j}, \qquad \frac{\partial T}{\partial \dot{q}_j} = \sum_k m_k \mathbf{v}_k \cdot \frac{\partial \mathbf{v}_k}{\partial \dot{q}_j}, \tag{2.16}$$

and notice that they are exactly the two sums participating in Eq. (2.14). As a result, we obtain a system of J *Lagrange equations*[7],

$$\frac{d}{dt}\frac{\partial T}{\partial \dot{q}_j} - \frac{\partial T}{\partial q_j} - \mathcal{F}_j = 0, \quad \text{for } j = 1, 2, \ldots, J. \tag{2.17}$$

Their big advantage over the initial Newton's law equation (2.4) is that the Lagrange equations do not include the constraining forces $\mathbf{N}_k$ (and that there are only J of them—typically many fewer than $3N$).

This is as far as we can go for arbitrary forces. However, if all the forces may be expressed in a form similar to, but somewhat more general than Eq. (1.31), $\mathbf{F}_k = -\nabla_k U(\mathbf{r}_1, \mathbf{r}_2, \ldots, \mathbf{r}_N, t)$, where U is the effective potential energy of the system[8], and sign ∇_k denotes the spatial differentiation over coordinates of the kth particle, we may recast Eq. (2.8) into a simpler form:

$$\mathcal{F}_j \equiv \sum_k \mathbf{F}_k \cdot \frac{\partial \mathbf{r}_k}{\partial q_j} = -\sum_k \left(\frac{\partial U}{\partial x_k} \cdot \frac{\partial x_k}{\partial q_j} + \frac{\partial U}{\partial y_k} \cdot \frac{\partial y_k}{\partial q_j} + \frac{\partial U}{\partial z_i} \cdot \frac{\partial z_i}{\partial q_j} \right) \equiv -\frac{\partial U}{\partial q_j}. \tag{2.18}$$

Since we assume that U depends only on particle coordinates (and possibly time), but not velocities, $\partial U / \partial \dot{q}_j = 0$, with the substitution of Eq. (2.18), the Lagrange equation (2.17) may be represented in the so-called *canonical form*:

$$\frac{d}{dt}\frac{\partial L}{\partial \dot{q}_j} - \frac{\partial L}{\partial q_j} = 0, \tag{2.19a}$$

where L is called the *Lagrangian function* (or just the 'Lagrangian'), defined as

$$L \equiv T - U. \tag{2.19b}$$

[7] They were derived in 1788 by J-L Lagrange, who pioneered the whole field of analytical mechanics, not to mention his key contributions to number theory and celestial mechanics.

[8] Note that due to the possible time dependence of U, equation (2.17) does not mean that forces $\mathbf{F}_k$ have to be conservative—see the next section for more discussion. With this understanding, I will still use for the function U the convenient name of 'potential energy'.

(It is crucial to distinguish this function from the mechanical energy (1.26), $E = T + U$.)

Using the Lagrangian formalism in practice, the reader should always remember that, first, each system has only *one* Lagrange function L, but is described by $J \geqslant 1$ Lagrange equations of motion (with j taking values 1, 2,..., J), and second, that differentiating the functions T and L, we have to consider the generalized velocities as its independent arguments, ignoring the fact they are actually the time derivatives of the generalized coordinates.

Note also that according to Eq. (2.18), for a system under the effect of a fixed generalized external force $\mathscr{F}_j$, we have to use in all these relations not the internal potential energy U_{int} of the system, but its Gibbs energy $U_G \equiv U_{\text{int}} - \mathscr{F}_j q_j$—see the discussion in section 1.4.

2.2 Three simple examples

As the first, simplest example, consider a particle constrained to move along one axis (say, x):

$$T = \frac{m}{2}\dot{x}^2, \qquad U = U(x, t). \tag{2.20}$$

In this case, it is natural to consider x as the (only) generalized coordinate, and $\dot{x}$ as the generalized velocity, so that

$$L \equiv T - U = \frac{m}{2}\dot{x}^2 - U(x, t). \tag{2.21}$$

Considering $\dot{x}$ an independent variable, we obtain $\partial L/\partial \dot{x} = m\dot{x}$ and $\partial L/\partial x = -\partial U/\partial x$, so that Eq. (2.19) (the only Lagrange equation in this case of the single degree of freedom!) yields

$$\frac{d}{dt}(m\dot{x}) - \left(-\frac{\partial U}{\partial x}\right) = 0, \tag{2.22}$$

evidently the same result as the x-component of Newton's second law with $F_x = -\partial U/\partial x$. This is a good sanity check, but we see that the Lagrange formalism does not provide too much advantage in this particular case.

This advantage is, however, evident for our testbed problem—see figure 2.1. Indeed, taking the polar angle θ for the (only) generalized coordinate, we see that in this case the kinetic energy depends not only on the generalized velocity, but also on the generalized coordinate[9]:

$$\begin{aligned} T &= \frac{m}{2}R^2(\dot{\theta}^2 + \omega^2 \sin^2\theta), \qquad U = -mgz + \text{const} = -mgR\cos\theta + \text{const}, \\ L &\equiv T - U = \frac{m}{2}R^2(\dot{\theta}^2 + \omega^2 \sin^2\theta) + mgR\cos\theta + \text{const}. \end{aligned} \tag{2.23}$$

[9] This expression for $T \equiv (m/2)(\dot{x}^2 + \dot{y}^2 + \dot{z}^2)$ may be readily obtained either by the formal differentiation of Eq. (2.2) over time, or just by noticing that the velocity vector has two perpendicular components: one (of magnitude $R\dot{\theta}$) along the ring and another (of magnitude $\omega\rho = \omega R \sin\theta$) normal to the ring plane.

Here it is particularly important to remember that when substantiating the Lagrange equation, θ and $\dot{\theta}$ have to be treated as independent arguments of L, so that

$$\frac{\partial L}{\partial \dot{\theta}} = mR^2\dot{\theta}, \qquad \frac{\partial L}{\partial \theta} = mR^2\omega^2 \sin\theta \cos\theta - mgR \sin\theta, \tag{2.24}$$

giving us the following equation of motion:

$$\frac{d}{dt}(mR^2\dot{\theta}) - (mR^2\omega^2 \sin\theta \cos\theta - mgR \sin\theta) = 0. \tag{2.25}$$

As a sanity check, at $\omega = 0$, Eq. (2.25) is reduced to Eq. (1.18) of the usual pendulum:

$$\ddot{\theta} + \Omega^2 \sin\theta = 0, \quad \text{where } \Omega \equiv \left(\frac{g}{R}\right)^{1/2}. \tag{2.26}$$

We will explore the full dynamic equation (2.25) in more detail later, but please note how simple its derivation was in comparison to writing the 3D Newton law and then excluding the reaction force.

Next, although the Lagrangian formalism was derived from the Newton law for mechanical systems, the resulting equations (2.19) are applicable to other dynamic systems, in particular those for which the kinetic and potential energies may be readily expressed via some generalized coordinates. As the simplest example, consider the well-known connection (figure 2.4) of a capacitor with capacitance C to an inductive coil with self-inductance $\mathscr{L}$.[10] (Electrical engineers frequently call it the *LC tank circuit*.)

As the reader (hopefully) knows, at relatively low frequencies we may use the so-called lumped-circuit approximation, in which the total energy of the system is the sum of two components—the electric energy E_e localized inside the capacitor and the magnetic energy E_m localized inside the inductance coil:

$$E_e = \frac{Q^2}{2C}, \quad E_m = \frac{\mathscr{L} I^2}{2}. \tag{2.27}$$

Since the electric current I through the coil and the electric charge Q on the capacitor are connected by the charge continuity equation $dQ/dt = I$ (as is evident from figure 2.4), it is natural to declare the charge the generalized coordinate of the system, and the current, its generalized velocity. With this choice, the electrostatic energy $E_e(Q)$ should may be treated as the potential energy U of the system, and the magnetic energy $E_m(I)$, as its kinetic energy T. With this attribution, we obtain

$$\frac{\partial T}{\partial \dot{q}} \equiv \frac{\partial E_m}{\partial I} = \mathscr{L} I \equiv \mathscr{L} \dot{Q}, \qquad \frac{\partial T}{\partial q} \equiv \frac{\partial E_m}{\partial Q} = 0, \qquad \frac{\partial U}{\partial q} \equiv \frac{\partial E_e}{\partial Q} = \frac{Q}{C}, \tag{2.28}$$

[10] A fancy font it used here to avoid any chance of confusion between the inductance and the Lagrange function.

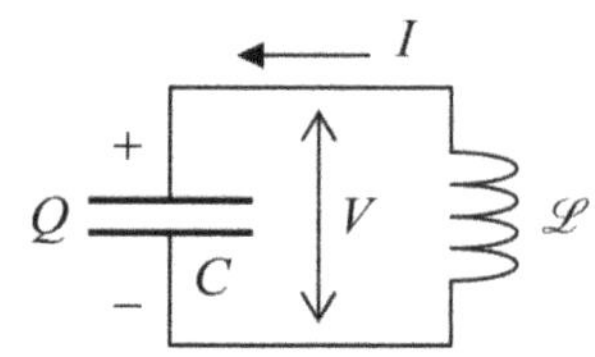

Figure 2.4. *LC* tank circuit.

so that the Lagrange equation (2.19) becomes

$$\frac{d}{dt}(\mathscr{L}\dot{Q}) - \left(-\frac{Q}{C}\right) = 0. \qquad (2.29)$$

Note, however, that the above choice of the generalized coordinate and velocity is not unique. Instead, one can use as the generalized coordinate the magnetic flux Φ through the inductive coil, related to the common voltage V across the circuit (figure 2.4) by Faraday's induction law $V = -d\Phi/dt$. With this choice, $(-V)$ becomes the generalized velocity, $E_m = \Phi^2/2\mathscr{L}$ should be understood as the *potential* energy, and $E_e = CV^2/2$ treated as the *kinetic* energy. It is straightforward to verify that for this choice, the resulting Lagrange equation of motion is equivalent to Eq. (2.29). If both parameters of the circuit, $\mathscr{L}$ and C, are constant in time, Eq. (2.29) is just the harmonic oscillator equation, and describes sinusoidal oscillations with frequency

$$\omega_0 = \frac{1}{(\mathscr{L}C)^{1/2}}. \qquad (2.30)$$

This is of course a well-known result, which may be derived in the more standard way—by equating the voltage drops across the capacitor ($V = Q/C$) and the inductor ($V = -\mathscr{L}dI/dt \equiv -\mathscr{L}d^2Q/dt^2$). However, the Lagrangian approach is much more convenient for more complex systems, for example, for the general description of the electromagnetic field and its interaction with charged particles[11].

2.3 Hamiltonian function and energy

The canonical form (2.19) of the Lagrange equation has been derived using Eq. (2.18), which is formally similar to Eq. (1.22) for a potential force. Does this mean that the system described by Eq. (2.19) always conserves energy? Not necessarily, because the 'potential energy', U, that participates in Eq. (2.18) may depend not only on the generalized coordinates, but on time as well. Let us start the analysis of this issue with the introduction of two new (and very important!) notions: the *generalized momentum* corresponding to each generalized coordinate q_j,

$$p_j \equiv \frac{\partial L}{\partial \dot{q}_j}, \qquad (2.31)$$

[11] See, e.g. *Part EM* section 9.8.

and the *Hamiltonian function*[12]

$$H \equiv \sum_j \frac{\partial L}{\partial \dot{q}_j}\dot{q}_j - L \equiv \sum_j p_j \dot{q}_j - L. \tag{2.32}$$

In order to see whether the Hamiltonian function is conserved during the motion, let us differentiate both sides of its definition (2.32) over time:

$$\frac{dH}{dt} = \sum_j \left[\frac{d}{dt}\left(\frac{\partial L}{\partial \dot{q}_j}\right)\dot{q}_j + \frac{\partial L}{\partial \dot{q}_j}\ddot{q}_j\right] - \frac{dL}{dt}. \tag{2.33}$$

If we want to make use of the Lagrange equation (2.19), the last derivative has to be calculated considering L as a function of independent arguments q_j, $\dot{q}_j$, and t, so that

$$\frac{dL}{dt} = \sum_j \left(\frac{\partial L}{\partial q_j}\dot{q}_j + \frac{\partial L}{\partial \dot{q}_j}\ddot{q}_j\right) + \frac{\partial L}{\partial t}, \tag{2.34}$$

where the last term is the derivative of L as an *explicit* function of time. We see that the last term in the square brackets of Eq. (2.33) immediately cancels with the last term in the parentheses of Eq. (2.34). Moreover, using the Lagrange equation (2.19) for the first term in the square brackets of Eq. (2.33), we see that it cancels with the first term in the parentheses of Eq. (2.34). Thus we arrive at a very simple and important result:

$$\frac{dH}{dt} = -\frac{\partial L}{\partial t}. \tag{2.35}$$

The most important corollary of this formula is that if the Lagrangian function does not depend on time *explicitly* ($\partial L/\partial t = 0$), the Hamiltonian function is an integral of motion:

$$H = \text{const}. \tag{2.36}$$

Let us see how it works, using the first two examples discussed in the previous section. For a 1D particle, the definition (2.31) of the generalized momentum yields

$$p_x \equiv \frac{\partial L}{\partial v} = mv, \tag{2.37}$$

[12] Named after W R Hamilton, who developed his approach to analytical mechanics in 1833 on the basis of Lagrangian mechanics. This function is sometimes called just the 'Hamiltonian', but it is advisable to use the full term 'Hamiltonian function' in classical mechanics in order to distinguish it from the *Hamiltonian operator* used in quantum mechanics. (Their relation will be discussed in section 10.1 below.)

so that it coincides with the usual momentum—or rather with its x-component. According to Eq. (2.32), the Hamiltonian function for this case (with just one degree of freedom) is

$$H \equiv p_x - L = m\dot{x}^2 - \left(\frac{m}{2}\dot{x}^2 - U\right) = \frac{m}{2}\dot{x}^2 + U, \tag{2.38}$$

i.e. it coincides with the particle's mechanical energy $E = T + U$. Since the Lagrangian does not depend on time explicitly, both H and E are conserved.

However, it is not always that simple! Indeed, let us return again to our testbed problem (figure 2.1). In this case, the generalized momentum corresponding to the generalized coordinate θ is

$$p_\theta \equiv \frac{\partial L}{\partial \dot{\theta}} = mR^2\dot{\theta}, \tag{2.39}$$

and Eq. (2.32) yields:

$$\begin{aligned} H &\equiv p_\theta\dot{\theta} - L = mR^2\dot{\theta}^2 - \left[\frac{m}{2}R^2(\dot{\theta}^2 + \omega^2 \sin^2\theta) + mgR\cos\theta\right] + \text{const} \\ &= \frac{m}{2}R^2(\dot{\theta}^2 - \omega^2 \sin^2\theta) - mgR\cos\theta + \text{const}. \end{aligned} \tag{2.40}$$

This means that (as soon as $\omega \neq 0$), the Hamiltonian function *differs* from the mechanical energy

$$E \equiv T + U = \frac{m}{2}R^2(\dot{\theta}^2 + \omega^2 \sin^2\theta) - mgR\cos\theta + \text{const}. \tag{2.41}$$

The difference, $E–H = mR^2\omega^2 \sin^2\theta$ (besides an inconsequential constant), may change at bead's motion along the ring, so that although H *is* an integral of motion (since $\partial L/\partial t = 0$), the energy is *not* conserved.

In this context, let us find out when these two functions, E and H, do coincide. In mathematics, there is a notion of a *homogeneous function* $f(x_1, x_2, \ldots)$ *of degree* λ, defined in the following way: for an arbitrary constant a,

$$f(ax_1, ax_2, \ldots) = a^\lambda f(x_1, x_2, \ldots). \tag{2.42}$$

Such functions obey the following *Euler theorem*[13]:

$$\sum_j \frac{\partial f}{\partial x_j} x_j = \lambda f, \tag{2.43}$$

that may be readily proven by differentiating both parts of Eq. (2.42) over a and then setting this parameter to the particular value $a = 1$. Now, consider

[13] This is just one of many theorems bearing the name of the mathematics genius L Euler (1707–83).

the case when the kinetic energy is a quadratic form of all generalized velocities $\dot{q}_j$:

$$T = \sum_{j,j'} t_{jj'}(q_1, q_2, \ldots, t)\dot{q}_j\dot{q}_{j'}, \tag{2.44}$$

with no other terms. It is evident that such T satisfies the definition of a homogeneous function of the velocities with $\lambda = 2$,[14] so that the Euler theorem (2.43) gives

$$\sum_j \frac{\partial T}{\partial \dot{q}_j}\dot{q}_j = 2T. \tag{2.45}$$

But since U is independent of the generalized velocities, $\partial L/\partial \dot{q}_j = \partial T/\partial \dot{q}_j$, and the left-hand side of Eq. (2.45) is exactly the first term in the definition (2.32) of the Hamiltonian function, so that in this case

$$H = 2T - L = 2T - (T - U) = T + U = E. \tag{2.46}$$

So, for the kinetic energy of the type (2.44), for example a free particle with the kinetic energy considered as a function of its Cartesian velocities,

$$T = \frac{m}{2}\left(v_x^2 + v_y^2 + v_z^2\right), \tag{2.47}$$

the notions of the Hamiltonian function and mechanical energy are identical. Indeed, some textbooks, very regretfully, do not distinguish these notions at all! However, as we have seen from our bead-on-the-rotating-ring example, this is not always true. For that problem, the kinetic energy, in addition to the term proportional to $\dot{\theta}^2$, has another, velocity-independent term—see the first of Eqs. (2.23)—and hence is *not* a quadratic-homogeneous function of the angular velocity.

Thus, Eq. (2.36) expresses a new conservation law, generally different from that of the mechanical energy conservation.

2.4 Other conservation laws

Looking at the Lagrange equation (2.19), we immediately see that if $L \equiv T-U$ is independent of some generalized coordinate q_j, $\partial L/\partial q_j = 0$,[15] then the corresponding generalized momentum is an integral of motion[16]:

$$p_j \equiv \frac{\partial L}{\partial \dot{q}_j} = \text{const}. \tag{2.48}$$

[14] Such functions are called *quadratic-homogeneous*.

[15] Such coordinates are frequently called *cyclic*, because in some cases (such as in the second example considered below) they represent periodic coordinates such as angles. However, this terminology is misleading, because some 'cyclic' coordinates (e.g. x in our first example) have nothing to do with rotation.

[16] This fact may be considered a particular case of a more general mathematical statement called the *Noether theorem* (named after its author, A E Nöther, sometimes called the 'greatest woman mathematician who ever lived'). For a discussion see, e.g. [5] section 13.7.

For example, for a 1D particle with the Lagrangian function (2.21), the momentum p_x is conserved if the potential energy is constant (the x-component of force is zero)—of course. As a less obvious example, let us consider a 2D motion of a particle in the field of central forces. If we use polar coordinates r and φ in the role of the generalized coordinates, the Lagrangian function[17].

$$L \equiv T - U = \frac{m}{2}(\dot{r}^2 + r^2\dot{\varphi}^2) - U(r), \qquad (2.49)$$

is independent of φ and hence the corresponding generalized momentum,

$$p_\varphi \equiv \frac{\partial L}{\partial \dot{\varphi}} = mr^2\dot{\varphi}, \qquad (2.50)$$

is conserved. This is just a particular (2D) case of angular-momentum conservation—see Eq. (1.24). Indeed, for the 2D motion within the $[x, y]$-plane, the angular-momentum vector,

$$\mathbf{L} \equiv \mathbf{r} \times \mathbf{p} = \begin{vmatrix} \mathbf{n}_x & \mathbf{n}_y & \mathbf{n}_z \\ x & y & z \\ m\dot{x} & m\dot{y} & m\dot{z} \end{vmatrix}, \qquad (2.51)$$

has only one component different from zero, namely the one normal to the motion plane:

$$L_z = x(m\dot{y}) - y(m\dot{x}). \qquad (2.52)$$

Differentiating the well-known relations between the polar and Cartesian coordinates,

$$x = r\cos\varphi, \quad y = r\sin\varphi, \qquad (2.53)$$

over time, and plugging the result into (2.52), we see that

$$L_z = mr^2\dot{\varphi} \equiv p_\varphi. \qquad (2.54)$$

Thus the Lagrangian formalism provides a powerful way of searching for non-evident integrals of motion. On the other hand, if such a conserved quantity is evident or known *a priori*, it is helpful for the selection of the most appropriate generalized coordinates, giving the simplest Lagrange equations. For example, in the last problem, if we knew in advance that p_φ had to be conserved, this could provide a motivation for using angle φ as one of the generalized coordinates.

2.5 Problems

In each of problems 2.1–2.11, for the given system:

(i) introduce a set of convenient generalized coordinate(s) q_j of the system;
(ii) write down Lagrangian L as a function of q_j, $\dot{q}_j$, and (if appropriate) time;

[17] Note that here $\dot{r}^2$ is just the square of the scalar derivative $\dot{r}$, rather than the square of vector $\dot{\mathbf{r}} = \mathbf{v}$.

(iii) write down the Lagrangian equation(s) of motion;
(iv) calculate the Hamiltonian function H and find out whether it is conserved;
(v) calculate energy E; is $E = H$?; is the energy conserved?;
(vi) are there any other evident integrals of motion?

Problem 2.1. A double pendulum—see the figure below. Consider only the motion in the vertical plane containing the suspension point.

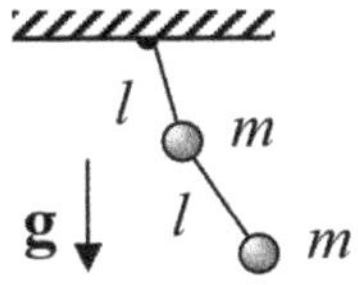

Problem 2.2. A stretchable pendulum (i.e. a mass hung on an elastic cord that exerts force $F = -\kappa(l - l_0)$, where κ and l_0 are positive constants), confined to a vertical plane.

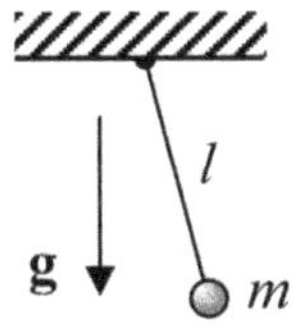

Problem 2.3. A fixed-length pendulum hanging from a horizontal support whose motion law $x_0(t)$ is fixed. (No vertical plane constraint here.)

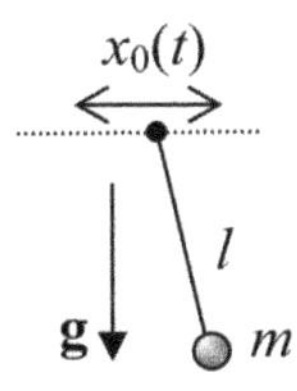

Problem 2.4. A pendulum of mass m, hung on another point mass m', which may slide, without friction, along a straight horizontal rail—see the figure below. Its motion is confined to the vertical plane that contains the rail.

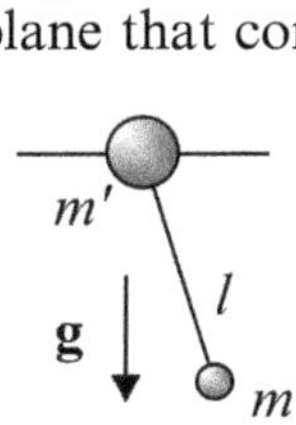

Problem 2.5. A point-mass pendulum of length l is attached to the rim of a disk of radius R, which is rotated in a vertical plane with a constant angular velocity ω—see the figure below. (Consider only the motion within the disk's plane.)

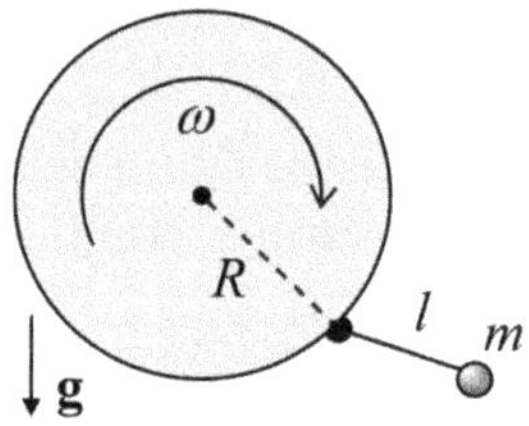

Problem 2.6. A bead of mass m, sliding without friction along a light string stretched by a fixed force $\mathscr{T}$, between two horizontally displaced points—see the figure to the right. Here, in contrast to the similar problem 1.10, the string tension T may be comparable to the bead's weight mg, and the motion is *not* restricted to the vertical plane.

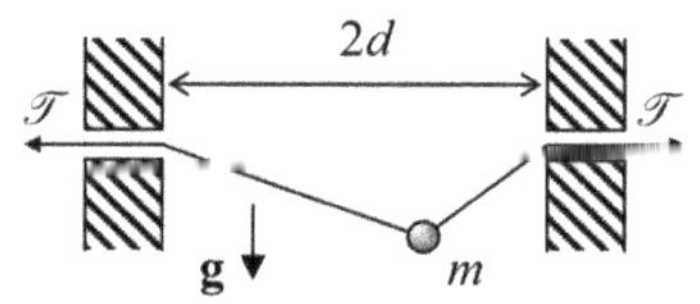

Problem 2.7. A bead of mass m, sliding without friction along a light string of a fixed length $2l$, which is hung between two points, horizontally displaced by distance $2d < 2l$—see the figure below. As in the previous problem, the motion is not restricted to the vertical plane.

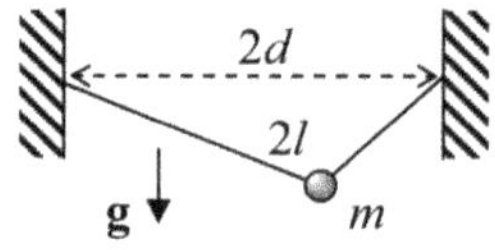

Problem 2.8. A block of mass m that can slide, without friction, along the inclined plane surface of a heavy wedge with mass m'. The wedge is free to move, also without friction, along a horizontal surface—see the figure below. (Both motions are within the vertical plane containing the steepest slope line.)

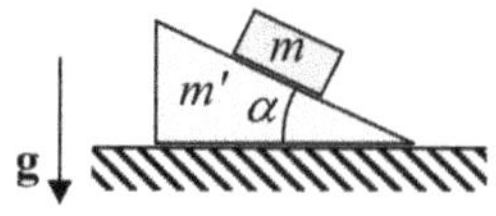

Problem 2.9. The two-pendulum system that was the subject of problem 1.8—see the figure below.

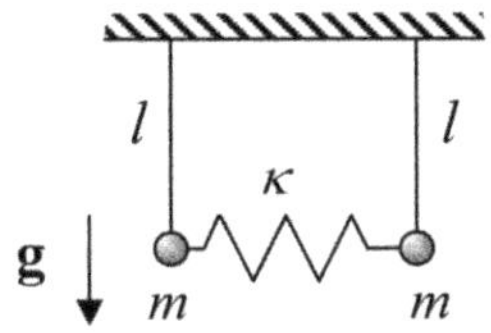

Problem 2.10. A system of two similar, inductively coupled *LC* circuits—see the figure below.

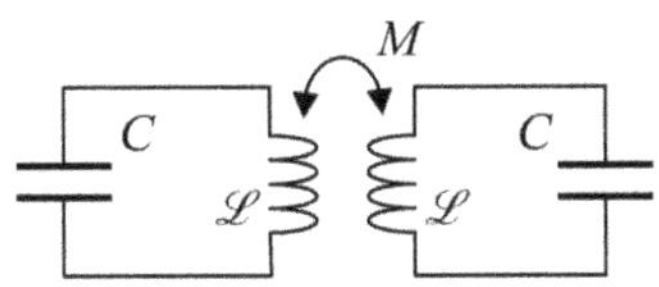

Problem 2.11.* A small Josephson junction—a system of two superconductors (S), weakly coupled by Cooper-pair tunneling through a thin insulating layer (I) that separates them—see the figure below.

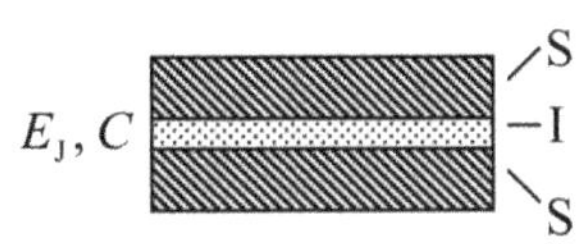

Hints:

(i) At not very high frequencies (whose quantum $\hbar\omega$ is lower than the binding energy 2Δ of the Cooper pairs), the Josephson effect in a sufficiently small junction may be described by the following coupling energy:

$$U(\varphi) = -E_J \cos\varphi + \text{const},$$

where constant E_J describes the coupling strength, and the variable φ (called the *Josephson-phase difference*) is connected to the voltage V across the junction by the famous *frequency-to-voltage relation*

$$\frac{d\varphi}{dt} = \frac{2e}{\hbar}V,$$

where $e \approx 1.6 \times 10^{-19}$ C is the fundamental electric charge and $\hbar \approx 1.054 \times 10^{-34}$ J·s is the Planck constant[18].

(ii) The junction (as any system of two close conductors) has a substantial electric capacitance C.

[18] More discussion of the Josephson effect and the physical sense of the variable φ may be found, for example, in *Part EM* section 6.4 and *Part QM* sections 2.3 and 2.8 of this lecture note series, but the current problem may be solved without this additional information.

References

[1] Gelfand I and Fomin S 2000 *Calculus of Variations* (New York: Dover)
[2] Elsgolc L 2007 *Calculus of Variations* (New York: Dover)
[3] Arfken G *et al* 2012 *Mathematical Methods for Physicists* 7th edn (New York: Academic)
[4] Weinstock R 2007 *Calculus of Variations* (New York: Dover)
[5] Goldstein H *et al* 2002 *Classical Mechanics* 3rd edn (Reading, MA: Addison-Wesley)

IOP Publishing

Chapter 3

A few simple problems

The objective of this chapter is the solution of a few simple but very important problems of particle dynamics, which may be reduced to 1D motion. They notably include the famous 'planetary' problem of two particles interacting via a spherically symmetric potential, and the classical particle scattering problem. In the process of solution, several methods that will be essential for the analysis of more complex systems are also discussed.

3.1 One-dimensional and 1D-reducible systems

If a particle is confined to motion along a straight line (say, axis x), its position, of course, is completely defined by this coordinate. In this case, as we already know, the particle's Lagrangian is given by Eq. (2.21):

$$L = T(\dot{x}) - U(x, t), \quad T(\dot{x}) = \frac{m}{2}\dot{x}^2, \tag{3.1}$$

so that the Lagrange equation of motion, given by Eq. (2.22),

$$m\ddot{x} = -\frac{\partial U(x, t)}{\partial x} \tag{3.2}$$

is just the x-component of Newton's second law.

It is convenient to discuss the dynamics of such *really 1D systems* as a part of a more general class of *effectively 1D systems*, whose position, due to holonomic constraints and/or conservation laws, is also fully determined by one generalized coordinate q, and whose Lagrangians may be presented in a form similar to Eq. (3.1):

$$L = T_{\text{ef}}(\dot{q}) - U_{\text{ef}}(q, t), \quad T_{\text{ef}} = \frac{m_{\text{ef}}}{2}\dot{q}^2, \tag{3.3}$$

doi:10.1088/978-0-7503-1398-8ch3

where m_{ef} is some constant which may be considered as the *effective mass* of the system, and the function U_{ef} its *effective potential energy*. In this case the Lagrange equation (2.19), describing the system's dynamics, has a form similar to Eq. (3.2):

$$m_{ef}\ddot{q} = -\frac{\partial U_{ef}(q, t)}{\partial q}. \tag{3.4}$$

As an example, let us return again to our testbed system shown in figure 2.1. We have already seen that for that system, having one degree of freedom, the genuine kinetic energy T expressed by the first of Eqs. (2.23) is *not* a quadratically homogeneous function of the generalized velocity. However, the system's Lagrangian function Eq. (2.23) still may be represented in the form (3.3),

$$L = \frac{m}{2}R^2\dot{\theta}^2 + \frac{m}{2}R^2\omega^2 \sin^2\theta + mgR\cos\theta + \text{const} = T_{ef} - U_{ef}, \tag{3.5}$$

if we take

$$T_{ef} \equiv \frac{m}{2}R^2\dot{\theta}^2, \quad U_{ef} \equiv -\frac{m}{2}R^2\omega^2 \sin^2\theta - mgR\cos\theta + \text{const}. \tag{3.6}$$

In this new partitioning of the function L, which is legitimate because U_{ef} depends only on the generalized coordinate θ, but not on the corresponding generalized velocity, T_{ef} includes only a part of the full kinetic energy T of the bead, while U_{ef} includes not only its real potential energy U in the gravity field, but also an additional term related to ring rotation. (As we will see in section 4.6, this term may be interpreted as the effective potential energy due to the inertial centrifugal 'force'.)

Returning to the general case of effectively 1D systems with Lagrangians of the type (3.3), let us calculate their Hamiltonian function, using its definition (2.32):

$$H = \frac{\partial L}{\partial \dot{q}}\dot{q} - L = m_{ef}\dot{q}^2 - (T_{ef} - U_{ef}) = T_{ef} + U_{ef}. \tag{3.7}$$

So, H is expressed via T_{ef} and U_{ef} exactly as the mechanical energy E is expressed via genuine T and U.

3.2 Equilibrium and stability

Autonomous systems are defined as the dynamic systems whose equations of motion do not depend on time explicitly. For the 1D (and the effectively 1D) systems obeying Eq. (3.4), this means that their function U_{ef}, and hence the Lagrangian function (3.5), should not depend on time explicitly. According to Eq. (2.35), in such systems the Hamiltonian function (3.7), i.e. the sum $T_{ef} + U_{ef}$, is an integral of motion. However, be careful! This may not be true for the system's mechanical energy E; for example, as we already know from section 2.2, for our testbed problem with the generalized coordinate $q = \theta$ (figure 2.1), $E \neq H$.

According to Eq. (3.4), an autonomous system, at appropriate initial conditions, may stay in equilibrium at one or several *stationary* (alternatively called *fixed*) *points*

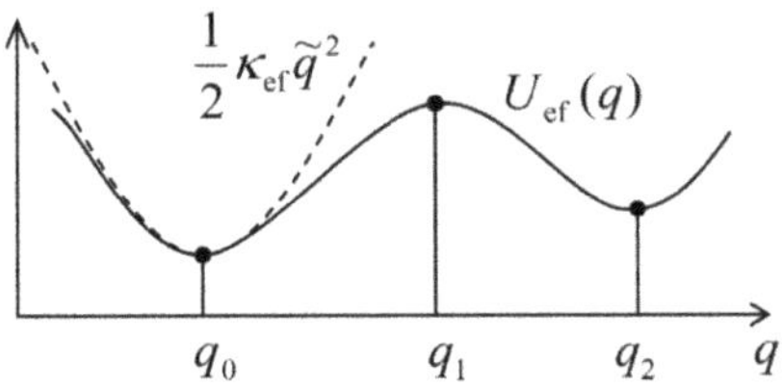

Figure 3.1. The effective potential energy profile near stable (q_0, q_2) and unstable (q_1) fixed points, and its quadratic approximation (3.10) near point q_0 (schematically).

q_n, corresponding to either the minimum or maximum of the effective potential energy (see figure 3.1):

$$\frac{dU_{ef}}{dq}(q_n) = 0. \tag{3.8}$$

In order to explore the *stability* of such fixed points, let us analyze the dynamics of small deviations

$$\tilde{q}(t) \equiv q(t) - q_n \tag{3.9}$$

from the equilibrium. For that, let us expand the function U_{ef}(q) in the Taylor series at a fixed point:

$$U_{ef}(q) = U_{ef}(q_n) + \frac{dU_{ef}}{dq}(q_n)\,\tilde{q} + \frac{1}{2}\frac{d^2U_{ef}}{dq^2}(q_n)\,\tilde{q}^2 + \dots . \tag{3.10}$$

The first term on the right-hand side, $U_{ef}(q_n)$, is arbitrary and does not affect motion. The next term, linear in deviation $\tilde{q}$, is equal to zero—see the fixed-point definition (3.8). Hence the fixed-point stability is determined by the next term, quadratic in $\tilde{q}$, more exactly by its coefficient

$$\kappa_{ef} \equiv \frac{d^2U_{ef}}{dq^2}(q_n), \tag{3.11}$$

which plays the role of the effective spring constant. Indeed, neglecting the higher terms of the Taylor expansion (3.10)[1], we see that Eq. (3.4) takes the familiar form:

$$m_{ef}\ddot{\tilde{q}} + \kappa_{ef}\tilde{q} = 0. \tag{3.12}$$

I am confident that the reader of these notes knows everything about this equation, but since we will soon run into similar but more complex equations, let us review the formal procedure of its solution. From the mathematical standpoint, Eq. (3.12) is an ordinary, linear differential equation of the second order, with

[1] Those terms may be important only in the very special case when κ_{ef} is exactly zero, i.e. when a fixed point is an *inflection point* of the function $U_{ef}(q)$.

constant coefficients. The theory of such equations tells us that its general solution (for any initial conditions) may be presented as

$$\tilde{q}(t) = c_+ e^{\lambda_+ t} + c_- e^{\lambda_- t}, \tag{3.13}$$

where the constants $c_\pm$ are determined by initial conditions, while the so-called *characteristic exponents* $\lambda_\pm$ are completely defined by the equation itself. In order to calculate these exponents, it is sufficient to plug just one partial solution, $e^{\lambda t}$, into the equation. In our simple case (3.12), this yields the following *characteristic equation*:

$$m_{\text{ef}}\lambda^2 + \kappa_{\text{ef}} = 0. \tag{3.14}$$

If the ratio $\kappa_{\text{ef}}/m_{\text{ef}}$ is positive, i.e. the fixed point corresponds to the minimum of potential energy (e.g. points q_0 and q_2 in figure 3.1), the characteristic equation yields

$$\lambda_\pm = \pm i\omega_0, \qquad \text{with } \omega_0 \equiv \left(\frac{\kappa_{\text{ef}}}{m_{\text{ef}}}\right)^{1/2}, \tag{3.15}$$

(where i is the imaginary unity, $i^2 = -1$), so that Eq. (3.13) describes sinusoidal oscillations of the system[2],

$$\tilde{q}(t) = c_+ e^{+i\omega_0 t} + c_- e^{-i\omega_0 t} \equiv c_c \cos\omega_0 t + c_s \sin\omega_0 t, \tag{3.16}$$

with frequency ω_0, about the fixed point which is thereby *stable*[3]. On the other hand, at the potential energy maximum ($\kappa_{\text{ef}} < 0$, e.g. at point q_1 in figure 3.1), we obtain

$$\lambda_\pm = \pm\lambda, \quad \lambda \equiv \left(\frac{|\kappa_{\text{ef}}|}{m_{\text{ef}}}\right)^{1/2}, \quad \tilde{q}(t) = c_+ e^{+\lambda t} + c_- e^{-\lambda t}. \tag{3.17}$$

Since the solution has an exponentially growing part[4], the fixed point is *unstable*.

Note that the *quadratic* expansion of function $U_{\text{ef}}(q)$, given by the truncation of Eq. (3.10) to the three displayed terms, is equivalent to a *linear* Taylor expansion of the effective force:

$$F_{\text{ef}} \equiv -\frac{dU_{\text{ef}}}{dq}\bigg|_{q=q_n} \approx -\kappa_{\text{ef}}\tilde{q}, \tag{3.18}$$

immediately resulting in the linear equation (3.12). Hence, in order to analyze the stability of a fixed point q_n, it is sufficient to *linearize* the equation of motion in small

[2] The reader should not be scared of the first form of Eq. (3.16), i.e. of the representation of a real variable (the deviation from equilibrium) via a sum of two complex functions. Indeed, any real initial conditions give $c_-^* = c_+$, so that the sum is real for any t. An even simpler way to deal with such complex representations of real functions will be discussed in the beginning of the next chapter, and then used throughout the series.

[3] This particular type of stability, when the deviation from the equilibrium oscillates with a constant amplitude A, neither growing nor decreasing in time, is sometimes called the *orbital* (or 'neutral', or 'indifferent') *stability*.

[4] Mathematically, the growing part vanishes at some special (exact) initial conditions which give $c_+ = 0$. However, the futility of this argument for real physical systems should be obvious for anybody who had ever tried to balance a pencil on its sharp point.

deviations from the point, and study possible solutions of the resulting linear equation. This linearization procedure is typically simpler to carry out than the quadratic expansion (3.10).

As an example, let us return to our testbed problem (figure 2.1) whose function $U_{\rm ef}$ we already know—see the second of Eqs. (3.6). With it, the equation of motion (3.4) becomes

$$mR^2\ddot{\theta} = -\frac{dU_{\rm ef}}{d\theta} = mR^2(\omega^2 \cos\theta - \Omega^2)\sin\theta, \quad \text{i.e. } \ddot{\theta} = (\omega^2\cos\theta - \Omega^2)\sin\theta, \tag{3.19}$$

where $\Omega \equiv (g/R)^{1/2}$ is the frequency of small oscillations of the system at $\omega = 0$—see Eq. (2.26)[5]. From the requirement (3.8), we see that on any 2π long segment of angle θ,[6] the system may have four fixed points; for example, on the closed segment $[-\pi, +\pi]$ these points are

$$\theta_0 = 0, \quad \theta_1 = \pm\pi, \quad \theta_{2,3} = \pm\cos^{-1}\frac{\Omega^2}{\omega^2}. \tag{3.20}$$

The last two fixed points, corresponding to the bead shifted to either side of the rotating ring, exist only if the angular velocity ω of the rotation exceeds Ω. (In the limit of very fast rotation, $\omega \gg \Omega$, the last of Eqs. (3.20) yields $\theta_{2,3} \to \pm\pi/2$, i.e. the stationary positions approach the horizontal diameter of the ring—in accordance with our physical intuition.)

In order to analyze the fixed-point stability, we may again use Eq. (3.9), in the form $\theta = \theta_n + \tilde{\theta}$, plug it into Eq. (3.19), and Taylor-expand the trigonometric functions of θ up to the first term in $\tilde{\theta}$:

$$\ddot{\tilde{\theta}} = [\omega^2(\cos\theta_n - \sin\theta_n\,\tilde{\theta}) - \Omega^2]\,(\sin\theta_n + \cos\theta_n\,\tilde{\theta}). \tag{3.21}$$

Generally, this equation may be linearized further by purging its right-hand side of the term proportional to $\tilde{\theta}^2$; however in this simple case, Eq. (3.21) is already convenient for analysis. In particular, for the fixed point $\theta_0 = 0$ (corresponding to the bead position at the bottom of the ring), we have $\cos\theta_0 = 1$ and $\sin\theta_0 = 0$, so that Eq. (3.21) is reduced to a linear differential equation

$$\ddot{\tilde{\theta}} = (\omega^2 - \Omega^2)\,\tilde{\theta}, \tag{3.22}$$

whose characteristic equation is similar to Eq. (3.14) and yields

$$\lambda^2 = \omega^2 - \Omega^2, \quad \text{for } \theta \approx \theta_0. \tag{3.23a}$$

This result shows that if $\omega < \Omega$, when both roots λ are imaginary, this fixed point is orbitally stable. However, the roots become real, $\lambda_\pm = (\omega^2 - \Omega^2)^{1/2}$, with one of them

[5] Note that Eq. (3.19) coincides with Eq. (2.25). This is a good sanity check illustrating that the procedure (3.5)–(3.6) of moving of a term from the potential to kinetic energy within the Lagrangian function is indeed legitimate.

[6] For this particular problem, the values of θ that differ by a multiple of 2π are physically equivalent.

positive, so that the fixed point becomes unstable beyond this threshold, i.e. as soon as fixed points $\theta_{2,3}$ exist. Absolutely similar calculations for other fixed points yield

$$\lambda^2 = \Omega^2 + \omega^2 > 0, \quad \text{for } \theta \approx \theta_1, \tag{3.23b}$$

$$\lambda^2 = \Omega^2 - \omega^2, \quad \text{for } \theta \approx \theta_{2,3}. \tag{3.23c}$$

These results show that the fixed point θ_1 (the bead on the top of the ring) is always unstable—just as we could foresee, while the side fixed points $\theta_{2,3}$ are orbitally stable as soon as they exist (at $\omega > \Omega$).

Thus, our fixed-point analysis may be summarized in a simple way: an increase of the ring rotation speed ω beyond a certain threshold value, equal to Ω (2.26), causes the bead to move to one of the ring sides, oscillating about one of the fixed points $\theta_{2,3}$. Together with the rotation about the vertical axis, this motion yields quite a complex (generally, open) spatial trajectory as observed from a lab frame, so it is fascinating that we could analyze it qualitatively in such a simple way.

Later in this course we will repeatedly use the linearization of the equations of motion for the analysis of the stability of more complex systems, including those with energy dissipation.

3.3 Hamiltonian 1D systems

The autonomous systems that are described by time-independent Lagrangians are frequently called *Hamiltonian,* because their Hamiltonian function H (again, not necessarily equal to the genuine mechanical energy E!) is conserved. In our current 1D case, described by Eq. (3.3),

$$H = \frac{m_{\text{ef}}}{2}\dot{q}^2 + U_{\text{ef}}(q) = \text{const}. \tag{3.24}$$

From the mathematical standpoint, this conservation law is just the first integral of motion. Solving Eq. (3.24) for $\dot{q}$, we obtain the first-order differential equation,

$$\frac{dq}{dt} = \pm\left\{\frac{2}{m_{\text{ef}}}[H - U_{\text{ef}}(q)]\right\}^{1/2}, \qquad \text{i.e. } \pm\left(\frac{m_{\text{ef}}}{2}\right)^{1/2}\frac{dq}{[H - U_{\text{ef}}(q)]^{1/2}} = dt, \tag{3.25}$$

which may be readily integrated:

$$\pm\left(\frac{m_{\text{ef}}}{2}\right)^{1/2}\int_{q(t_0)}^{q(t)}\frac{dq'}{[H - U_{\text{ef}}(q')]^{1/2}} = t - t_0. \tag{3.26}$$

Since the constant H (as well as the proper sign before the integral—see below) is fixed by the initial conditions, Eq. (3.26) gives the reciprocal form, $t = t(q)$, of the desired law of system motion, $q(t)$. Of course, for any particular problem the integral in Eq. (3.26) still has to be worked out, either analytically or numerically, but even the latter procedure is typically much easier than the numerical integration of the initial, second-order differential equation of motion, because with addition of many

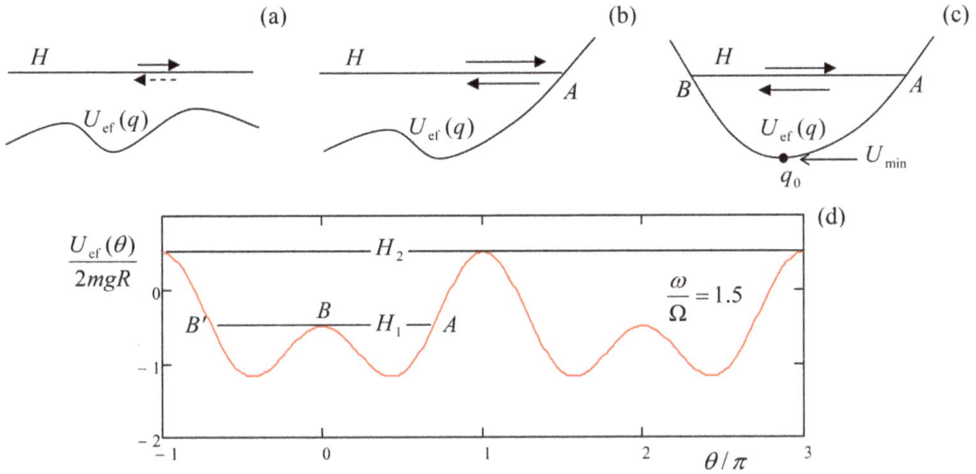

Figure 3.2. Graphical representations of Eq. (3.25) for three different cases: (a) an unbound motion, with the velocity sign conserved, (b) a reflection from a 'classical turning point', accompanied with the velocity sign change, and (c) bound, periodic motion between two turning points (schematically). (d) The effective potential energy Eq. (3.6) of the bead on the rotating ring (figure 2.1) for a particular case with $\omega > \Omega$.

values (to which the numerical integration is reduced[7]) the rounding errors are effectively averaged out.

Moreover, Eq. (3.25) also allows a general classification of 1D system motion. Indeed:

(i) If $H > U_{ef}(q)$ in the whole range of interest, the effective kinetic energy T_{ef} (3.3) is always positive. Hence the derivative dq/dt cannot change its sign, so that the effective velocity retains the sign it had initially. This is an unbound motion in one direction—see figure 3.2a.

(ii) Now let the particle approach a *classical turning point* A where $H = U_{ef}(x)$—see figure 3.2b[8]. According to (3.25), at that point the particle velocity vanishes, while its acceleration, according to Eq. (3.4), is still finite. Evidently, this corresponds to the particle reflection from this *potential wall*, with the change of the velocity sign.

(iii) If, after the reflection from point A, the particle runs into another classical turning point B figure 3.2c, the reflection process is repeated again and again, so that the particle is bound to a periodic motion between two turning points.

The last case of *periodic oscillations* presents a large conceptual and practical interest, and the whole next chapter will be devoted to a detailed analysis of this phenomenon and numerous associated effects. Here I will only note that for an

[7] See, e.g. Eqs. (A.26) and (A.27).

[8] This terminology comes from quantum mechanics which shows that actually a particle (or rather its wavefunction) can, to a certain extent, penetrate into the 'classically forbidden range' where $H < U_{ef}(x)$.

autonomous Hamiltonian system, described by Eq. (3.4), relation (3.26) immediately enables us to calculate the oscillation period:

$$\mathcal{T} = 2\left(\frac{m_{\rm ef}}{2}\right)^{1/2} \int_B^A \frac{dq}{[H - U_{\rm ef}(q)]^{1/2}}, \tag{3.27}$$

where the additional upfront factor 2 accounts for two time intervals: of the motion from B to A and back figure 3.2c. Indeed, according to Eq. (3.25), in each classically allowed point q the velocity magnitude is the same, so that these time intervals are equal to each other.

(Note that the dependence of points A and B on the 'energy' H is not necessarily continuous. For example, for our testbed problem (whose effective potential energy is plotted in figure 3.2d for a particular value of $\omega > \Omega$) a gradual increase of H leads to a sudden jump, at $H = H_1$, of point B to position B', corresponding to a sudden switch from oscillations about one fixed point $\theta_{2,3}$ to oscillations about two adjacent fixed points—before the beginning of a persistent rotation along the ring at $H > H_2$.)

Now let us consider a particular, but very important, limit of Eq. (3.27). As figure 3.2c shows, if H is reduced to approach $U_{\min}$, the periodic oscillations take place at the very bottom of the 'potential well', about a stable fixed point q_0. Hence, if the potential energy profile is smooth enough, we may limit the Taylor expansion (3.10) to the displayed quadratic term. Plugging it into Eq. (3.27), and using the mirror symmetry of this particular problem about the fixed point q_0, we obtain:

$$\mathcal{T} = 4\left(\frac{m_{\rm ef}}{2}\right)^{1/2} \int_0^A \frac{d\tilde{q}}{\left[H - \left(U_{\min} + \kappa_{\rm ef}\tilde{q}^2/2\right)\right]^{1/2}} = \frac{4}{\omega_0} I,$$
$$\text{with } I \equiv \int_0^1 \frac{d\xi}{(1 - \xi^2)^{1/2}}, \tag{3.28}$$

where $\xi \equiv \tilde{q}/A$, with $A \equiv (2/\kappa_{\rm ef})^{1/2}[H - U_{\min}]^{1/2}$ being the classical turning point, i.e. the oscillation amplitude, and ω_0 the frequency given by Eq. (3.15). Taking into account that the elementary integral I in that equation equals $\pi/2$,[9] we finally obtain

$$\mathcal{T} = \frac{2\pi}{\omega_0}, \tag{3.29}$$

as it should be for harmonic oscillations (3.16). Note that the oscillation period does not depend on the oscillation amplitude A, i.e. on the difference $(H - U_{\min})$, while it is sufficiently small.

3.4 Planetary problems

Leaving a more detailed study of oscillations for the next chapter, let us now discuss the so-called *planetary systems*[10] whose description, somewhat surprisingly, may be

[9] Indeed, introducing a new variable ζ by relation $\xi \equiv \sin\zeta$, we obtain $d\xi = \cos\zeta\, d\zeta = (1-\xi^2)^{1/2}\, d\zeta$, so that the function under the integral is just $d\zeta$.

[10] This name is very conditional, because this group of problems includes, for example, charged particle scattering (see section 3.5).

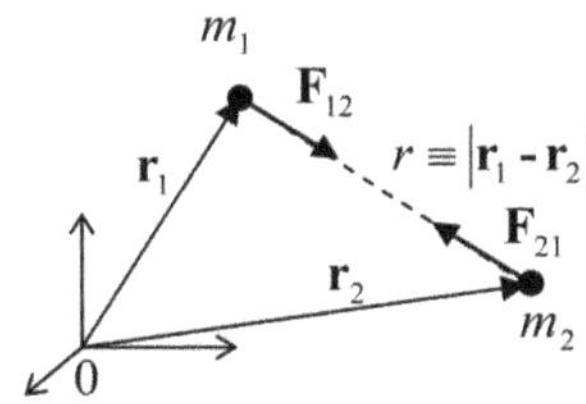

Figure 3.3. Vectors in the 'planetary' problem.

also reduced to an effectively 1D problem. Consider two particles that interact via a conservative, central force $\mathbf{F}_{21} = -\mathbf{F}_{12} = \mathbf{n}_r F(r)$, where r and $\mathbf{n}_r$ are, respectively, the magnitude and direction of the *distance vector* $\mathbf{r} \equiv \mathbf{r}_1 - \mathbf{r}_2$ connecting the two particles (figure 3.3).

Generally, two particles moving without constrains in 3D space have 3 + 3 = 6 degrees of freedom, which may described, e.g. by their Cartesian coordinates $\{x_1, y_1, z_1, x_2, y_2, z_2\}$. However, for this particular form of interaction, the following series of tricks allows the number of essential degrees of freedom to be reduced to just one.

First, the central, conservative force of particle interaction may be described by time-independent potential energy $U(r)$, such that $F(r) = -\partial U(r)/\partial r$[11]. Hence the Lagrangian of the system is

$$L \equiv T - U(r) = \frac{m_1}{2}\dot{\mathbf{r}}_1^2 + \frac{m_2}{2}\dot{\mathbf{r}}_2^2 - U(r). \tag{3.30}$$

Let us perform the transfer from the initial six scalar coordinates of the particles to six generalized coordinates: three Cartesian components of the distance vector

$$\mathbf{r} \equiv \mathbf{r}_1 - \mathbf{r}_2, \tag{3.31}$$

and three components of the following vector:

$$\mathbf{R} \equiv \frac{m_1\mathbf{r}_1 + m_2\mathbf{r}_2}{M}, \quad \text{with } M \equiv m_1 + m_2, \tag{3.32}$$

which defines the position of the *center-of-mass* of the system, with the total mass M. Solving the system of two linear equations (3.31) and (3.32) for $\mathbf{r}_1$ and $\mathbf{r}_2$, we obtain

$$\mathbf{r}_1 = \mathbf{R} + \frac{m_2}{M}\mathbf{r}, \quad \mathbf{r}_2 = \mathbf{R} - \frac{m_1}{M}\mathbf{r}. \tag{3.33}$$

Plugging these relations into Eq. (3.30), we see that it is reduced it to

$$L = \frac{M}{2}\dot{\mathbf{R}}^2 + \frac{m}{2}\dot{\mathbf{r}}^2 - U(r), \tag{3.34}$$

[11] See, e.g. Eq. (A.66) with $\partial/\partial\theta = \partial/\partial\varphi = 0$.

where m is the so-called *reduced mass*:

$$m \equiv \frac{m_1 m_2}{M}, \quad \text{so that } \frac{1}{m} \equiv \frac{1}{m_1} + \frac{1}{m_2}. \tag{3.35}$$

Note that according to Eq. (3.35), the reduced mass is lower than that of the lightest component of the two-body system. If one of $m_{1,2}$ is much less that its counterpart (as it is in most star–planet or planet–satellite systems), then with a good precision $m \approx \min\,[m_1, m_2]$.

Since the Lagrangian function (3.34) depends only on $\dot{\mathbf{R}}$ rather than $\mathbf{R}$ itself, according to our discussion in section 2.4, all Cartesian components of R are cyclic coordinates, and the corresponding generalized momenta are conserved:

$$P_j \equiv \frac{\partial L}{\partial \dot{R}_j} = M\dot{R}_j = \text{const}, \quad j = 1, 2, 3. \tag{3.36}$$

Physically, this is just the conservation law for the full momentum $\mathbf{P} \equiv M\mathbf{R}$ of our system, due to the absence of external forces. Actually, in the axiomatics used in section 1.3 this law is postulated—see Eq. (1.10)—but now we may attribute the momentum $\mathbf{P}$ to a certain geometric point, the center-of-mass $\mathbf{R}$. In particular, since according to Eq. (3.36), the center moves with a constant velocity in the inertial reference frame used to write Eq. (3.30), we may create a new inertial frame with the origin at point $\mathbf{R}$. In this new frame, $\mathbf{R} \equiv 0$, so that the vector $\mathbf{r}$ (and hence the scalar r) remain the same as in the old frame (because the frame transfer vector adds equally to $\mathbf{r}_1$ and $\mathbf{r}_2$, and cancels in $\mathbf{r} = \mathbf{r}_1 - \mathbf{r}_2$), and the Lagrangian function (3.34) is now reduced to

$$L = \frac{m}{2}\dot{\mathbf{r}}^2 - U(r). \tag{3.37}$$

Thus our initial problem has been reduced to just three degrees of freedom—three scalar components of the vector $\mathbf{r}$. Moreover, Eq. (3.37) shows that the dynamics of the vector $\mathbf{r}$ of our initial, two-particle system is identical to that of the radius-vector of a *single particle* with the effective mass m, moving in the central potential field $U(r)$.

Two more degrees of freedom may be excluded from the planetary problem by noticing that according to Eq. (1.35), the angular momentum $\mathbf{L} = \mathbf{r} \times \mathbf{p}$ of our effective particle is also conserved, both in magnitude and direction. Since the direction of $\mathbf{L}$ is, by definition, perpendicular to both $\mathbf{r}$ and $\mathbf{v} = \mathbf{p}/m$, this means that the particle's motion is confined to the plane whose orientation is determined by the initial directions of vectors $\mathbf{r}$ and $\mathbf{v}$. Hence we can completely describe a particle's position by just two coordinates in that plane, for example by the distance r to the origin, and the polar angle φ. In these coordinates, Eq. (3.37) takes a form identical to Eq. (2.49):

$$L = \frac{m}{2}(\dot{r}^2 + r^2\dot{\varphi}^2) - U(r). \tag{3.38}$$

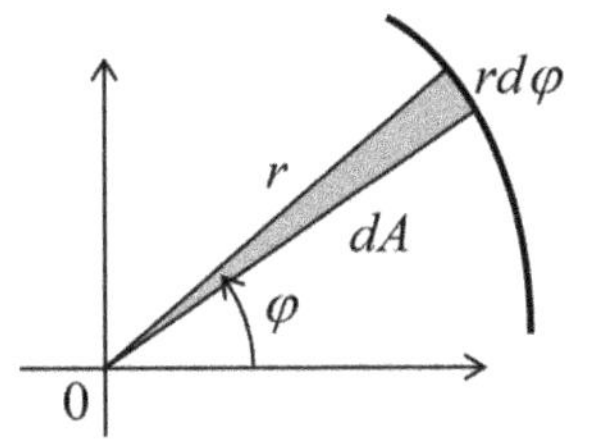

Figure 3.4. The area differential dA in the polar coordinates.

Moreover, the latter coordinate, polar angle φ, may also be eliminated by using the conservation of the magnitude of angular momentum, in the form of Eq. (2.50)[12]:

$$L_z = mr^2\dot{\varphi} = \text{const.} \tag{3.39}$$

A direct corollary of this conservation is the so-called *second Kepler law*[13]: the radius-vector **r** sweeps equal areas A in equal times. Indeed, in the linear approximation in $dA \ll A$, the area differential dA equals the area of a narrow right triangle with the base being the arc differential $rd\varphi$, and the height equal to r—see figure 3.4. As a result, according to Eq. (3.39), the time derivative of the area,

$$\frac{dA}{dt} = \frac{r(rd\varphi)/2}{dt} \equiv \frac{1}{2}r^2\dot{\varphi} = \frac{L_z}{2m}, \tag{3.40}$$

remains constant. Integration of this equation over an arbitrary (not necessarily small!) time interval proves the second Kepler law.

Now note that since $\partial L/\partial t = 0$, the Hamiltonian function H is also conserved, and since, according to Eq. (3.38), the kinetic energy of the system is a quadratic-homogeneous function of the generalized velocities $\dot{r}$ and $\dot{\varphi}$, $H = E$, so that the system energy E,

$$E = \frac{m}{2}\dot{r}^2 + \frac{m}{2}r^2\dot{\varphi}^2 + U(r), \tag{3.41}$$

is also a first integral of motion[14]. But according to Eq. (3.39), the second term of Eq. (3.41) may be represented as

$$\frac{m}{2}r^2\dot{\varphi}^2 = \frac{L_z^2}{2mr^2}, \tag{3.42}$$

[12] Here the index z stands for the coordinate perpendicular to the motion plane. Since other components of the angular momentum are equal to zero, the index is not really necessary, but I will still use it, just to make a clear distinction between the angular momentum L_z and the Lagrangian function L.

[13] This is one of three laws deduced almost exactly 400 years ago by J Kepler, from the extremely detailed astronomical data collected by T Brahe (1546–1601). In turn, the set of three Kepler laws became the main basis for Sir Isaac Newton's discovery, in the 1680s, of the law of gravity (1.15). This relentless march of physics...

[14] One may claim that this fact should have been evident from the very beginning, because the effective particle of mass m moves in a potential field $U(r)$, which conserves energy.

so that the energy Eq. (3.41) may be expressed as that of a 1D particle moving along axis r,

$$E = \frac{m}{2}\dot{r}^2 + U_{\text{ef}}(r), \tag{3.43}$$

in the following effective potential[15]:

$$U_{\text{ef}}(r) \equiv U(r) + \frac{L_z^2}{2mr^2}. \tag{3.44}$$

So the planetary motion problem has been reduced to the dynamics of an effectively 1D system[16].

Now we may proceed just as we did in section 3.3, with due respect for the very specific effective potential Eq. (3.44) which, in particular, diverges at $r \to 0$—apart from the very special case of an exactly radial motion, $L_z = 0$. In particular, we may solve Eq. (3.43) for dr/dt to obtain

$$dt = \left(\frac{m}{2}\right)^{1/2} \frac{dr}{[E - U_{\text{ef}}(r)]^{1/2}}. \tag{3.45}$$

This formula allows us not only to obtain a direct relation between time t and distance r, similar to Eq. (3.26),

$$t = \pm\left(\frac{m}{2}\right)^{1/2} \int \frac{dr}{[E - U_{\text{ef}}(r)]^{1/2}} = \pm\left(\frac{m}{2}\right)^{1/2} \int \frac{dr}{[E - U(r) - L_z^2/2mr^2]^{1/2}}, \tag{3.46}$$

but also perform a similar calculation of angle φ. Indeed, integrating Eq. (3.39),

$$\varphi \equiv \int \dot{\varphi}\, dt = \frac{L_z}{m} \int \frac{dt}{r^2}, \tag{3.47}$$

and plugging dt from Eq. (3.45), we obtain an explicit expression for the particle trajectory $\varphi(r)$:

$$\varphi = \pm \frac{L_z}{(2m)^{1/2}} \int \frac{dr}{r^2[E - U_{\text{ef}}(r)]^{1/2}} = \pm \frac{L_z}{(2m)^{1/2}} \int \frac{dr}{r^2[E - U(r) - L_z^2/2mr^2]^{1/2}}. \tag{3.48}$$

Note that according to Eq. (3.39), the derivative $d\varphi/dt$ does *not* change sign at the reflection from any classical turning point $r \neq 0$, so that, in contrast to Eq. (3.46), the sign on the right-hand side of Eq. (3.48) is uniquely determined by the initial conditions and cannot change during the motion.

Let us use these results, valid for any interaction law $U(r)$, for the planetary motion's classification. The following cases should be distinguished. (Following a

[15] The physical sense of this potential, as well as that given by Eq. (3.6), will be discussed in section 4.6 below.

[16] Note that this reduction has been carried out using a different method to that used for our testbed problem (shown in figure 2.1) in section 3.2 above. (The reader is encouraged to analyze this difference.) In order to emphasize this fact, I will keep writing E instead of H here, although for the planetary problem we are discussing, these two notions coincide.

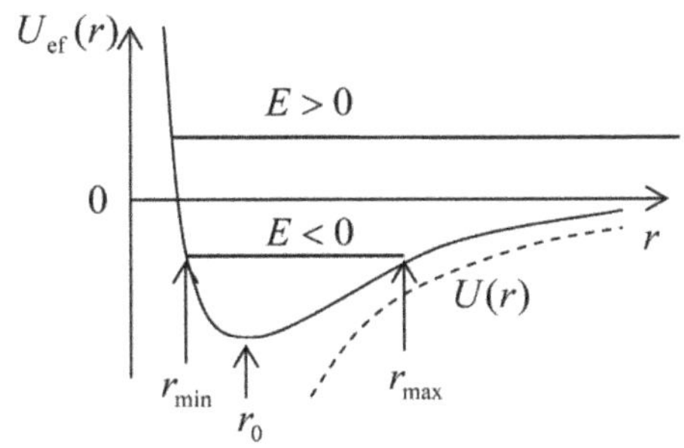

Figure 3.5. Effective potential profile of and two types of motion in an attractive central field.

good tradition, in what follows I will select the arbitrary constant in the potential energy so as to provide $U \to 0$, and hence $U_{ef} \to 0$ at $r \to \infty$.)

If the particle interaction is *attractive* (as it always is in the case of gravity), and the divergence of the attractive potential at $r \to 0$ is faster than $1/r^2$, then $U_{ef}(r) \to -\infty$ at $r \to 0$, so that at appropriate initial conditions ($E < 0$) the particle may drop on the center even if $L_z \neq 0$—an event called the *capture*. On the other hand, with $U(r)$ either converging, or diverging slower than $1/r^2$, at $r \to 0$, the effective energy profile $U_{ef}(r)$ has the shape shown schematically in figure 3.5. This is true, in particular, for the very important case

$$U(r) = -\frac{\alpha}{r}, \quad \text{with } \alpha > 0, \tag{3.49}$$

which describes, in particular, the *Coulomb* (electrostatic) *interaction* of two particles with electric charges of the opposite sign, and Newton's law of gravity (1.15). This particular case will be analyzed below, and now let us return to the analysis of an arbitrary attractive potential $U(\mathbf{r}) < 0$ leading to the effective potential shown in figure 3.5, when the angular-momentum term dominates at small distances r.

According to the analysis in section 3.3, such potential profile, with a minimum at some distance r_0, may sustain two types of motion, depending on the energy E (which is of course determined by the initial conditions):

(i) If $E > 0$, there is only one classical turning point where $E = U_{ef}$, so that distance r either grows with time from the very beginning, or (if the initial value of $\dot{r}$ was negative) first decreases and then, after the reflection from the increasing potential U_{ef}, starts to grow indefinitely. The latter case, of course, describes the *scattering* of the effective particle by the attractive center.

(ii) In the opposite case, if the energy is within the range

$$U_{ef}(r_0) \leqslant E < 0, \tag{3.50}$$

the system moves periodically between two classical turning points r_{min} and r_{max}. These oscillations of distance r correspond to the bound orbital motion of our effective particle about the attracting center[17].

[17] In the opposite case when the interaction is *repulsive*, $U(r) > 0$, the addition of the positive angular energy term only increases the trend, and only the scattering scenario is possible.

Let us start with a discussion of the bound motion, with energy within the range (3.50). If energy has its minimal possible value,

$$E = U_{\rm ef}(r_0) = \min\,[U_{\rm ef}(r)], \tag{3.51}$$

the distance cannot change, $r = r_0 = \text{const}$, so the orbit is circular, with the radius r_0 satisfying the condition $dU_{\rm ef}/dr = 0$. Using Eq. (3.44) we see that the condition for r_0 may be written as

$$\frac{L_z^2}{mr_0^3} = \frac{dU}{dr}\bigg|_{r=r_0}. \tag{3.52}$$

Since in a circular motion velocity $\mathbf{v}$ is perpendicular to the radius-vector $\mathbf{r}$, L_z is just mr_0v, the left-hand side of Eq. (3.52) equals mv^2/r_0, while its right-hand side is just the magnitude of the attractive force, so that this equation expresses the well-known second Newton law for circular motion. Plugging this result into Eq. (3.47), we obtain a linear law of angle change, $\varphi = \omega t + \text{const}$, with the angular velocity

$$\omega = \frac{L_z}{mr_0^2} = \frac{v}{r_0}, \tag{3.53}$$

and hence the rotation period $\mathcal{T}_\varphi \equiv 2\pi/\omega$ obeys the elementary relation

$$\mathcal{T}_\varphi = \frac{2\pi r_0}{v}. \tag{3.54}$$

Now let the energy be above its minimum value (but still negative). Using Eq. (3.46) just as in section 3.3, we see that distance r now oscillates with period

$$\mathcal{T}_r = (2m)^{1/2}\int_{r_{\min}}^{r_{\max}} \frac{dr}{[E - U(r) - L_z^2/2mr^2]^{1/2}}. \tag{3.55}$$

This period is, in general, different from the $\mathcal{T}_\varphi$ that corresponds to the 2π-change of the angle. Indeed, according to Eq. (3.48), the change of angle φ between two sequential points of the nearest approach,

$$|\Delta\varphi| = 2\frac{L_z}{(2m)^{1/2}}\int_{r_{\min}}^{r_{\max}} \frac{dr}{r^2[E - U(r) - L_z^2/2mr^2]^{1/2}}, \tag{3.56}$$

is generally different from 2π. Hence, the general trajectory of the bound motion has a spiral shape—see, e.g. the illustration in figure 3.6.

The situation is special, however, for a very important particular case, namely that of the *Coulomb potential* described by Eq. (3.49). Indeed, plugging this potential into Eq. (3.48), we obtain

$$\varphi = \pm\frac{L_z}{(2m)^{1/2}}\int \frac{dr}{r^2\left(E + \alpha/r - L_z^2/2mr^2\right)^{1/2}}. \tag{3.57}$$

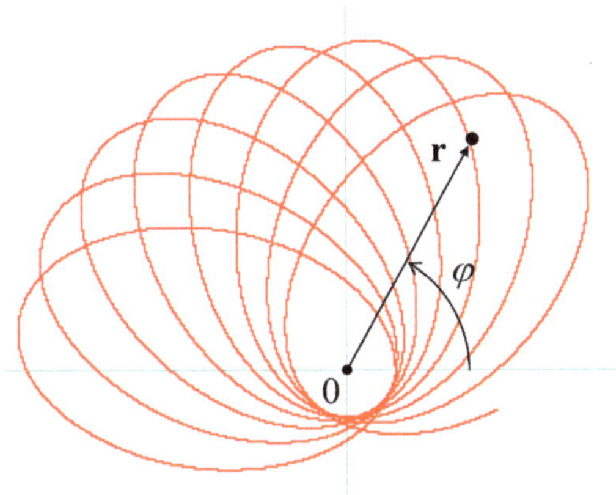

Figure 3.6. A typical open orbit of a particle moving in a non-Coulomb central field.

This is a table integral[18], giving

$$\varphi = \pm\cos^{-1}\frac{L_z^2/m\alpha r - 1}{(1 + 2EL_z^2/m\alpha^2)^{1/2}} + \text{const.} \tag{3.58}$$

The reciprocal function, $r(\varphi)$, is 2π-periodic,

$$r = \frac{p}{1 + e\cos(\varphi + \text{const})}, \tag{3.59}$$

so that at $E < 0$, the orbit is a closed line[19], characterized by the following parameters[20]:

$$p \equiv \frac{L_z^2}{m\alpha}, \qquad e \equiv \left(1 + \frac{2EL_z^2}{m\alpha^2}\right)^{1/2}. \tag{3.60}$$

The physical meaning of these parameters is very simple. Indeed, according to the general equation (3.52), in the Coulomb potential, for which $dU/dr = \alpha/r^2$, we see that p is just the circular orbit radius[21] for the given L_z: $r_0 = L_z^2/m\alpha \equiv p$, so that

$$\min\left[U_{\text{ef}}(r)\right] \equiv U_{\text{ef}}(r_0) = -\frac{\alpha^2 m}{2L_z^2}. \tag{3.61}$$

[18] See, e.g. Eq. (A.30*a*).

[19] It may be proved that for the power-law interaction, $U \propto r^\nu$, the orbits are closed lines only if $\nu = -1$ (i.e. our current case of the Coulomb potential) or if $\nu = +2$ (the 3D harmonic oscillator)—the so-called *Bertrand theorem*.

[20] Let me hope that the difference between the parameter p and the particle momentum's magnitude is absolutely clear from the context, so that using the same (traditional) notation for both notions cannot lead to confusion.

[21] Mathematicians prefer a more solemn terminology: the parameter $2p$ is called the *latus rectum* of the ellipse.

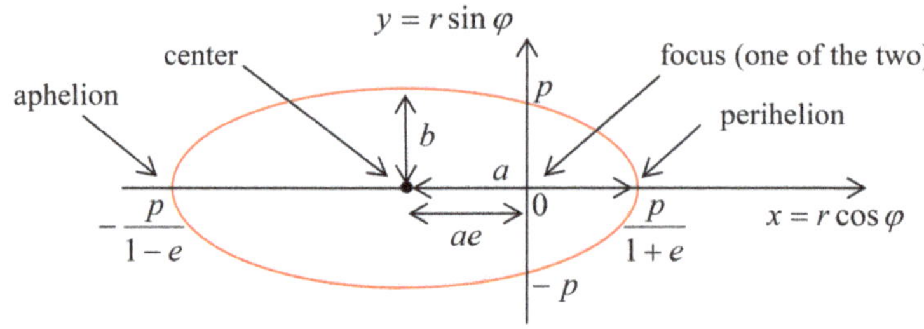

Figure 3.7. Ellipse, and its special points and dimensions.

Using this equality, the parameter e (called *eccentricity*) may be represented just as

$$e = \left\{1 - \frac{E}{\min\,[U_{\text{ef}}(r)]}\right\}^{1/2}. \tag{3.62}$$

Analytical geometry tells us that Eq. (3.59), with $e < 1$, is one of canonical forms for representation of an *ellipse,* with one of its two focuses located at the origin. This fact is known as the *first Kepler law*. Figure 3.7 shows the relation between the main dimensions of the ellipse and parameters p and e.[22]

In particular, the *major semi-axis* a and the *minor semi-axis* b are simply related to p and e, and hence, via (3.60), to the motion integrals E and L_z:

$$a = \frac{p}{1 - e^2} = \frac{\alpha}{2\,|\,E\,|}, \qquad b = \frac{p}{(1 - e^2)^{1/2}} = \frac{L_z}{(2m\,|\,E\,|)^{1/2}}. \tag{3.63}$$

As was mentioned above, at $E \to \min\,[U_{\text{ef}}(r)]$ the orbit is almost circular, with $r(\varphi) \cong r_0 = p$. In contrast, as E is increased to approach zero (its maximum value for the closed orbit), then $e \to 1$, so that the aphelion point $r_{\max} = p/(1-e)$ tends to infinity, i.e. the orbit becomes extremely extended—see the red lines in figure 3.8.

The above relations enable, in particular, a ready calculation of the rotation period $\mathcal{T} \equiv \mathcal{T}_r = \mathcal{T}_\varphi$. (In the case of a closed trajectory, $\mathcal{T}_r$ and $\mathcal{T}_\varphi$ coincide.) Indeed, it is well known that the ellipse area $A = \pi ab$. But according to the second Kepler law (3.40), $dA/dt = L_z/2m =$ const. Hence

$$\mathcal{T} = \frac{A}{dA/dt} = \frac{\pi ab}{L_z/2m}. \tag{3.64a}$$

Using Eqs. (3.60) and (3.63), this important result may be represented in several other forms:

$$\mathcal{T} = \frac{\pi p^2}{(1 - e^2)^{3/2}(L_z/2m)} = \pi\alpha\left(\frac{m}{2\,|\,E\,|^3}\right)^{1/2} = 2\pi a^{3/2}\left(\frac{m}{\alpha}\right)^{1/2}. \tag{3.64b}$$

[22] In this figure, the constant participating in Eqs. (3.58)–(3.59) is assumed to be zero. It is evident that a different choice of the constant corresponds just to a different origin of φ, i.e. a constant turn of the ellipse about the origin.

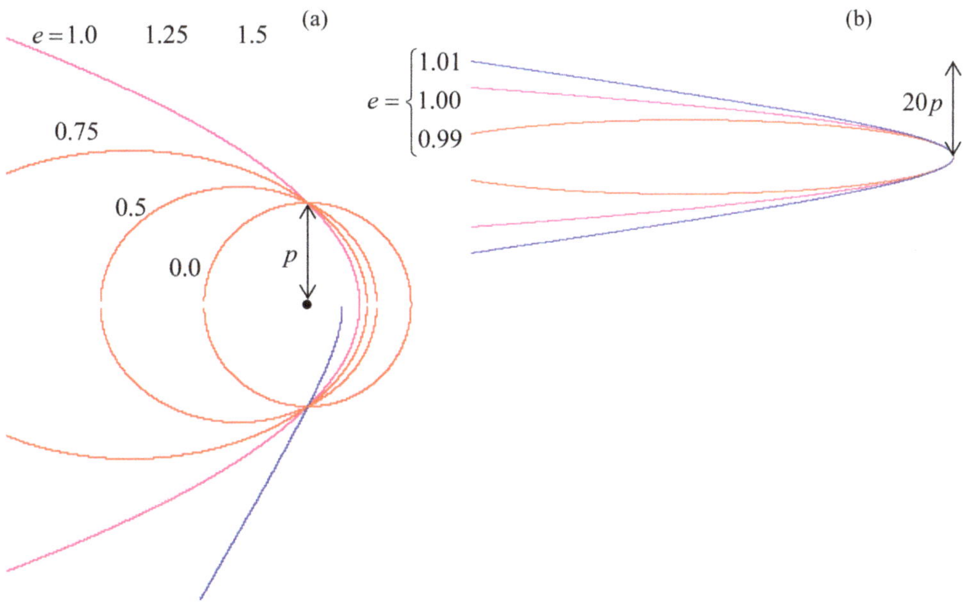

Figure 3.8. (a) Zoom-in and (b) zoom-out on the Coulomb-field trajectories corresponding to the same parameter p (i.e. the same L_z), but different values of the eccentricity parameter e, i.e. of the energy E—see Eq. (3.60): ellipses ($e < 1$, red lines), the parabola ($e = 1$, magenta line), and hyperbolas ($e > 1$, blue lines). Note that the transition from closed to open trajectories at $e = 1$ is dramatic only at very large distances, $r \gg p$.

Since for the Newtonian gravity (1.15), $\alpha = Gm_1m_2 = GmM$, at $m_1 \ll m_2$ (i.e. $m \ll M$) this constant is proportional to m, and the last form of Eq. (3.64b) yields the *third Kepler law*: periods of motion of different planets in the same central field, say that of our Sun, scale as $T \propto a^{3/2}$. Note that in contrast to the second Kepler law (which is valid for any central field), the first and third Kepler laws are potential-specific.

Now reviewing the above derivation of Eqs. (3.59)–(3.60), we see that they are also valid in the case of $E \geqslant 0$—see the top horizontal line in figure 3.5 and its discussion above, if we limit the results to the physically meaningful range $r \geqslant 0$. This means that if the energy is exactly zero, Eq. (3.59) (with $e = 1$) is still valid for all values of φ (except for one special point $\varphi = \pi$ where r becomes infinite) and describes a *parabolic* (i.e. open) trajectory—see the magenta lines in figure 3.8.

Moreover, if $E > 0$, Eq. (3.59) is still valid within a certain sector of angles φ,

$$\Delta\varphi = 2\cos^{-1}\frac{1}{e} \equiv 2\cos^{-1}\left(1 + \frac{2EL_z^2}{m\alpha^2}\right)^{-1/2} < \pi, \qquad \text{for } E > 0, \tag{3.65}$$

and describes an open, *hyperbolic* trajectory (see the blue lines in figure 3.8). As was mentioned earlier, such trajectories are typical, in particular, for particle scattering.

3.5 Elastic scattering

If $E > 0$, the motion is unbound for any interaction potential. In this case, the two most important parameters of the particle trajectory are the *scattering angle* θ and

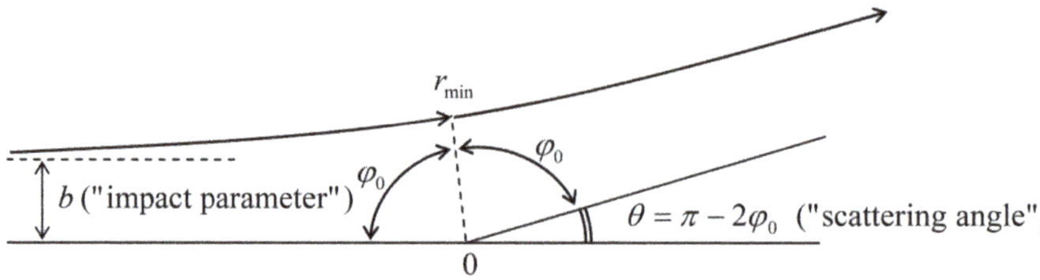

Figure 3.9. Main geometric parameters of the scattering problem.

the *impact parameter b* (figure 3.9), and the main task for theory is to find the relation between them in the given potential $U(r)$. For that, it is convenient to note that b is related to two conserved quantities, the particle's energy[23] E and its angular momentum L_z, in a simple way[24]:

$$L_z = b(2mE)^{1/2}. \tag{3.66}$$

Hence the angular contribution to the effective potential (3.44) may be represented as

$$\frac{L_z^2}{2mr^2} = E\frac{b^2}{r^2}. \tag{3.67}$$

Second, according to Eq. (3.48), the trajectory sections going from infinity to the nearest approach point ($r = r_{min}$), and from that point to infinity, have to be similar, and hence correspond to equal angle changes φ_0—see figure 3.9.

Hence we may apply the general equation (3.48) to just one of the sections, say $[r_{min}, \infty]$, to find the scattering angle:

$$\begin{aligned}\theta = \pi - 2\varphi_0 &= \pi - 2\frac{L_z}{(2m)^{1/2}}\int_{r_{min}}^{\infty}\frac{dr}{r^2\left[E - U(r) - L_z^2/2mr^2\right]^{1/2}}\\ &= \pi - 2\int_{r_{min}}^{\infty}\frac{bdr}{r^2[1 - U(r)/E - b^2/r^2]^{1/2}}.\end{aligned} \tag{3.68}$$

In particular, for the Coulomb potential (3.49), now with an arbitrary sign of α, we can apply the same table integral as in the previous section to obtain[25]

$$\theta = \left|\pi - 2\arccos\frac{\alpha/2Eb}{[1 + (\alpha/2Eb)^2]^{1/2}}\right|. \tag{3.69a}$$

[23] The energy conservation law is frequently emphasized by calling such a process *elastic scattering*.

[24] Indeed, at $r \gg b$, the definition $\mathbf{L} = \mathbf{r} \times (m\mathbf{v})$ yields $L_z = bmv_\infty$, where $v_\infty = (2E/m)^{1/2}$ is the initial (and hence the final) velocity of the particle.

[25] Alternatively, this result may be recovered directly from the first form of Eq. (3.65), with the eccentricity e expressed via the same dimensionless parameter $(2Eb/\alpha)$: $e = [1 + (2Eb/\alpha)^2]^{1/2} > 1$.

This result may be more conveniently rewritten as

$$\tan\frac{|\theta|}{2} = \frac{|\alpha|}{2Eb}. \tag{3.69b}$$

Very clearly, the scattering angle's magnitude increases with the potential strength α, and decreases as either the particle energy or the impact parameter (or both) are increased.

The general result (3.68) and the Coulomb-specific relations (3.69) represent a formally complete solution of the scattering problem. However, in a typical experiment on elementary particle scattering, the impact parameter b of a single particle is random and unknown. In this case, our results may be used to obtain the statistics of the scattering angle θ, in particular the so-called *differential cross-section*[26]

$$\frac{d\sigma}{d\Omega} \equiv \frac{1}{n}\frac{dN}{d\Omega}, \tag{3.70}$$

where n is the average number of incident particles per unit area, and dN is the average number of particles scattered into a small solid angle range $d\Omega$. For a spherically symmetric scattering center, which provides an axially symmetric scattering pattern, $d\sigma/d\Omega$ may be calculated by counting the number of incident particles that have the impact parameters within a small range db:

$$dN = n\, 2\pi\, bdb, \tag{3.71}$$

and hence are scattered into the corresponding small solid angle range $d\Omega = 2\pi \sin\theta\, d\theta$. Plugging these relations into Eq. (3.70), we obtain the following general geometric relation:

$$\frac{d\sigma}{d\Omega} = \frac{b}{\sin\theta}\left|\frac{db}{d\theta}\right|. \tag{3.72}$$

In particular, for the Coulomb potential (3.49), a straightforward differentiation of Eq. (3.68) yields the so-called *Rutherford scattering formula*

$$\frac{d\sigma}{d\Omega} = \left(\frac{\alpha}{4E}\right)^2 \frac{1}{\sin^4(\theta/2)}. \tag{3.73}$$

This result, which shows a very strong scattering to small angles (so strong that the integral that expresses the full cross-section σ is formally diverging at $\theta \to 0$)[27], and

[26] This terminology stems from the fact that an integral of $d\sigma/d\Omega$ over the full solid angle, called the *full cross-section* σ, has the dimension of area: $\sigma = N/n$, where N is the total number of scattered particles.

[27] This divergence, which persists at the quantum-mechanical treatment of the problem, is due to particles with large values of b, and disappears at an account, for example, of any finite concentration of the scattering centers.

very weak *backscattering* (to angles $\theta \approx \pi$) was historically extremely significant: in the early 1910s, its good agreement with α-particle scattering experiments carried out by Rutherford's group gave a strong justification for 'planetary' models of atoms, with electrons moving about very small nuclei.

Note that elementary particle scattering is frequently accompanied by electromagnetic radiation and/or other processes leading to the loss of the initial mechanical energy of the system, leading to *inelastic scattering*, that may give significantly different results. (In particular, a capture of an incoming particle becomes possible even for a Coulomb attracting center.) Also, quantum-mechanical effects may be important at scattering, so that the above results should be used with caution.

3.6 Problems

Problem 3.1. For the system considered in problem 2.6 (a bead sliding along a string with fixed tension $\mathscr{T}$, see the figure below), analyze small oscillations of the bead near equilibrium.

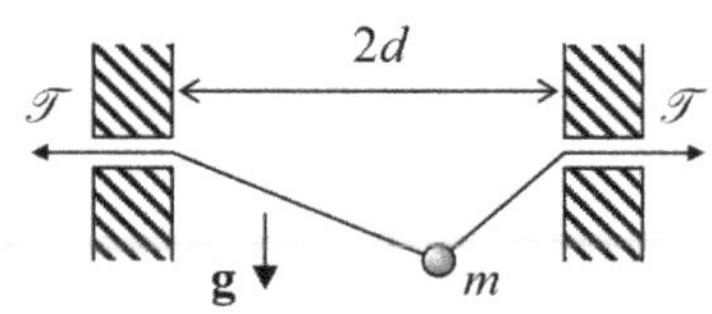

Problem 3.2. For the system considered in problem 2.7 (a bead sliding along a string of a fixed length $2l$, see the figure below), analyze small oscillations near the equilibrium.

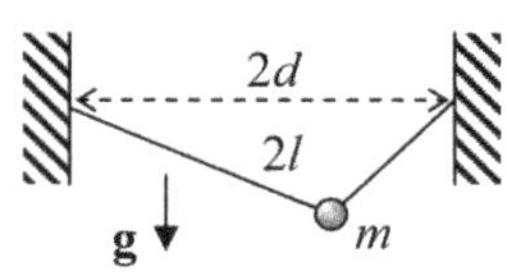

Problem 3.3. Calculate the functional dependence of period $\mathcal{T}$ of oscillations of a 1D particle of mass m in potential $U(q) = \alpha q^{2n}$ (where $\alpha > 0$, and n is a positive integer) on energy E. Explore the limit $n \to \infty$.

Problem 3.4. Explain why the term $mr^2\dot{\varphi}^2/2$, recast in accordance with Eq. (3.42), cannot be merged with $U(r)$ in Eq. (3.38), to form an effective 1D potential energy $U(r) - L_z^2/2mr^2$, with the second term's sign opposite to that given by Eq. (3.44). We have done an apparently similar thing for our testbed, bead-on-rotating-ring problem at the very end of section 3.1—see Eq. (3.6); why cannot the same trick work for the planetary problem? In addition to a formal explanation, discuss the physics behind this difference.

Problem 3.5. A system consisting of two equal masses m on a light rod of length l (frequently called a *dumbbell*) can slide without friction along a vertical ring of radius R, rotated about its vertical diameter with constant angular velocity ω—see the figure below. Derive the condition of stability of the lower horizontal position of the dumbbell.

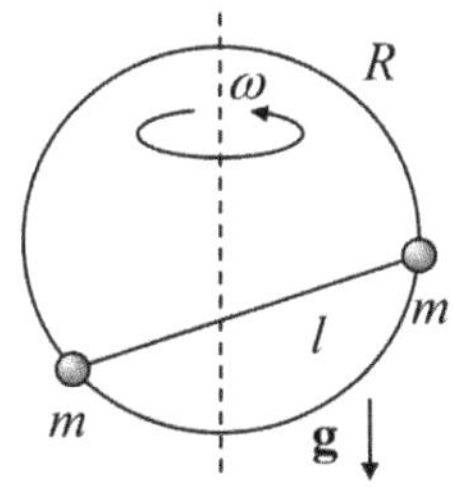

Problem 3.6. Analyze the dynamics of the so-called *spherical pendulum*—a point mass hung, in a uniform gravity field **g**, on a light cord of length l, with no motion confinement to a vertical plane. In particular:

(i) find the integrals of motion and reduce the problem to a 1D problem;
(ii) calculate the time period of the possible circular motion around the vertical axis; and
(iii) explore small deviations from the circular motion. (Are the pendulum orbits closed?)[28].

Problem 3.7. The orbits of Mars and Earth around the Sun may be well approximated as circles, with a radius ratio of 3/2. Use this fact, and the Earth year duration (which you should know), to calculate the time of travel to Mars expending the least energy (neglecting the planets' size and the effects of their gravitational fields on the spacecraft).

Problem 3.8. Derive the first-order and second-order differential equations for the reciprocal distance $u \equiv 1/r$ as a function of φ, describing the trajectory of a particle's motion in a central potential $U(r)$. Spell out the latter equation for the particular case of the Coulomb potential (3.49) and discuss the result.

Problem 3.9. For the motion of a particle in the Coulomb attractive field (with potential $U(r) = -\alpha/r$, with $\alpha > 0$), calculate and sketch the so-called *hodograph*[29]—the trajectory followed by the head of the velocity vector **v**, provided that its tail is kept at the origin.

Problem 3.10. For motion in the central potential

$$U(r) = -\frac{\alpha}{r} + \frac{\beta}{r^2},$$

(i) find the orbit $r(\varphi)$, for positive α and β, and all possible ranges of energy E;
(ii) prove that in the limit $\beta \to 0$, and for energy $E < 0$, the orbit may be represented as a slowly rotating ellipse; and

[28] Solving this problem is very good preparation for the analysis of symmetric top rotation in the next chapter.
[29] The use of this notion for the characterization of motion may be traced back at least to an 1846 treatise by W Hamilton. Currently, it is most often used in applied fluid mechanics, in particular meteorology.

(iii) express the angular velocity of this slow orbit rotation via parameters α and β of the potential, and the particle's mass m, its energy E, and the angular momentum L_z.

Problem 3.11. A particle is moving in the field of an attractive central force, with potential

$$U(r) = -\frac{\alpha}{r^n}, \qquad \text{where } \alpha n > 0.$$

For what values of n is a circular orbit stable?

Problem 3.12. Determine the condition for a particle of mass m, moving under the effect of a central attractive force

$$\mathbf{F} = -\alpha \frac{\mathbf{r}}{r^3} \exp\left\{-\frac{r}{R}\right\},$$

where α and R are positive constants, to have a stable circular orbit.

Problem 3.13. A particle of mass m, with angular momentum L_z, moves in the field of an attractive central force with a distance-independent magnitude F. If particle's energy E is slightly higher than the value $E_{\min}$ corresponding to the circular orbit of the particle, what is the time period of its radial oscillations? Compare the period with that of the circular orbit at $E = E_{\min}$.

Problem 3.14. For particle scattering by a repulsive Coulomb field, calculate the minimum approach distance $r_{\min}$ and the velocity $v_{\min}$ at that point, and analyze their dependence on the impact parameter b (see figure 3.9) and the initial velocity v_∞ of the particle.

Problem 3.15. A particle is launched from afar, with impact parameter b, toward an attracting center with

$$U(r) = -\frac{\alpha}{r^n}, \qquad \text{with } n > 2, \quad \alpha > 0.$$

(i) Express the minimum distance between the particle and the center via b, if the initial kinetic energy E of the particle is barely sufficient to escape capture by the attracting center.
(ii) Calculate the capture's full cross-section; explore the limit $n \to 2$.

Problem 3.16. A meteorite with initial velocity v_∞ approaches an atmosphere-free planet of mass M and radius R.

(i) Find the condition on the impact parameter b for the meteorite to hit the planet's surface.
(ii) If the meteorite barely avoids the collision, what is its scattering angle?

Problem 3.17. Calculate the differential and full cross-sections of the classical, elastic scattering of small particles by a hard sphere of radius R.

Problem 3.18. The most famous[30] confirmation of Einstein's theory of general relativity came from the observation, by A Eddington and his associates, of starlight's deflection by the Sun during the May 1919 solar eclipse. Considering light photons as classical particles propagating with the speed of light, $v_0 \to c \approx 3.00 \times 10^8$ m s^{-1}, and the astronomic data for the Sun's mass, $M_S \approx 2.0 \times 10^{30}$ kg, and radius, $R_S \approx 7.0 \times 10^8$ m, calculate the non-relativistic mechanics prediction for the angular deflection of the light rays grazing the Sun's surface.

[30] It was not the first confirmation, however. The first came four years earlier from Einstein himself, who showed that his theory may qualitatively explain the difference between the rate of Mercury orbit's precession, known from earlier observations, and the non-relativistic theory of this effect.

IOP Publishing

Classical Mechanics
Lecture notes
Konstantin K Likharev

Chapter 4

Rigid-body motion

This chapter discusses the motion of rigid bodies, with a heavy focus on its most non-trivial part: the rotation. Some by-product results of this analysis will enable us to discuss, at the end of the chapter, the description of motion of point particles in non-inertial reference frames.

4.1 Translation and rotation

Let us start a discussion of solid bodies from a (relatively) simple limit when the changes of distances $r_{kk'} \equiv |\mathbf{r}_k - \mathbf{r}_{k'}|$ between particles are negligibly small. Such an abstraction is called the (*absolutely*) *rigid body*, and is a reasonable approximation in many practical problems, including the motion of solids. In other words, this model neglects *deformations*—which will be the subject of the following chapters. The rigid-body approximation reduces the number of degrees of freedom of the system of N particles from $3N$ to just 6—for example, three Cartesian coordinates of one point (say, 0), and three angles of the system rotation about three mutually perpendicular axes passing through this point[1].

As follows from the discussion in sections 1.1–1.3, any purely *translational motion* of a rigid body, at which the velocity vectors $\mathbf{v}$ of all points are equal, is not more complex than that of a point particle. Indeed, according to Eqs. (1.8) and (1.30), in an inertial 'laboratory' reference frame such a body moves, upon the effect of the net external force $\mathbf{F}^{(\mathrm{ext})}$, exactly as a point particle. However, the rotation is a bit more tricky.

[1] An alternative way to arrive at the same number 6 is to consider three points of the body, which uniquely define its position. If movable independently, the points would have 9 degrees of freedom, but since three distances $r_{kk'}$ between them are now fixed, the resulting three constraints reduce the number of degrees of freedom to 6.

doi:10.1088/978-0-7503-1398-8ch4

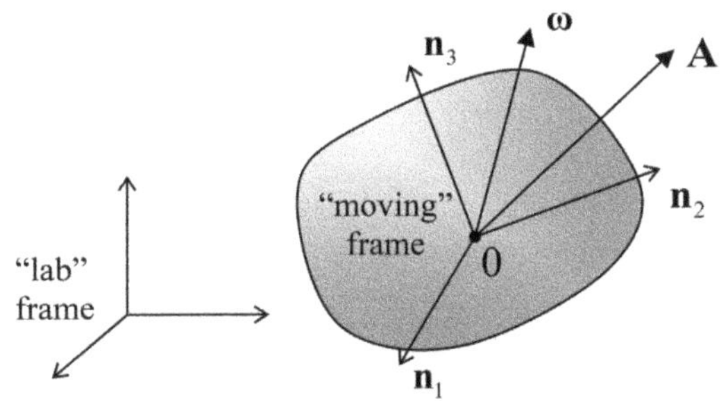

Figure 4.1. Deriving Eq. (4.8).

Let us start by showing that an arbitrary elementary displacement of a rigid body may be always considered as a sum of the translational motion discussed above, and what is called a *pure rotation*. For that, consider a 'moving' reference frame, firmly bound to the body, and an arbitrary vector **A**—see figure 4.1.

The vector may be represented by its Cartesian components A_j in that moving frame:

$$\mathbf{A} = \sum_{j=1}^{3} A_j \mathbf{n}_j. \tag{4.1}$$

Let us calculate its time derivative in a different ('lab') frame, taking into account that if the body rotates relative to this frame, then the directions of the unit vectors $\mathbf{n}_j$, as seen from the lab frame, change in time. Hence, we have to differentiate both operands in each product contributing to the sum (4.1):

$$\frac{d\mathbf{A}}{dt}\Big|_{\text{in lab}} = \sum_{j=1}^{3} \frac{dA_j}{dt}\mathbf{n}_j + \sum_{j=1}^{3} A_j \frac{d\mathbf{n}_j}{dt}. \tag{4.2}$$

In this expression, the first sum evidently describes the change of the vector **A** as observed from the moving frame. In the second sum, each of the infinitesimal vectors $d\mathbf{n}_j$ may be represented by its Cartesian components:

$$d\mathbf{n}_j = \sum_{j'=1}^{3} \mathbf{n}_{j'} d\varphi_{jj'}, \tag{4.3}$$

where $d\varphi_{jj'}$ are some scalar coefficients. In order to find more about them, let us scalar-multiply each side of Eq. (4.3) by an arbitrary unit vector $\mathbf{n}_{j''}$, and take into account the evident orthonormality condition:

$$\mathbf{n}_{j'} \cdot \mathbf{n}_{j''} = \delta_{j'j''}, \tag{4.4}$$

where $\delta_{j'j''}$ is the Kronecker symbol[2]. As a result, we obtain

[2] See, e.g. Eq. (A.82).

$$d\varphi_{jj''} = d\mathbf{n}_j \cdot \mathbf{n}_{j''}. \tag{4.5}$$

Now let us use Eq. (4.5) to calculate the first differential of Eq. (4.4):

$$\begin{aligned} d\mathbf{n}_{j'} \cdot \mathbf{n}_{j''} + \mathbf{n}_{j'} \cdot d\mathbf{n}_{j''} = d\varphi_{j'j''} + d\varphi_{j''j'} = 0; \\ \text{in particular,} \quad 2d\mathbf{n}_j \cdot \mathbf{n}_j = 2d\varphi_{jj} = 0. \end{aligned} \tag{4.6}$$

These relations, valid for any choice of indices j, j', and j'' of the set $\{1, 2, 3\}$, show that the matrix of elements $d\varphi_{jj'}$ is antisymmetric with respect to the swap of its indices; this means that there are not nine, but just three non-zero, independent coefficients $d\varphi_{jj'}$, all with $j \neq j'$. Hence it is natural to renumber them in a simpler way: $d\varphi_{jj'} = -d\varphi_{j'j} \equiv d\varphi_{j''}$, where indices j, j', and j'' follow in a 'correct' order—either $\{1,2,3\}$, or $\{2,3,1\}$, or $\{3,1,2\}$. It is easy to check (say, just by a component-by-component comparison) that in this new notation, Eq. (4.3) may be represented just as a vector product:

$$d\mathbf{n}_j = d\boldsymbol{\varphi} \times \mathbf{n}_j, \tag{4.7}$$

where $d\boldsymbol{\varphi}$ is the infinitesimal vector defined by its Cartesian components $d\varphi_j$ in the reference frame $\{\mathbf{n}_1, \mathbf{n}_2, \mathbf{n}_3\}$.

This relation is the basis of all rotation kinematics. Using it, Eq. (4.2) may be rewritten as

$$\begin{aligned} \left.\frac{d\mathbf{A}}{dt}\right|_{\text{in lab}} = \left.\frac{d\mathbf{A}}{dt}\right|_{\text{in mov}} + \sum_{j=1}^{3} A_j \frac{d\boldsymbol{\varphi}}{dt} \times \mathbf{n}_j = \left.\frac{d\mathbf{A}}{dt}\right|_{\text{in mov}} + \boldsymbol{\omega} \times \mathbf{A}, \\ \text{where } \boldsymbol{\omega} \equiv \frac{d\boldsymbol{\varphi}}{dt}. \end{aligned} \tag{4.8}$$

In order to interpret the physical sense of the vector $\boldsymbol{\omega}$, let us apply Eq. (4.8) to the particular case when $\mathbf{A}$ is the radius-vector $\mathbf{r}$ of a point of the body, and the lab frame is selected in a special way: its origin moves with the same velocity as that of the moving frame in the particular instant under consideration. In this case the first term on the right-hand side of Eq. (4.8) is zero, and we obtain

$$\left.\frac{d\mathbf{r}}{dt}\right|_{\text{in special lab frame}} = \boldsymbol{\omega} \times \mathbf{r}, \tag{4.9}$$

were the vector $\mathbf{r}$ is the same in both frames. According to the vector product definition, the particle velocity described by this formula has a direction perpendicular to vectors $\boldsymbol{\omega}$ and $\mathbf{r}$ (figure 4.2), and magnitude $\omega r \sin\theta$. As figure 4.2 shows, this expression may be rewritten as $\omega\rho$, where $\rho = r \sin\theta$ is the distance from the line that is parallel to vector $\boldsymbol{\omega}$ and passes through point 0. This is of course just the pure *rotation* about that line (called the *instantaneous axis of rotation*), with the angular velocity ω. Since, according to Eqs. (4.3) and (4.8), the *angular velocity vector* $\boldsymbol{\omega}$ is defined by the time evolution of the moving frame alone, it is the same for all points $\mathbf{r}$, i.e. for the rigid body as a whole. Note that nothing in our calculations forbids not

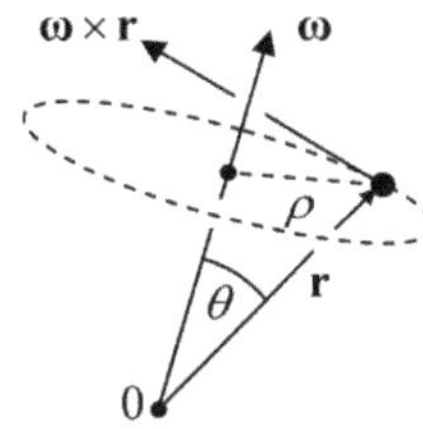

Figure 4.2. The instantaneous axis and the angular velocity of rotation.

only the magnitude but also the direction of vector $\boldsymbol{\omega}$, and thus of the instantaneous axis of rotation, from changing in time (and in many cases it does); hence the name.

Now let us generalize our result a step further, considering two laboratory reference frames that do not rotate versus each other: one arbitrary, and another one selected in the special way described above such that Eq. (4.9) is valid in it. Since the relative motion of these two reference frames is purely translational, we can use the simple velocity addition rule given by Eq. (1.6) to write

$$\mathbf{v}|_{\text{in lab}} = \mathbf{v}_0|_{\text{in lab}} + \mathbf{v}|_{\text{in special lab frame}} = \mathbf{v}_0|_{\text{in lab}} + \boldsymbol{\omega} \times \mathbf{r}, \tag{4.10}$$

where $\mathbf{r}$ is the radius-vector of a point is measured in the body-bound ('moving') frame 0.

4.2 Inertia tensor

Since the dynamics of each point of a rigid body is strongly constrained by conditions $r_{kk'}$ = const, this is one of the most important fields of application of the Lagrangian formalism discussed in chapter 2. To use this approach, the first thing we need to calculate is the kinetic energy of the body in an inertial reference frame. Since it is just the sum of the kinetic energies (1.19) of all its points, we can use Eq. (4.10) to write[3]:

$$\begin{aligned} T \equiv \sum \frac{m}{2}\mathbf{v}^2 &= \sum \frac{m}{2}(\mathbf{v}_0 + \boldsymbol{\omega} \times \mathbf{r})^2 \\ &= \sum \frac{m}{2} v_0^2 + \sum m\mathbf{v}_0 \cdot (\boldsymbol{\omega} \times \mathbf{r}) + \sum \frac{m}{2}(\boldsymbol{\omega} \times \mathbf{r})^2. \end{aligned} \tag{4.11}$$

Let us apply to the right-hand side of Eq. (4.11) two general vector analysis formulas, listed in appendix A: the so-called *operand rotation rule* (A.48) to the second term and Eq. (A.49*b*) to the third term. The result is

$$T = \sum \frac{m}{2} v_0^2 + \sum m\mathbf{r} \cdot (\mathbf{v}_0 \times \boldsymbol{\omega}) + \sum \frac{m}{2}[\omega^2 r^2 - (\boldsymbol{\omega} \cdot \mathbf{r})^2]. \tag{4.12}$$

This expression may be further simplified by making a specific choice of point 0 (from which the radius-vectors $\mathbf{r}$ of all particles are measured), namely if we use for

[3] Actually, all symbols for particle masses, coordinates and velocities should carry the particle's index, over which the summation is carried out. However, for the sake of notation simplicity, in this section this index is just implied.

this point the *center-of-mass* of the body. As was already mentioned in section 3.4 for the two-point case, the radius-vector $\mathbf{R}$ of this point is defined as

$$M\mathbf{R} \equiv \sum m\mathbf{r}, \quad M \equiv \sum m, \tag{4.13}$$

where M is just the total mass of the body. In the reference frame centered as this point, $\mathbf{R} = 0$, so that in the second sum in Eq. (4.12) vanishes, and the kinetic energy is a sum of just two terms:

$$T = T_{\text{tran}} + T_{\text{rot}}, \quad T_{\text{tran}} \equiv \frac{M}{2}V^2, \quad T_{\text{rot}} \equiv \sum \frac{m}{2}[\omega^2 r^2 - (\boldsymbol{\omega} \cdot \mathbf{r})^2], \tag{4.14}$$

where $\mathbf{V} \equiv d\mathbf{R}/dt$ is the center-of-mass velocity in our inertial reference frame, and all particle positions $\mathbf{r}$ have to be measured relatively the center-of-mass frame. Since the angular velocity vector $\boldsymbol{\omega}$ is common for all points of a rigid body, it is more convenient to rewrite the rotational energy in a form in which the summation over the components of this vector is clearly separated from the summation over the points of the body:

$$T_{\text{rot}} = \frac{1}{2}\sum_{j,j'=1}^{3} I_{jj'}\omega_j\omega_{j'}, \tag{4.15}$$

where the 3×3 matrix with elements

$$I_{jj'} \equiv \sum m\left(r^2\delta_{jj'} - r_j r_{j'}\right) \tag{4.16}$$

is called the *inertia tensor* of the body.

Actually, the term 'tensor' for this matrix has to be justified, because in physics this term implies a certain reference-frame-independent notion, whose elements have to obey certain rules at the transfer between reference frames. In order to show that the inertia tensor indeed deserves its title, let us calculate another key quantity, the total angular momentum $\mathbf{L}$ of the same body[4]. Summing up the angular momenta of each particle, defined by Eq. (1.31), and then using Eq. (4.10) again, in our inertial reference frame we obtain

$$\begin{aligned}\mathbf{L} \equiv \sum \mathbf{r} \times \mathbf{p} = \sum m\mathbf{r} \times \mathbf{v} = \sum m\mathbf{r} \times (\mathbf{v}_0 + \boldsymbol{\omega} \times \mathbf{r}) \\ = \sum m\mathbf{r} \times \mathbf{v}_0 + \sum m\mathbf{r} \times (\boldsymbol{\omega} \times \mathbf{r}).\end{aligned} \tag{4.17}$$

We see that the momentum may be represented as a sum of two terms. The first one,

$$\mathbf{L}_0 \equiv \sum m\mathbf{r} \times \mathbf{v}_0 = M\mathbf{R} \times \mathbf{v}_0, \tag{4.18}$$

describes possible rotation of the center-of-mass around the inertial frame's origin. This term evidently vanishes if the moving reference frame's origin 0 is positioned at

[4] Hopefully, there is a little chance of to confuse the angular momentum $\mathbf{L}$ (a vector) and its Cartesian components L_j (scalars with indices) on one hand, and the Lagrange function L (a scalar without an index) on the other hand.

the center-of-mass. In this case we are left with only the second term, which describes the pure rotation of the body about its center-of-mass:

$$\mathbf{L} = \mathbf{L}_{\text{rot}} \equiv \sum m\mathbf{r} \times (\boldsymbol{\omega} \times \mathbf{r}). \tag{4.19}$$

Using one more vector algebra formula, the 'bac minis cab' rule[5], we may rewrite this expression as

$$\mathbf{L} = \sum m[\boldsymbol{\omega} r^2 - \mathbf{r}(\mathbf{r} \cdot \boldsymbol{\omega})]. \tag{4.20}$$

Let us spell out an arbitrary Cartesian component of this vector:

$$L_j = \sum m\left[\omega_j r^2 - r_j \sum_{j'=1}^{3} r_{j'}\omega_{j'}\right] = \sum m \sum_{j'=1}^{3} \omega_{j'}\left(r^2\delta_{jj'} - r_j r_{j'}\right). \tag{4.21}$$

Changing the order of the summations, and comparing the result with Eq. (4.16), we see that the angular momentum may be conveniently expressed via the same matrix elements $I_{jj'}$ as the rotational kinetic energy:

$$L_j = \sum_{j'=1}^{3} I_{jj'}\omega_{j'}. \tag{4.22}$$

Since $\mathbf{L}$ and $\boldsymbol{\omega}$ are both legitimate vectors (meaning that they describe physical vectors independent on the reference frame choice), their connection, the matrix of elements $I_{jj'}$, is a legitimate tensor. This fact, and the symmetry of the tensor ($I_{jj'} = I_{j'j}$), which is evident from its definition (4.16), allow the tensor to be further simplified. In particular, mathematics tells us that by a certain choice of the axes' orientations, any symmetric tensor may be reduced to a diagonal form

$$I_{jj'} = I_j\delta_{jj'}, \tag{4.23}$$

where, in our case

$$I_j = \sum m\left(r^2 - r_j^2\right) = \sum m\left(r_{j'}^2 + r_{j''}^2\right) \equiv \sum m\rho_j^2, \tag{4.24}$$

ρ_j being the distance of the particle from the jth axis, i.e. the length of the perpendicular dropped from the point to that axis. The axes of such special coordinate systems are called the *principal axes*, while the diagonal elements I_j given by Eq. (4.24) are the *principal moments of inertia* of the body. In such a special reference frame, Eqs. (4.15) and (4.22) are reduced to very simple forms:

$$T_{\text{rot}} = \sum_{j=1}^{3} \frac{I_j}{2}\omega_j^2, \tag{4.25}$$

[5] See, e.g. Eq. (A.47).

$$L_j = I_j\omega_j. \tag{4.26}$$

Both these results recall the corresponding relations for the translational motion, $T_{\text{tran}} = MV^2/2$ and $\mathbf{P} = M\mathbf{V}$, with the angular velocity $\boldsymbol{\omega}$ replacing the linear velocity $\mathbf{V}$, and the tensor of inertia playing the role of scalar mass M. However, let me emphasize that even in the specially selected reference frame, with axes pointing in the principal directions, the analogy is incomplete, and rotation is generally more complex than translation, because the measures of inertia, I_j, are generally different for each principal axis.

Let me illustrate this fact on a simple but instructive system of three similar massive particles fixed in the vertices of an equilateral triangle (figure 4.3).

Due to the symmetry of the configuration, one of the principal axes has to pass through the center-of-mass 0, and be normal to the plane of the triangle. For the corresponding principal moment of inertia, Eq. (4.24) readily yields $I_3 = 3m\rho^2$. If we want to express the result in terms of the triangle side a, we may notice that due to the system's symmetry, the angle marked in figure 4.3 equals $\pi/6$, and from the shaded right triangle, $a/2 = \rho\cos(\pi/6) \equiv \rho\sqrt{3}/2$, giving $\rho = a/\sqrt{3}$, so that, finally, $I_3 = ma^2$.

Let us use this simple case to illustrate the following general *axis shift theorem*, which may be rather useful—particularly for more complex cases. For that, let us relate the inertia tensor components $I_{jj'}$ and $I'_{jj'}$, calculated in two reference frames—one with the origin in the center-of-mass 0, and another one (0′) displaced by a certain vector $\mathbf{d}$ (figure 4.4a), so that for an arbitrary point, $\mathbf{r}' = \mathbf{r} + \mathbf{d}$. Plugging this relation into Eq. (4.16), we obtain

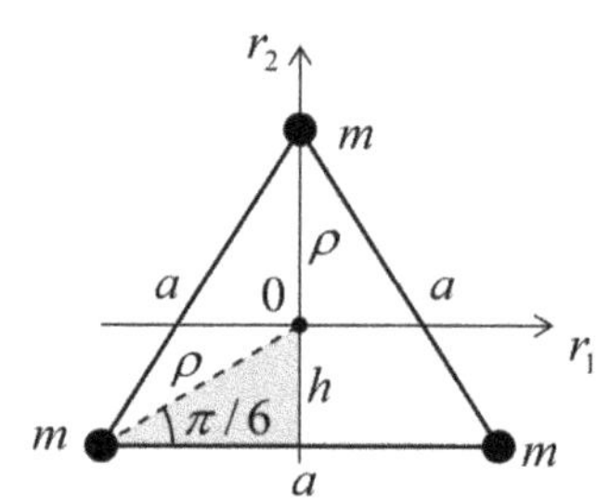

Figure 4.3. Principal moments of inertia: a simple case study.

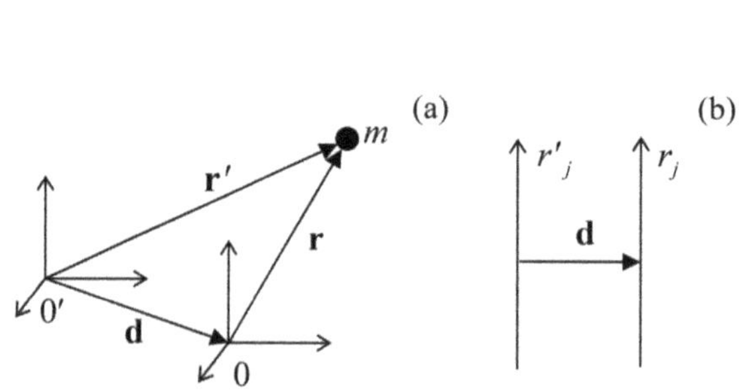

Figure 4.4. (a) A general reference frame shift from the center-of-mass and (b) a shift perpendicular to one of the principal axes.

$$\begin{aligned} I'_{jj'} &= \sum m\left[(\mathbf{r}+\mathbf{d})^2\delta_{jj'} - \left(r_j + d_j\right)\left(r_{j'} + d_{j'}\right)\right] \\ &= \sum m\left[(r^2 + 2\mathbf{r}\cdot\mathbf{d} + d^2)\delta_{jj'} - \left(r_j r_{j'} + r_j d_{j'} + r_{j'} d_j + d_j d_{j'}\right)\right]. \end{aligned} \tag{4.27}$$

Since in the center-of-mass frame, all sums $\sum m r_j$ equal zero, we may use Eq. (4.16) to finally obtain

$$I'_{jj'} = I_{jj'} + M\left(\delta_{jj'}d^2 - d_j d_{j'}\right). \tag{4.28}$$

In particular, this equation shows that if the shift vector $\mathbf{d}$ is perpendicular to one (say, jth) of the principal axes (figure 4.4b), i.e. $d_j = 0$, then Eq. (4.28) is reduced to a very simple formula:

$$I'_j = I_j + Md^2. \tag{4.29}$$

Now returning to the system shown in figure 4.3, let us perform such a shift to the new ('primed') axis passing through the location of one of the particles, still perpendicular to the particles' plane. Then the contribution of that particular mass to the primed moment of inertia vanishes, and $I'_3 = 2ma^2$. Now, returning to the center-of-mass and applying Eq. (4.29), we obtain $I_3 = I'_3 - M\rho^2 = 2ma^2 - (3m)(a/\sqrt{3})^2 = ma^2$, i.e. the same result as above.

The symmetry situation inside the triangle plane is somewhat less evident, so let us start with calculating the moments of inertia for the axes shown vertical and horizontal in figure 4.3. From Eq. (4.24) we readily obtain:

$$\begin{aligned} I_1 &= 2mh^2 + m\rho^2 = m\left[2\left(\frac{a}{2\sqrt{3}}\right)^2 + \left(\frac{a}{\sqrt{3}}\right)^2\right] = \frac{ma^2}{2}, \\ I_2 &= 2m\left(\frac{a}{2}\right)^2 = \frac{ma^2}{2}, \end{aligned} \tag{4.30}$$

where h is the distance from the center-of-mass and any side of the triangle: $h = \rho\sin(\pi/6) = \rho/2 = a/2\sqrt{3}$. We see that $I_1 = I_2$, and mathematics tells us that in this case *any* in-plane axis (passing through the center-of-mass 0) may be considered as principal, and has the same moment of inertia. A rigid body with this property, $I_1 = I_2 \neq I_3$, is called the *symmetric top*. (The last direction is called the *main principal axis* of the system.)

Despite the symmetric top's name, the situation may be even more symmetric in the so-called *spherical tops*, i.e. highly symmetric systems whose principal moments of inertia are all equal,

$$I_1 = I_2 = I_3 \equiv I. \tag{4.31}$$

Mathematics says that in this case the moment of inertia for rotation about *any* axis (but still passing through the center-of-mass) is equal to the same I. Hence Eqs. (4.25) and (4.26) are further simplified for any direction of vector $\boldsymbol{\omega}$:

$$T_{\text{rot}} = \frac{I}{2}\omega^2, \quad \mathbf{L} = I\boldsymbol{\omega}, \tag{4.32}$$

thus making the analogy of rotation and translation complete. (As will be discussed in the next section, the analogy is also complete if the rotation axis is fixed by external constraints.)

An evident example of a spherical top is a uniform sphere or spherical shell; a less obvious example is a uniform cube—with masses either concentrated in vertices, or uniformly spread over the faces, or uniformly distributed over the volume. Again, in this case *any* axis passing through the center-of-mass is principal, and has the same principal moment of inertia. For a sphere, this is natural; for a cube, rather surprising—but may be confirmed by a direct calculation.

4.3 Fixed-axis rotation

Now we are well equipped for a discussion of rigid body's rotational dynamics. The general equation of this dynamics is given by Eq. (1.38), which is valid for dynamics of any system of particles—either rigidly connected or not:

$$\dot{\mathbf{L}} = \boldsymbol{\tau}, \tag{4.33}$$

where $\boldsymbol{\tau}$ is the net torque of external forces. Let us start exploring this equation from the simplest case when the axis of rotation, i.e. the direction of vector $\boldsymbol{\omega}$, is fixed by some external constraints. Let us direct axis z along this vector; then $\omega_x = \omega_y = 0$. According to Eq. (4.22), in this case, the z-component of the angular momentum,

$$L_z = I_{zz}\omega_z, \tag{4.34}$$

where I_{zz}, though not necessarily one of the principal moments of inertia, still may be calculated using Eq. (4.24):

$$I_{zz} = \sum m\rho_z^2 = \sum m(x^2 + y^2), \tag{4.35}$$

with ρ_z being the distance of each particle from the rotation axis z. According to Eq. (4.15), the rotational kinetic energy in this case is just

$$T_{\text{rot}} = \frac{I_{zz}}{2}\omega_z^2. \tag{4.36}$$

Moreover, it is straightforward to use Eqs. (4.12), (4.17), and (4.28) to show that if the rotation axis is fixed, Eqs. (4.34)–(4.36) are valid even if the axis does not pass through the center-of-mass—if only distances ρ_z are now measured from that axis. (The proof is left for a reader's exercise.)

As a result, we may not care about other components of vector $\mathbf{L}$,[6] and use just one component of Eq. (4.33),

[6] Note that according to Eq. (4.22), other Cartesian components of the angular momentum, $L_x = I_{xz}\omega_z$ and $L_y = I_{yz}\omega_z$ may be different from zero, and even evolve in time. (Indeed, if axes x and y are fixed in the lab frame, I_{xz} and I_{yz} may change due to body's rotation.) The corresponding torques $\tau_x^{(\text{ext})}$ and $\tau_y^{(\text{ext})}$, which obey Eq. (4.33), are automatically provided by the external forces that keep the rotation axis fixed.

$$\dot{L}_z = \tau_z, \tag{4.37}$$

because it, when combined with (4.34), completely determines the dynamics of rotation:

$$I_{zz}\dot{\omega}_z = \tau_z, \quad \text{i.e. } I_{zz}\ddot{\theta}_z = \tau_z, \tag{4.38}$$

where θ_z is the angle of rotation about the axis, so that $\omega_z = \dot{\theta}$. The scalar relations (4.34), (4.36) and (4.38), describing rotation about a fixed axis, are completely similar to the corresponding formulas of 1D motion of a single particle, with ω_z corresponding to the usual ('linear') velocity, the angular-momentum component L_z to the linear momentum, and I_z to the particle's mass.

The resulting motion about the axis is also frequently similar to that of a single particle. As a simple example, let us consider what is called the *physical pendulum* (figure 4.5)—a rigid body free to rotate about a fixed horizontal axis that does not pass through the center-of-mass 0, in a uniform gravity field $\mathbf{g}$. Let us drop a perpendicular from point 0 to the rotation axis, and call the corresponding vector $\mathbf{l}$ (figure 4.5). Then the torque (relative to the rotation axis 0′) of the forces keeping the axis fixed is zero, and the only contribution to the net torque is due to gravity alone:

$$\begin{aligned} \boldsymbol{\tau}\big|_{\text{in }0'} &\equiv \sum \mathbf{r}\big|_{\text{in }0'} \times \mathbf{F} = \sum\left(\mathbf{l} + \mathbf{r}\big|_{\text{in }0}\right) \times m\mathbf{g} \\ &= \sum m(\mathbf{l} \times \mathbf{g}) + \sum m\,\mathbf{r}\big|_{\text{in }0} \times \mathbf{g} = M\mathbf{l} \times \mathbf{g}. \end{aligned} \tag{4.39}$$

(The last step used the facts that the point 0 is the center-of-mass, so that the second term on the right-hand side equals zero, and that the vectors $\mathbf{l}$ and $\mathbf{g}$ are the same for all particles of the body.)

This result shows that the torque is directed along the rotation axis, and its (only) component τ_z is equal to $-Mgl \sin\theta$, where θ is the angle between vectors $\mathbf{l}$ and $\mathbf{g}$, i.e. the angular deviation of the pendulum from the position of equilibrium. As a result, Eq. (4.38) takes the form,

$$I'\ddot{\theta} = -Mgl \sin\theta, \tag{4.40}$$

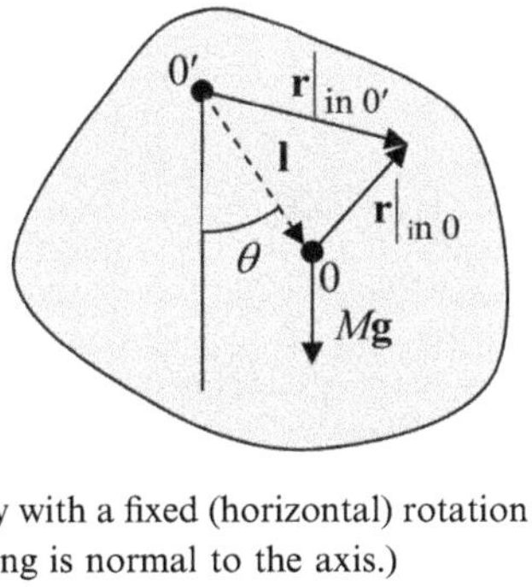

Figure 4.5. Physical pendulum: a body with a fixed (horizontal) rotation axis 0′ that does not pass through the center-of-mass 0. (The plane of drawing is normal to the axis.)

where I' is the moment of inertia for rotation about axis 0′ rather than about the center-of-mass. This equation is identical to Eq. (1.18) for the point-mass (sometimes called 'mathematical') pendulum, with a small-oscillation frequency

$$\Omega = \left(\frac{Mgl}{I'}\right)^{1/2}. \tag{4.41}$$

As a sanity check, in the simplest case when the linear size of the body is much smaller than the suspension length l, Eq. (4.35) yields $I' = Ml^2$, and Eq. (4.41) reduces to the well-known formula $\Omega = (g/l)^{1/2}$ for the point pendulum.

Now let us discuss the situations when a body not only rotates, but also moves as the whole. As we already know from our introductory chapter, the total momentum of the body,

$$\mathbf{P} \equiv \sum m\mathbf{v} = \sum m\dot{\mathbf{r}} = \frac{d}{dt}\sum m\mathbf{r}, \tag{4.42}$$

satisfies Newton's second law in the form (1.30). Using the definition (4.13) of the center-of-mass, the momentum may be represented as

$$\mathbf{P} = M\dot{\mathbf{R}} = M\mathbf{V}, \tag{4.43}$$

so that Eq. (1.30) may be rewritten as

$$M\dot{\mathbf{V}} = \mathbf{F}, \tag{4.44}$$

where **F** is the vector sum of all external forces. This equation shows that the center-of-mass of the body moves exactly as a point particle of mass M, under the effect of the net force **F**. In many cases this fact makes the translational dynamics of a rigid body absolutely similar to that of a point particle.

The situation becomes more complex if some of the forces contributing to the vector sum **F** depend on rotation of the same body, i.e. if its rotational and translational motions are coupled. Analysis of such coupled motion is rather straightforward if the *direction* of the rotation axis does not change in time, and hence Eqs. (4.35) and (4.36) are still valid. Possibly the simplest example is a round cylinder (say, a wheel) rolling on a surface without slippage (figure 4.6).

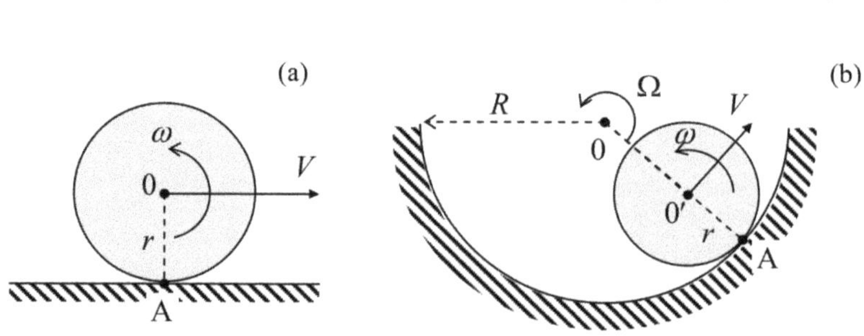

Figure 4.6. Round cylinder rolling over (a) a plane surface and (b) a concave surface.

The no-slippage condition may be represented as the requirement of the net velocity of the particular wheel point A that touches the surface to equal zero—in the reference frame connected to the surface. For the simplest case of plane surface (figure 4.6a), this condition may be spelled out using Eq. (4.10), giving the following relation between the angular velocity ω of the wheel and the linear velocity V of its center:

$$V + r\omega = 0. \tag{4.45}$$

Such kinematic relations are essentially holonomic constraints, which reduce the number of degrees of freedom of the system. For example, without the no-slippage condition (4.45), the wheel on a plane surface has to be considered as a system with two degrees of freedom, making its total kinetic energy (4.14) a function of two independent generalized velocities, say V and ω:

$$T = T_{\text{tran}} + T_{\text{rot}} = \frac{M}{2}V^2 + \frac{I}{2}\omega^2. \tag{4.46}$$

Using Eq. (4.45) we may eliminate, for example, the linear velocity and reduce Eq. (4.46) to

$$T = \frac{M}{2}(\omega r)^2 + \frac{I}{2}\omega^2 = \frac{I_{\text{ef}}}{2}\omega^2, \quad \text{where } I_{\text{ef}} \equiv I + Mr^2. \tag{4.47}$$

This result may be interpreted as the kinetic energy of pure rotation of the wheel about the instantaneous rotation axis A, with I_{ef} being the moment of inertia about that axis, satisfying Eq. (4.29).

Kinematic relations are not always as simple as Eq. (4.45). For example, if the wheel is rolling on a concave surface (figure 4.6b), we need relate the angular velocities of wheel's rotation about its axis 0 (say, ω) and that (say, Ω) of its axis' rotation about the center $0'$ of curvature of the surface. A popular error here is to write $\Omega = -(r/R)\omega$ (WRONG!). A prudent way to derive the correct relation is to note that Eq. (4.45) holds for this situation as well, and on the other hand the same linear velocity of wheel's center may be expressed as $V = (R-r)\Omega$. Combining these formulas, we obtain the correct relation

$$\Omega = -\frac{r}{R-r}\omega. \tag{4.48}$$

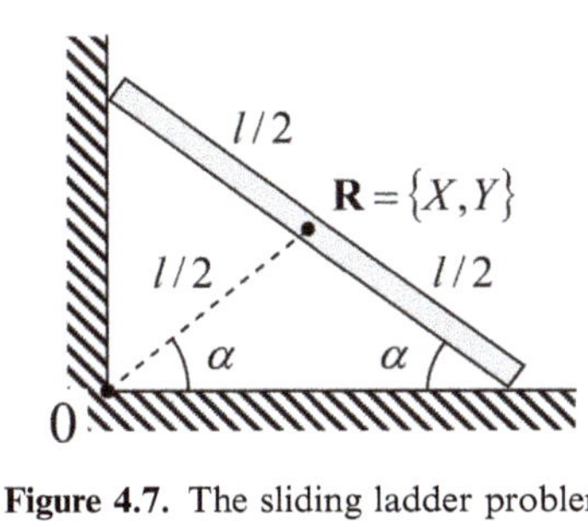

Figure 4.7. The sliding ladder problem.

Another famous example of the relation between the translational and rotational motion is given by the 'sliding ladder' problem (figure 4.7). Let us analyze it for the simplest case of negligible friction, and a ladder thickness that is small in comparison with its length l.

In order to use the Lagrangian formalism, we may write the kinetic energy of the ladder as the sum (4.14) of the translational and rotational parts:

$$T = \frac{M}{2}(\dot{X}^2 + \dot{Y}^2) + \frac{I}{2}\dot{\alpha}^2, \tag{4.49}$$

where X and Y are the Cartesian coordinates of its center-of-mass in an inertial reference frame, and I is the moment of inertia for rotation about the z-axis passing through the center-of-mass. (For the uniformly distributed mass, an elementary integration of Eq. (4.35) yields $I = Ml^2/12$.) In the reference frame with the center in the corner 0, both X and Y may be simply expressed via angle α:

$$X = \frac{l}{2}\cos\alpha, \qquad Y = \frac{l}{2}\sin\alpha. \tag{4.50}$$

(The easiest way to obtain these relations is to notice that the dashed line in figure 4.7 has the same slope α, and length $l/2$.) Plugging these expressions into Eq. (4.49), we obtain

$$T = \frac{I_{\text{ef}}}{2}\dot{\alpha}^2, \qquad I_{\text{ef}} \equiv I + M\left(\frac{l}{2}\right)^2 = \frac{1}{3}Ml^2. \tag{4.51}$$

Since the potential energy of the ladder in the gravity field may be also expressed via the same angle,

$$U = MgY = Mg\frac{l}{2}\sin\alpha, \tag{4.52}$$

α may be conveniently used as the (only) generalized coordinate of the system. Even without writing the Lagrangian equation of motion for that coordinate explicitly, we may notice that since the Lagrangian function $L \equiv T - U$ does not depend on time explicitly, and the kinetic energy (4.51) is a quadratic-homogeneous function of the generalized velocity $\dot{\alpha}$, the full mechanical energy,

$$E \equiv T + U = \frac{I_{\text{ef}}}{2}\dot{\alpha}^2 + Mg\frac{l}{2}\sin\alpha = \frac{Mgl}{2}\left(\frac{l\dot{\alpha}^2}{3g} + \sin\alpha\right), \tag{4.53}$$

is conserved, giving us the first integral of motion. Moreover, Eq. (4.53) shows that the system's energy (and hence dynamics) is identical to that of a physical pendulum with an unstable fixed point $\alpha_1 = \pi/2$, stable fixed point at $\alpha_2 = -\pi/2$, and frequency

$$\Omega = \left(\frac{3g}{2l}\right)^{1/2} \tag{4.54}$$

of small oscillations near the latter point. (Of course, this fixed point cannot be reached in the simple geometry shown in figure 4.7, where the ladder's fall on the floor would change its equations of motion. Moreover, even before that, the left end of the ladder may detach from the wall. The analysis of this issue is left for a reader's exercise.)

4.4 Free rotation

Now let us proceed to more complex case when the rotation axis is *not* fixed. A good illustration of the complexity arising is this case comes from the simplest case of a rigid body left alone, i.e. not subjected to external forces and hence its potential energy U is constant. Since in this case, according to Eq. (4.44), the center-of-mass moves (as measured from any inertial reference frame) with a constant velocity, we can always use a convenient inertial reference frame with the origin at that point. From the point of view of such a frame, the body's motion is a pure rotation, and $T_{\text{tran}} = 0$. Hence, the system's Lagrangian equals just the rotational energy (4.15), which is, first, a quadratic-homogeneous function of components ω_j (which may be taken for generalized velocities), and, second, does not depend on time explicitly. As we know from chapter 2, in this case the mechanical energy, here equal to T_{rot} alone, is conserved. For the components of vector $\boldsymbol{\omega}$ in the principal axes, this means that

$$T_{\text{rot}} = \sum_{j=1}^{3} \frac{I_j}{2} \omega_j^2 = \text{const.} \tag{4.55}$$

Next, as Eq. (4.33) shows, in the absence of external forces the angular momentum $\mathbf{L}$ of the body is conserved as well. However, though we can certainly use Eq. (4.26) to represent this fact as

$$\mathbf{L} = \sum_{j=1}^{3} I_j \omega_j \mathbf{n}_j = \text{const}, \tag{4.56}$$

where $\mathbf{n}_j$ are the principal axes of inertia, this does not mean that components ω_j of the angular velocity vector $\boldsymbol{\omega}$ are constant, because the principal axes are fixed relative to the rigid body, and hence may rotate with it.

Before exploring these complications, let us briefly mention two conceptually trivial, but practically very important, particular cases. The first is a spherical top ($I_1 = I_2 = I_3 = I$). In this case Eqs. (4.55) and (4.56) imply that all components of vector $\boldsymbol{\omega} = \mathbf{L}/I$, i.e. both the magnitude and the direction of the angular velocity, are conserved for any initial spin. In other words, the body conserves its rotation speed and axis direction, as measured in an inertial frame.

The most obvious example is a spherical planet. For example, our Mother Earth, rotating about its axis with angular velocity $\omega = 2\pi/(1 \text{ day}) \approx 7.3 \times 10^{-5} \text{ s}^{-1}$, keeps its axis at a nearly constant angle of $23°27'$ to the *ecliptic pole*, i.e. the axis normal to the plane of its motion around the Sun. (In section 4.6 below, we will discuss some very slow motions of this axis, due to gravity effects.)

Spherical tops are also used in the most accurate gyroscopes, usually with gas-jet or magnetic suspension in a vacuum. If done carefully, such systems may have

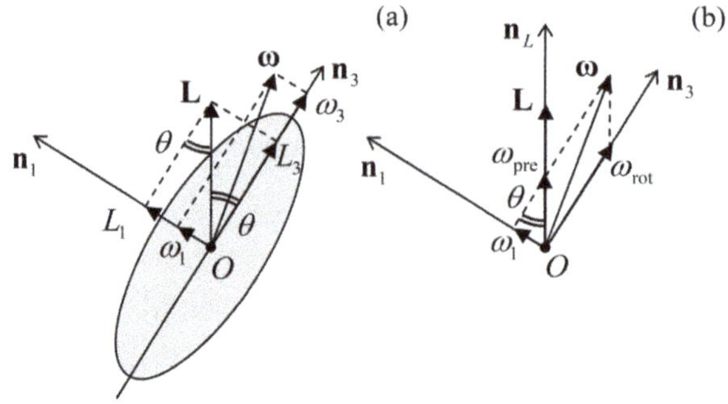

Figure 4.8. Free rotation of a symmetric top: (a) the general configuration of vectors and (b) calculating the free-precession frequency.

spectacular stability. For example, the gyroscope system of the Gravity Probe B satellite experiment, flown in 2004–5, was based on quartz spheres—round with a precision of about 10 nm and covered with superconducting thin films (which have enabled their magnetic suspension and monitoring). The whole system was stable enough to measure that the so-called *geodetic effect* in general relativity (essentially, the space curving by Earth's mass), resulting in the axis precession by just 6.6 arc seconds per year, i.e. with a precession frequency of just $\sim 10^{-11}\ \text{s}^{-1}$, which agrees with theory with a record ~0.3% accuracy[7].

The second simple case is that of the symmetric top ($I_1 = I_2 \neq I_3$), with the initial vector **L** aligned with the main principal axis. In this case, $\boldsymbol{\omega} = \mathbf{L}/I_3 = \text{const}$, so that the rotation axis is conserved[8]. Such tops, typically in the shape of a *flywheel* (heavy, flat rotor), and supported by a three-ring *gimbal* system (which allows for torque-free rotation in all three directions)[9], are broadly used in more common gyroscopes—core parts of automatic guidance systems, for example, in ships, airplanes, missiles, etc. Even if the ship's hull wobbles or drifts, the suspended gyroscope sustains its direction relative to the Earth (which is sufficiently inertial for these applications)[10].

However, in the general case with no such special initial alignment, the dynamics of symmetric tops is more complex. In this case, the vector **L** is still conserved, including its direction, but the vector $\boldsymbol{\omega}$ is not. Indeed, let us direct axis $\mathbf{n}_2$ perpendicular to the common plane of vectors **L** and the current instantaneous direction $\mathbf{n}_3$ of the main principal axis (in figure 4.8 below, the plane of drawing); then, in that particular instant, $L_2 = 0$. Now let us recall that in a symmetric top, axis $\mathbf{n}_2$ is a principal one.

[7] However, the main goal of this expensive (~$750M) project, an accurate measurement of a subtler relativistic effect, the so-called *frame-dragging drift* (also called 'the Schiff precession'), predicted to be about 0.04 arc seconds per year, has not been achieved.

[8] This is also true for an asymmetric top, i.e. an arbitrary body (with, say, $I_1 < I_2 < I_3$), but in this case the alignment of vector **L** with axis $\mathbf{n}_2$, corresponding to the intermediate moment of inertia, is unstable.

[9] See, for example, a very nice animation available online at http://en.wikipedia.org/wiki/Gimbal.

[10] The much more compact (and much less accurate) gyroscopes used, for example, in smartphones and tablet computers, are based on a different effect of rotation on the oscillator frequency, and are implemented as micro-electromechanical systems (MEMS) on the silicon chip surface—see, e.g. [1] chapter 22.

According to Eq. (4.26) with $j = 2$, the corresponding component ω_2 has to be equal to L_2/I_2, so it is equal to zero. This means that the vector $\boldsymbol{\omega}$ lies in this plane (the common plane of vectors $\mathbf{L}$ and $\mathbf{n}_3$) as well—see figure 4.8a.

Now consider any point located on the main axis $\mathbf{n}_3$, and hence on the plane $[\mathbf{n}_3, \mathbf{L}]$. Since $\boldsymbol{\omega}$ is the instantaneous axis of rotation, according to Eq. (4.9), the instantaneous velocity $\mathbf{v} = \boldsymbol{\omega} \times \mathbf{r}$ of this point is directed normally to that plane. Since this is true for each point of the main axis (apart from one, with $\mathbf{r} = 0$, i.e. the center-of-mass, which does not move), this axis as a whole has to move perpendicular to the common plane of vectors $\mathbf{L}$, $\boldsymbol{\omega}$, and $\mathbf{n}_3$. Since such conclusion is valid for any moment of time, it means that vectors $\boldsymbol{\omega}$ and $\mathbf{n}_3$ rotate about the space-fixed vector $\mathbf{L}$ together, with some angular velocity ω_{pre}, at each moment staying in one plane. This effect is usually called the *free precession* (or 'torque-free', or 'regular' precession) and has to be clearly distinguished it from the completely different effect of the *torque-induced precession* which will be discussed in the next section.

In order to calculate ω_{pre}, let us represent the instant vector $\boldsymbol{\omega}$ as a sum of not its Cartesian coordinates (as in figure 4.8a), but rather of two non-orthogonal vectors directed along $\mathbf{n}_3$ and $\mathbf{L}$ (figure 4.8b):

$$\boldsymbol{\omega} = \omega_{\text{rot}}\mathbf{n}_3 + \omega_{\text{pre}}\mathbf{n}_L, \qquad \mathbf{n}_L \equiv \frac{\mathbf{L}}{L}. \tag{4.57}$$

Figure 4.8b shows that ω_{rot} has the meaning of the angular velocity of body rotation of the body about its main principal axis, while ω_{pre} is the angular velocity of rotation of that axis about the constant direction of vector $\mathbf{L}$, i.e. the frequency of precession—exactly what we are trying to find. Now ω_{pre} may be readily calculated from the comparison of two panels of figure 4.8, by noticing that the same angle θ between vectors $\mathbf{L}$ and $\mathbf{n}_3$ participates in two relations:

$$\sin\theta = \frac{L_1}{L} = \frac{\omega_1}{\omega_{\text{pre}}}. \tag{4.58}$$

Since the axis $\mathbf{n}_1$ is a principal one, we may use Eq. (4.26) for $j = 1$, i.e. $L_1 = I_1\omega_1$, to eliminate ω_1 from Eq. (4.58), and get a very simple formula

$$\omega_{\text{pre}} = \frac{L}{I_1}. \tag{4.59}$$

This result shows that the precession *frequency* is constant and independent of the alignment of the vector $\mathbf{L}$ with the main principal axis $\mathbf{n}_3$, while the *amplitude* of this motion (characterized by the angle θ) does depend on the alignment, and vanishes if $\mathbf{L}$ is parallel to $\mathbf{n}_3$[11]. Note also that if all principal moments of inertia are of the same order, ω_{pre} is of the same order as the total angular velocity $\omega = |\boldsymbol{\omega}|$ of rotation.

[11] For our Earth, the free-precession amplitude is so small (corresponding to less than 10 m of linear displacement on the Earth surface) that this effect is of the same order as other, irregular motions of the rotation axis, resulting from the turbulent fluid flow effects in the planet's interior and its atmosphere.

Now let us briefly discuss the free precession in the general case of an 'asymmetric top', i.e. a body with arbitrary $I_1 \neq I_2 \neq I_3$. In this case the effect is more complex because here not only the *direction* but also the *magnitude* of the instantaneous angular velocity $\boldsymbol{\omega}$ may evolve in time. If we are only interested in the relation between the instantaneous values of ω_j and L_j, i.e. the 'trajectories' of the vectors $\boldsymbol{\omega}$ and $\mathbf{L}$ as observed from the reference frame $\{\mathbf{n}_1, \mathbf{n}_2, \mathbf{n}_3\}$ of the principal axes of the body (rather than the explicit law of their time evolution), they may be found directly from the conservation laws. (Let me emphasize again that the vector $\mathbf{L}$, being constant in an inertial frame, generally evolves in the frame rotating with the body.) Indeed, Eq. (4.55) may be understood as the equation of an ellipsoid in Cartesian coordinates $\{\omega_1, \omega_2, \omega_3\}$, so that for a free body, the vector $\boldsymbol{\omega}$ has to stay on the surface of that ellipsoid[12]. On the other hand, since the reference-frame rotation preserves the length of any vector, the *magnitude* (but not the direction!) of the vector $\mathbf{L}$ is also an integral of motion in the moving frame, and we can write

$$L^2 \equiv \sum_{j=1}^{3} L_j^2 = \sum_{j=1}^{3} I_j^2 \omega_j^2 = \text{const.} \tag{4.60}$$

Hence the trajectory of the vector $\boldsymbol{\omega}$ follows the closed curve formed by the intersection of two ellipsoids, (4.55) and (4.60). It is evident that this trajectory is generally 'taco-edge-shaped', i.e. more complex than a plane circle, but never very complex either.

The same argument may be repeated for the vector $\mathbf{L}$, for whom the first form of Eq. (4.60) descries a sphere, and Eq. (4.55), another ellipsoid:

$$T_{\text{rot}} = \sum_{j=1}^{3} \frac{1}{2I_j} L_j^2 = \text{const.} \tag{4.61}$$

On the other hand, if we are interested in the trajectory of the vector $\boldsymbol{\omega}$ as observed from an inertial frame (in which the vector $\mathbf{L}$ stays still), we may note that the general relation (4.15) for the same rotational energy T_{rot} may also be rewritten as

$$T_{\text{rot}} = \frac{1}{2} \sum_{j=1}^{3} \omega_j \sum_{j'=1}^{3} I_{jj'} \omega_{j'}. \tag{4.62}$$

But according to Eq. (4.22), the second sum on the right-hand side is nothing more than L_j, so that

$$T_{\text{rot}} = \frac{1}{2} \sum_{j=1}^{3} \omega_j L_j = \frac{1}{2} \boldsymbol{\omega} \cdot \mathbf{L}. \tag{4.63}$$

This equation shows that for a free body ($T_{\text{rot}} = \text{const}$, $\mathbf{L} = \text{const}$), even if the vector $\boldsymbol{\omega}$ changes in time, its end point should stay within a plane perpendicular to the

---

[12] It is frequently called the *Poinsot's ellipsoid*, after L Pointset (1777–1859) who made several pioneering contributions to rigid-body mechanics.

angular momentum **L**. (Earlier, we have seen that for the particular case of the symmetric top—see figure 4.8b, but for an asymmetric top, the trajectory of the end point may not be circular.)

If we are interested not only in the trajectory of vector $\boldsymbol{\omega}$, but also its explicit evolution in time, it may be calculated using the general Eq. (4.33) represented in principal components ω_j. For that, we have to recall that Eq. (4.33) is only valid in an inertial reference frame, while the frame $\{\mathbf{n}_1, \mathbf{n}_2, \mathbf{n}_3\}$ may rotate with the body and hence is generally not inertial. We may handle this problem by applying, to the vector **L**, the general relation (4.8):

$$\frac{d\mathbf{L}}{dt}\Big|_{\text{in lab}} = \frac{d\mathbf{L}}{dt}\Big|_{\text{in mov}} + \boldsymbol{\omega} \times \mathbf{L}. \tag{4.64}$$

Combining it with Eq. (4.33), in the moving frame we obtain

$$\frac{d\mathbf{L}}{dt} + \boldsymbol{\omega} \times \mathbf{L} = \boldsymbol{\tau}, \tag{4.65}$$

where $\boldsymbol{\tau}$ is the external torque. In particular, for the principal-axis components L_j, related to components ω_j by Eq. (4.26), Eq. (4.65) is reduced to a set of three scalar *Euler equations*

$$I_j\dot{\omega}_j + \left(I_{j''} - I_{j'}\right)\omega_{j'}\omega_{j''} = \tau_j, \tag{4.66}$$

where the set of indices $\{j, j', j''\}$ has to follow the usual 'right' order—e.g. $\{1, 2, 3\}$, etc[13].

In order to obtain a feeling of how do the Euler equations work, let us return to the particular case of a free symmetric top ($\tau_1 = \tau_2 = \tau_3 = 0$, $I_1 = I_2 \neq I_3$). In this case, $I_1 - I_2 = 0$, so that Eq. (4.66) with $j = 3$ yields $\omega_3 = \text{const}$, while the equations for $j = 1$ and $j = 2$ take the following simple forms:

$$\dot{\omega}_1 = -\Omega_{\text{pre}}\omega_2, \qquad \dot{\omega}_2 = \Omega_{\text{pre}}\omega_1, \tag{4.67}$$

where Ω_{pre} is a constant determined by both the system parameters and the initial conditions:

$$\Omega_{\text{pre}} \equiv \omega_3\frac{I_3 - I_1}{I_1}. \tag{4.68}$$

Obviously, the system of equations (4.67) has a sinusoidal solution with frequency Ω_{pre}, and describe a uniform rotation of vector $\boldsymbol{\omega}$, with that frequency, about the main axis $\mathbf{n}_3$. This is just another representation of the torque-free precession analyzed above, this time as observed from the rotating body. Evidently, Ω_{pre} is substantially different from the frequency ω_{pre} (4.59) of the precession as observed

[13] These equations are of course valid in the simplest case of the fixed rotation axis as well. For example, if $\boldsymbol{\omega} = \mathbf{n}_z\omega$, i.e. $\omega_x = \omega_y = 0$, Eq. (4.66) is reduced to Eq. (4.38).

from the lab frame; for example, the former frequency vanishes for the spherical top (with $I_1 = I_2 = I_3$), while the latter frequency tends to the rotation frequency.

Unfortunately, for the rotation of an asymmetric top (i.e. an arbitrary rigid body), when no component ω_j is conserved, the Euler equations (4.66) are strongly nonlinear even in the absence of any external torque, and a discussion of their solutions would take more time than I can afford[14].

4.5 Torque-induced precession

The dynamics of rotation becomes even more complex in the presence of external forces. Let us consider the most important and counter-intuitive effect of *torque-induced precession*, for the simplest case of an axially symmetric body (which is a particular case of the symmetric top, $I_1 = I_2 \neq I_3$), supported at some point A of its symmetry axis that does not coincide with the center-of-mass 0 (figure 4.9). The uniform gravity field $\mathbf{g}$ creates bulk-distributed forces that, as we know from the analysis of the physical pendulum in section 4.3, are equivalent to a single force $M\mathbf{g}$ applied in the center-of-mass—in figure 4.9, point 0. The torque of this force relative to the support point A is

$$\boldsymbol{\tau} = \mathbf{r}_0 \Big|_{\text{in A}} \times M\mathbf{g} = Ml\,\mathbf{n}_3 \times \mathbf{g}. \tag{4.69}$$

Hence the general equation (4.33) of the angular-momentum evolution (valid in any inertial frame, for example the one with an origin in point A) becomes

$$\dot{\mathbf{L}} = Ml\,\mathbf{n}_3 \times \mathbf{g}. \tag{4.70}$$

Despite the apparent simplicity of this (exact!) equation, its analysis is straightforward only in the case when the top is launched spinning about its symmetry axis $\mathbf{n}_3$ with a very high angular velocity ω_{rot}. In this case, we may neglect the contribution to $\mathbf{L}$ due to a small precession velocity ω_{pre} (still to be calculated), and use Eq. (4.26) to write

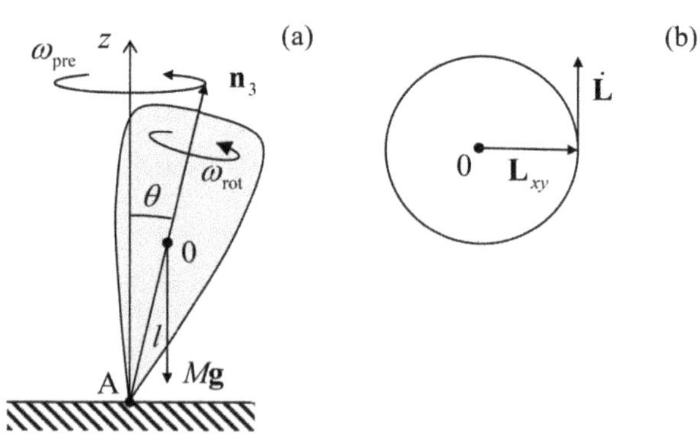

Figure 4.9. Symmetric top in the gravity field: (a) a side view at the system and (b) the top view at the evolution of the horizontal component of the angular-momentum vector.

[14] Such discussion may be found, e.g. in section 37 of [2].

$$\mathbf{L} = I_3\boldsymbol{\omega} = I_3\omega_{\text{rot}}\mathbf{n}_3. \tag{4.71}$$

Then (4.70) shows that the vector $\dot{\mathbf{L}}$is perpendicular to both $\mathbf{n}_3$ (and hence **L**) and **g**, i.e. lies within the horizontal plane, and is perpendicular to the horizontal component $\mathbf{L}_{xy}$ of the vector **L**—see figure 4.9b. Since, according to Eq. (4.70), the magnitude of this vector is constant, $|\dot{\mathbf{L}}| = mgl \sin \theta$, the vector **L** (and hence the body's main axis) rotates about the vertical axis with angular velocity

$$\omega_{\text{pre}} = \frac{|\dot{\mathbf{L}}|}{L_{xy}} = \frac{Mgl \sin \theta}{L \sin \theta} = \frac{Mgl}{L} = \frac{Mgl}{I_3\omega_{\text{rot}}}. \tag{4.72}$$

Thus, very counter-intuitively, the fast-rotating top 'does not want to' follow the external, vertical force and, in addition to fast spinning about the symmetry axis $\mathbf{n}_3$, also performs a revolution, called the *torque-induced precession*, about the vertical axis. Note that, similarly to the free-precession frequency (4.59), the torque-induced precession frequency (4.72) does not depend on the initial (and sustained) angle θ. However, the torque-induced precession frequency is inversely (rather than directly) proportional to ω_{rot}. This fact makes the above simple theory valid in many practical cases. Indeed, Eq. (4.71) may be used only if the contribution of the precession into **L** is relatively small: $I\omega_{\text{pre}} \ll I_3\omega_{\text{rot}}$, where I is a certain effective moment of inertia for the precession (to be calculated below). Using our result (4.72), this condition may be rewritten as

$$\omega_{\text{rot}} \gg \left(\frac{MglI}{I_3^2}\right)^{1/2}. \tag{4.73}$$

According to Eq. (4.16), for a body of not too extreme proportions, i.e. with all linear dimensions of the order of the same length scale l, all inertia moments are of the order of Ml^2, so that the right-hand side of Eq. (4.73) is of the order of $(g/l)^{1/2}$, i.e. comparable to the frequency of oscillations of the same body as the physical pendulum, in the absence of its fast rotation.

In order to develop a qualitative theory that would be valid beyond such approximate treatment, the Euler equation (4.66) may be used, but are not very convenient. A better approach, suggested by the same L Euler, is to introduce a set of three independent angles between the principal axes $\{\mathbf{n}_1, \mathbf{n}_2, \mathbf{n}_3\}$ bound to the rigid body, and the axes $\{\mathbf{n}_x, \mathbf{n}_y, \mathbf{n}_z\}$ of an inertial reference frame (figure 4.10), and then express the basic equation (4.33) of rotation, via these angles. There are several possible options for the definition of such angles; figure 4.10 shows the set of *Euler angels*, most convenient for analyses of fast rotation[15]. As one can see, the first Euler angle, θ, is the usual polar angle measured from axis $\mathbf{n}_z$ to axis $\mathbf{n}_3$. The second one is the azimuthal angle φ, measured from axis $\mathbf{n}_x$ to the so-called *line of nodes* formed by the intersection of planes $[\mathbf{n}_x, \mathbf{n}_y]$ and $[\mathbf{n}_1, \mathbf{n}_2]$. The last Euler angle, ψ, is measured within plane $[\mathbf{n}_1, \mathbf{n}_2]$, from the line of nodes to axis $\mathbf{n}_1$. For example, in the simple

[15] Of the several choices more convenient in the absence of fast rotation, the most common is the set of so-called *Tait–Bryan angles* (called the *yaw*, *pitch*, and *roll*), which are broadly used for aircraft and maritime navigation.

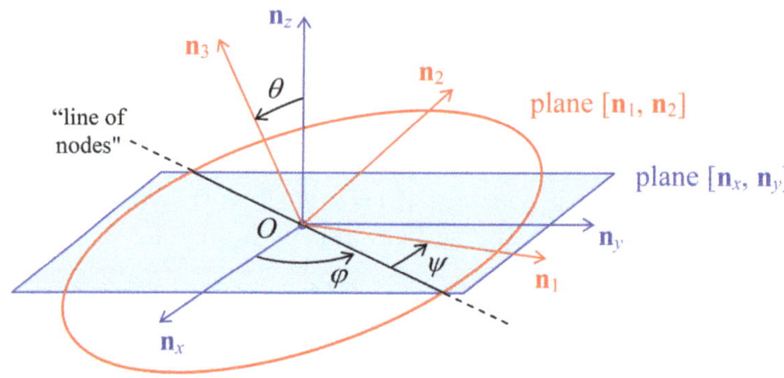

Figure 4.10. Definition of the Euler angles.

picture of the force-induced precession of a symmetric top, discussed above, the angle θ is constant, the angle ψ changes very rapidly, with the rotation velocity ω_{rot}, while the angle φ evolves with the precession frequency ω_{pre} (4.72).

Now we can express the principal-axes components of the instantaneous angular velocity vector, ω_1, ω_2, and ω_3, as measured in the lab reference frame, in terms of the Euler angles. It may readily be done by calculating, from figure 4.10, the contributions to the change of Euler angles to the rotation about each principal axis, and then adding them up:

$$\begin{aligned}\omega_1 &= \dot{\varphi} \sin\theta \sin\psi + \dot{\theta} \cos\psi,\\ \omega_2 &= \dot{\varphi} \sin\theta \cos\psi - \dot{\theta} \sin\psi,\\ \omega_3 &= \dot{\varphi} \cos\theta + \dot{\psi}.\end{aligned} \tag{4.74}$$

These relations enable the expression of the kinetic energy of rotation (4.25) and the angular-momentum components (4.26) via the generalized coordinates θ, φ, and ψ and their time derivatives (i.e. the corresponding generalized velocities), and then using the powerful Lagrangian formalism to derive their equations of motion. This is especially simple to do in the case of symmetric tops (with $I_1 = I_2$), because plugging Eqs. (4.74) into Eq. (4.25) we obtain an expression,

$$T_{\mathrm{rot}} = \frac{I_1}{2}(\dot{\theta}^2 + \dot{\varphi}^2 \sin^2\theta) + \frac{I_3}{2}(\dot{\varphi}\cos\theta + \dot{\psi})^2, \tag{4.75}$$

which does not include explicitly either φ or ψ. (This reflects the fact that for a symmetric top we can always select the axis $\mathbf{n}_1$ to coincide with the line of nodes, and hence take $\psi = 0$ at the considered moment of time. Note that this trick does *not* mean we can take $\dot{\psi} = 0$, because the axis $\mathbf{n}_1$, as observed from the inertial reference frame, moves!) Now we should not forget that at the torque-induced precession, the center-of-mass moves as well (see figure 4.9), so that according to Eq. (4.14), the total kinetic energy of the body is the sum of two terms,

$$T = T_{\mathrm{rot}} + T_{\mathrm{tran}}, \qquad T_{\mathrm{tran}} = \frac{M}{2}V^2 = \frac{M}{2}l^2(\dot{\theta}^2 + \dot{\varphi}^2 \sin^2\theta), \tag{4.76}$$

while the potential energy is just

$$U = Mgl\cos\theta + \text{const.} \tag{4.77}$$

Now we could readily write the Lagrangian equations of motion for the Euler angles, but it is better to immediately notice that according to Eqs. (4.75)–(4.77), the Lagrangian function, $T-U$, does not depend explicitly on the 'cyclic' coordinates φ and ψ, so that the corresponding generalized momenta (2.31) are conserved:

$$p_\varphi \equiv \frac{\partial T}{\partial \dot\varphi} = I_A\dot\varphi\sin^2\theta + I_3(\dot\varphi\cos\theta + \dot\psi)\cos\theta = \text{const}, \tag{4.78}$$

$$p_\psi \equiv \frac{\partial T}{\partial \dot\psi} = I_3(\dot\varphi\cos\theta + \dot\psi) = \text{const}, \tag{4.79}$$

where $I_A \equiv I_1 + Ml^2$. (According to Eq. (4.29), I_A is just the body's moment of inertia for rotation about a horizontal axis passing through the support point A.) According to the last equation of Eq. (4.74), p_ψ is just L_3, i.e. the angular momentum's component along the *rotating* axis $\mathbf{n}_3$. On the other hand, by its very definition (4.78), p_φ is L_z, the same vector $\mathbf{L}$'s component along the *static* axis z. (Actually, we could foresee in advance the conservation of both these components of $\mathbf{L}$, because the vector (4.69) of the external torque is perpendicular to both $\mathbf{n}_3$ and $\mathbf{n}_z$.) Using this notation, and solving the simple system of linear equations (4.78)–(4.79) for the angle derivatives, we obtain

$$\dot\varphi = \frac{L_z - L_3\cos\theta}{I_A\sin^2\theta}, \qquad \dot\psi = \frac{L_3}{I_3} - \frac{L_z - L_3\cos\theta}{I_A\sin^2\theta}\cos\theta. \tag{4.80}$$

One more conserved quantity in this problem is the full mechanical energy[16]

$$E \equiv T + U = \frac{I_A}{2}(\dot\theta^2 + \dot\varphi^2\sin^2\theta) + \frac{I_3}{2}(\dot\varphi\cos\theta + \dot\psi)^2 + Mgl\cos\theta. \tag{4.81}$$

Plugging Eqs. (4.80) into Eq. (4.81), we obtain a first-order differential equation for angle θ, which may be represented in the following physically transparent form:

$$\begin{aligned} &\frac{I_A}{2}\dot\theta^2 + U_{ef}(\theta) = E, \\ &U_{ef}(\theta) \equiv \frac{(L_z - L_3\cos\theta)^2}{2I_A\sin^2\theta} + \frac{L_3^2}{2I_3} + Mgl\cos\theta + \text{const.} \end{aligned} \tag{4.82}$$

Thus, similarly to the planetary problems considered in section 3.4, the torque-induced precession of a symmetric top has been reduced (without any approximations!) to a 1D problem of motion of one of its degrees of freedom, the polar angle θ, in an effective potential $U_{ef}(\theta)$. According to Eq. (4.82), very similarly to Eq. (3.44) for the planetary problem, this potential is the sum of the actual potential energy U, given by Eq. (4.77), and a contribution from the kinetic energy of motion along two

[16] Indeed, since the Lagrangian does not depend on time explicitly, $H = \text{const}$, and since the full kinetic energy T (4.75)–(4.76) is a quadratic-homogeneous function of the generalized velocities, $E = H$.

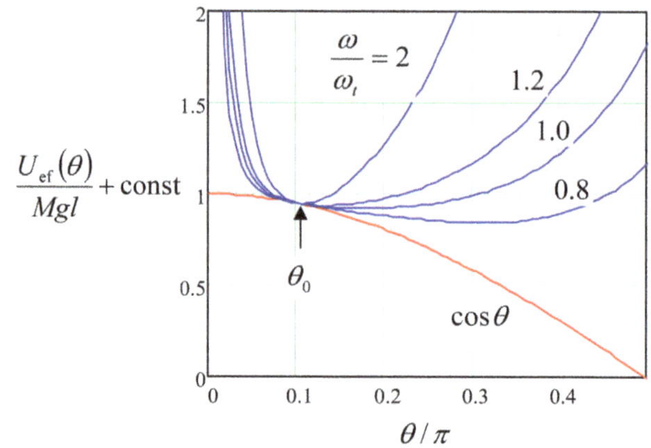

Figure 4.11. The effective potential energy U_{ef} of the symmetric top, given by Eq. (4.82), as a function of the polar angle θ, for a particular value (0.95) of the ratio $r \equiv L_z/L_3$ (so that at $\omega_{rot} \gg \omega_{th} \gg 1$, $\theta_0 = \cos^{-1}r \approx 0.1011\pi$), and several values of the ratio ω_{rot}/ω_{th}.

other angles. In the absence of rotation about axes $\mathbf{n}_z$ and $\mathbf{n}_3$ (i.e. $L_z = L_3 = 0$), Eq. (4.82) is reduced to the first integral of equation (4.40) of motion of a physical pendulum. If the rotation is present, then (apart from the case of very special initial conditions when $\theta(0) = 0$ *and* $L_z = L_3$)[17], the first contribution to $U_{ef}(\theta)$ diverges at $\theta \to 0$ and π, so that the effective potential energy has a minimum at some non-zero value θ_0 of the polar angle θ—see figure 4.11.

If the initial angle θ (0) equals this θ_0, i.e. if the initial effective energy is equal to its minimum value $U_{ef}(\theta_0)$, the polar angle remains constant through the motion: $\theta(t) = \theta_0$. This corresponds to the pure torque-induced precession whose angular velocity is given by the first of Eqs. (4.80):

$$\omega_{pre} \equiv \dot{\varphi} = \frac{L_z - L_3 \cos\theta_0}{I_A \sin^2\theta_0}. \tag{4.83}$$

The condition for finding θ_0, $dU_{ef}/d\theta = 0$, is a transcendental algebraic equation that cannot be solved analytically for arbitrary parameters. However, in the high spinning speed limit (4.73) this is possible. Indeed, in this limit the Mgl contribution to U_{ef} is small, and we may analyze its effect by successive approximations. In the 0th approximation, i.e. at $Mgl = 0$, the minimum of U_{ef} is evidently achieved at $\cos\theta_0 = L_z/L_3$, turning the precession frequency (4.83) to zero. In the next, first approximation, we may require that at $\theta = \theta_0$, the derivative of the first term of Eq. (4.82) for U_{ef} over $\cos\theta$, equal to $-L_z(L_z-L_3\cos\theta)/I_A \sin^2\theta$,[18] is cancelled with that

[17] In that simple case the body continues to rotate about the vertical symmetry axis: $\theta(t) = 0$. Note, however, that such motion is stable only if the spinning speed is sufficiently high—see (4.85) below.

[18] Indeed, the derivative of the fraction $1/2I_A \sin^2\theta$, taken at the point $\cos\theta = L_z/L_3$, is multiplied by the nominator, $(L_z-L_3\cos\theta)^2$, which turns to zero at this point.

of the gravity-induced term, equal to Mgl. This immediately yields $\omega_{\text{pre}} = (L_z - L_3 \cos\theta_0)/I_A \sin^2\theta_0 = Mgl/L_3$, so that identifying ω_{rot} with $\omega_3 \equiv L_3/I_3$ (see figure 4.8), we recover the simple expression (4.72).

The second important result that may be readily obtained from Eq. (4.82) is the exact expression for the threshold value of the spinning speed for a vertically rotating top ($\theta = 0$, $L_z = L_3$). Indeed, in the limit $\theta \to 0$ this expression may be readily simplified:

$$U_{\text{ef}}(\theta) \approx \text{const} + \left(\frac{L_3^2}{8I_A} - \frac{Mgl}{2}\right)\theta^2. \tag{4.84}$$

This formula shows that if $\omega_{\text{rot}} \equiv L_3/I_3$ is higher than the following threshold value,

$$\omega_{\text{th}} \equiv 2\left(\frac{MglI_A}{I_3^2}\right)^{1/2}, \tag{4.85}$$

then the coefficient at θ^2 in Eq. (4.84) is positive, so that U_{ef} has a stable minimum at $\theta_0 = 0$. On the other hand, if ω_3 is decreased below ω_{th}, the fixed point becomes unstable, so that the top falls down. As the plots in figure 4.11 show, the result (4.85) for the threshold frequency works very well even for non-zero, but small values of the precession angle θ_0. Note that if we take $I = I_A$ in the condition (4.73) of the approximate treatment, it acquires a very simple sense: $\omega_{\text{rot}} \gg \omega_{\text{th}}$.

Finally, Eqs. (4.82) give a natural description of one more phenomenon. If the initial energy is larger than $U_{\text{ef}}(\theta_0)$, the angle θ oscillates between two classical turning points on both sides of the fixed point θ_0. The law and frequency of these oscillations may be found exactly as in section 3.3—see Eqs. (3.27) and (3.28). At $\omega_3 \gg \omega_{\text{th}}$, this motion is a fast rotation of the symmetry axis $\mathbf{n}_3$ of the body about its average position performing the slow precession. Historically, these oscillations are called *nutations*, but their physics is similar to that of the free precession that was analyzed in the previous section, and the order of magnitude of their frequency is still given by Eq. (4.59).

It may be proved that small energy dissipation (not taken into account in our analysis) leads first to a decay of nutations, then to a slower drift of the precession angle θ_0 to zero and, finally, to a gradual decay of the spinning speed ω_{rot} until it reaches the threshold (4.85) and the top falls down.

4.6 Non-inertial reference frames

To complete this chapter, let us use the results of our discussion of the rotation kinematics in section 4.1 to complete the discussion of the transfer between two reference frames, started in the introductory chapter 1. As figure 4.12 (which reproduces figure 1.2 in a more convenient notation) shows, even if the 'moving' frame 0 rotates relative to the 'lab' frame 0′, the radius-vectors observed from these

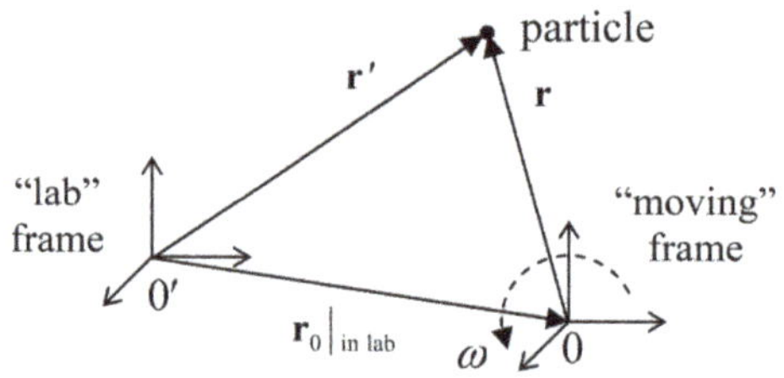

Figure 4.12. The general case of transfer between two reference frames.

two frames are still related, at any moment of time, by the simple Eq. (1.5). In our new notation:

$$\mathbf{r}'\big|_{\text{in lab}} = \mathbf{r}_0\big|_{\text{in lab}} + \mathbf{r}\big|_{\text{in lab}}. \tag{4.86}$$

However, as was mentioned in section 4.1, for velocities the general addition rule is already more complex. In order to find it, let us differentiate Eq. (4.86) over time:

$$\frac{d}{dt}\mathbf{r}'\big|_{\text{in lab}} = \frac{d}{dt}\mathbf{r}_0\big|_{\text{in lab}} + \frac{d}{dt}\mathbf{r}\big|_{\text{in lab}}. \tag{4.87}$$

The left-hand side of this relation is evidently the particle's velocity as measured in the lab frame, and the first term on the right-hand side is the velocity of point 0, as measured in the same lab frame. The last term is more complex: due to the possible mutual rotation of frames 0 and 0′, that term may not be zero even if the particle does not move relative to the rotating frame 0—see figure 4.12.

Fortunately, we have already derived the general Eq. (4.8) to analyze situations exactly like this one. Taking $\mathbf{A} = \mathbf{r}$, we may apply it to the last term of Eq. (4.87), to obtain

$$\mathbf{v}\big|_{\text{in lab}} = \mathbf{v}_0\big|_{\text{in lab}} + (\mathbf{v} + \boldsymbol{\omega} \times \mathbf{r}), \tag{4.88}$$

where $\boldsymbol{\omega}$ is the instantaneous angular velocity of an imaginary rigid body connected to the moving reference frame (or we may say, of this frame as such), and $\mathbf{v}$ is $d\mathbf{r}/dt$, as measured in the *moving* frame 0. (Here and later in this section, all vectors without indices imply their observation from the moving frame.) Relation (4.88), on one hand, is a natural generalization of Eq. (4.10) for $\mathbf{v} \neq 0$; on the other hand, if $\boldsymbol{\omega} = 0$, it is reduced to the simple equation (1.8) for the translational motion of frame 0.

Now, in order to calculate acceleration, me may just repeat the trick: differentiate Eq. (4.88) over time, and then use Eq. (4.8) again, now for the vector $\mathbf{A} = \mathbf{v} + \boldsymbol{\omega} \times \mathbf{r}$. The result is

$$\mathbf{a}\big|_{\text{in lab}} \equiv \mathbf{a}_0\big|_{\text{in lab}} + \frac{d}{dt}(\mathbf{v} + \boldsymbol{\omega} \times \mathbf{r}) + \boldsymbol{\omega} \times (\mathbf{v} + \boldsymbol{\omega} \times \mathbf{r}). \tag{4.89}$$

Carrying out the differentiation in the second term, we finally obtain the desired equation,

$$\mathbf{a}\big|_{\text{in lab}} \equiv \mathbf{a}_0\big|_{\text{in lab}} + \mathbf{a} + \dot{\boldsymbol{\omega}} \times \mathbf{r} + 2\boldsymbol{\omega} \times \mathbf{v} + \boldsymbol{\omega} \times (\boldsymbol{\omega} \times \mathbf{r}), \tag{4.90}$$

where **a** is particle's acceleration, as measured in the moving frame. This result is a natural generalization of the simple Eq. (1.9) to the rotating-frame case.

Now let the lab frame 0′ be inertial; then Newton's second law for a particle of mass m is

$$m\mathbf{a}\big|_{\text{in lab}} = \mathbf{F}, \tag{4.91}$$

where **F** is the vector sum of all forces exerted on the particle. This is simple and clear, however, in many cases it is much more convenient to work in a non-inertial reference frame; for example, in describing most phenomena on the Earth's surface, it is rather inconvenient to use a reference frame resting on the Sun (or the galactic center, etc). In order to understand what we should pay for the convenience of using the moving frame, we may combine Eqs. (4.90) and (4.91) to write

$$m\mathbf{a} = \mathbf{F} - m\mathbf{a}_0\big|_{\text{in lab}} - m\boldsymbol{\omega} \times (\boldsymbol{\omega} \times \mathbf{r}) - 2m\boldsymbol{\omega} \times \mathbf{v} - m\dot{\boldsymbol{\omega}} \times \mathbf{r}. \tag{4.92}$$

This result means that if we want to use a Newton's second law analog in a non-inertial reference frame, we have to add to the actual net force **F** acting on a particle, four pseudo-force terms, called *inertial forces*, all proportional to the particle's mass. Let us analyze them one by one, always remembering that these are just mathematical terms, not actual forces. (In particular, it would be futile to seek a Newton's third law counterpart for an inertial force.)

The first term, $-m\mathbf{a}_0|_{\text{in lab}}$, is the only one not related to rotation, and is well known from undergraduate mechanics. (Let me hope that the reader remembers all the weight-in-a-moving-elevator problems.) However, despite its simplicity, this term has more subtle consequences. As an example, let us consider, semi-qualitatively, the motion of a planet such as our Earth orbiting a star and also rotating about its own axis—see figure 4.13.

The bulk-distributed gravity forces, acting on a planet from its star, are not quite uniform, because they obey the $1/r^2$ gravity law (1.15), and hence are equivalent to a single force applied to a point A slightly offset from the planet's center-of-mass 0, toward the star. For a spherically symmetric planet, the direction from 0 to A would be exactly aligned with the direction toward the star. However, real planets are not

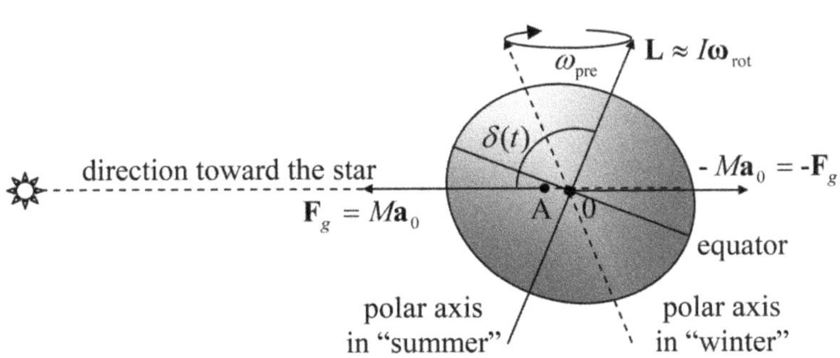

Figure 4.13. The axial precession of a planet (with the equatorial bulge and the 0A offset strongly exaggerated).

absolutely rigid, so that, due to the centrifugal 'force' (to be discussed imminently), their rotation about their own axis makes them slightly ellipsoidal—see figure 4.13. (For our Earth, this *equatorial bulge* is about 10 km.) As a result, the net gravity force does create a small torque relative to the center-of-mass 0. On the other hand, repeating all the arguments of this section for a body (rather than a point), we may see that, in the reference frame moving with the planet, the inertial force $-M\mathbf{a}_0$ (which is of course equal to the total gravity force and directed from the star) is applied exactly to the center-of-mass and hence does not create a torque about it. As a result, this pair of forces creates a torque $\boldsymbol{\tau}$ perpendicular to both the direction toward the star and the vector connecting points 0 and A. (In figure 4.13, the torque vector is perpendicular to the plane of drawing.) If the angle δ between the planet's 'polar' axis of rotation and the direction towards the star was fixed, then, as we have seen in the previous section, this torque would induce a slow axis precession about that direction.

However, as a result of the orbital motion, the angle δ oscillates in time much faster (once a year) between values $(\pi/2 + \varepsilon)$ and $(\pi/2 - \varepsilon)$, where ε is the axis tilt, i.e. angle between the polar axis (direction of vectors $\mathbf{L}$ and $\boldsymbol{\omega}_{\text{rot}}$) and the normal to the *ecliptic plane* of the planet's orbit. (For the Earth, $\varepsilon \approx 23.4°$.) A straightforward averaging over these fast oscillations[19] shows that the torque leads to the polar axis precession about the axis *perpendicular* to the ecliptic plane, keeping the angle ε constant—see figure 4.13. For the Earth, the period, $T_{\text{pre}} = 2\pi/\omega_{\text{pre}}$, of this *precession of the equinoxes* (or 'precession of the equator'), corrected to the substantial effect of Moon's gravity, is close to 26 000 years.

Returning to Eq. (4.92), the direction of the second term on its right-hand side,

$$\mathbf{F}_c \equiv -m\boldsymbol{\omega} \times (\boldsymbol{\omega} \times \mathbf{r}), \tag{4.93}$$

called the *centrifugal force*, is always perpendicular to, and directed out of the instantaneous rotation axis—see figure 4.14. Indeed, the vector $\boldsymbol{\omega} \times \mathbf{r}$ is perpendicular to both $\boldsymbol{\omega}$ and $\mathbf{r}$ (in figure 4.14, normal to the picture plane and directed from the reader) and has the magnitude $\omega r \sin\theta = \omega\rho$, where ρ is the distance of the particle from the rotation axis. Hence the outer vector product, with the account of the minus sign, is normal to the rotation axis $\boldsymbol{\omega}$, directed from the axis, and equal to

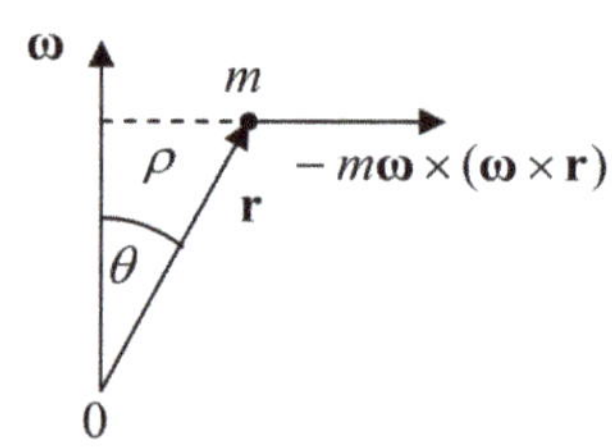

Figure 4.14. The centrifugal force.

[19] Details of this calculation may be found, e.g. in section 5.8 of the textbook [3].

$\omega^2 r \sin\theta = \omega^2\rho$. The 'centrifugal force' is of course just the result of the fact that the centripetal acceleration $\omega^2\rho$, explicit in the inertial reference frame, disappears in the rotating frame. For a typical location of the Earth ($\rho \sim R_E \approx 6 \times 10^6$ m), with its angular velocity $\omega_E \approx 10^{-4}$ s^{-1}, the acceleration is rather considerable, of the order of 3 cm s^{-2}, i.e. $\sim 0.003g$, and is responsible, in particular, for the largest part of the equatorial bulge mentioned above.

As an example of using the centrifugal force concept, let us return again to our 'testbed' problem on the bead sliding along a rotating ring—see figure 2.1. In the non-inertial reference frame attached to the ring, we have to add, to the actual forces $m\mathbf{g}$ and $\mathbf{N}$ acting on the bead, the horizontal centrifugal force[20] directed from the rotation axis, with magnitude $m\omega^2\rho$. In the notations of figure 2.1, its component tangential to the ring equals $m\omega^2\rho\cos\theta = m\omega^2 R\sin\theta\cos\theta$, and hence the Cartesian component of Eq. (4.92) along this direction is $ma = -mg\sin\theta + m\omega^2 R\sin\theta\cos\theta$. With $a = R\ddot{\theta}$, this gives us an equation of motion equivalent to Eq. (2.25), which had been derived in section 2.2 (in the inertial frame) using the Lagrangian formalism.

The third term on the right-hand side of Eq. (4.92)

$$\mathbf{F}_C \equiv -2m\boldsymbol{\omega} \times \mathbf{v}, \tag{4.94}$$

is the so-called *Coriolis force*[21], which is different from zero only if the particle moves in the rotating reference frame. Its physical sense may be understood by considering a projectile fired horizontally, say from the North Pole—see figure 4.15. From the point of view of the Earth-based observer, the projectile will be affected by an additional Coriolis force (4.94), directed westward, with magnitude $2m\omega_E v$, where $\mathbf{v}$ is the main, southward component of the velocity. This force would cause the westward acceleration $a = 2\omega_E v$, and the resulting westward deviation growing with time as $d = at^2/2 = \omega_E vt^2$. (This formula is exact only if d is much smaller than the distance $r = vt$ passed by the projectile.) On the other hand, from the point of

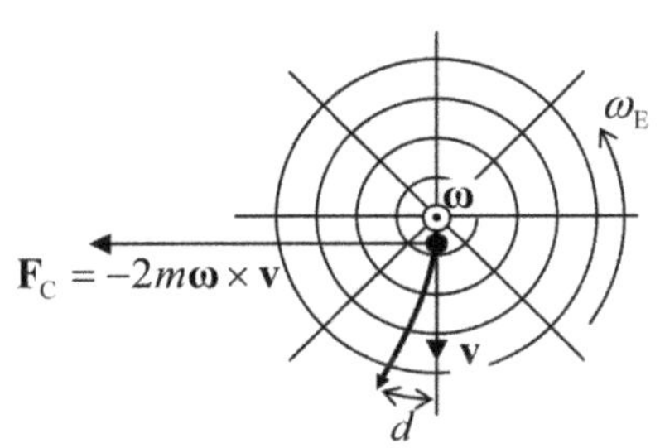

Figure 4.15. The trajectory of a projectile fired horizontally from the North Pole, from the point of view of an Earth-bound observer looking down. The circles show parallels, while the straight lines mark meridians.

[20] For this problem, all other inertial 'forces', apart from the besides the Coriolis force (see below), vanish, while the latter force is directed perpendicular to the ring and does not affect the bead's motion along it.

[21] Named after G-G Coriolis (1792–1843), who gave its quantitative theory (he is also credited for the first unambiguous definitions of the mechanical work and the kinetic energy), although the first qualitative descriptions of this effect were given by G Riccioli and C Dechales as early as in the mid-1600s.

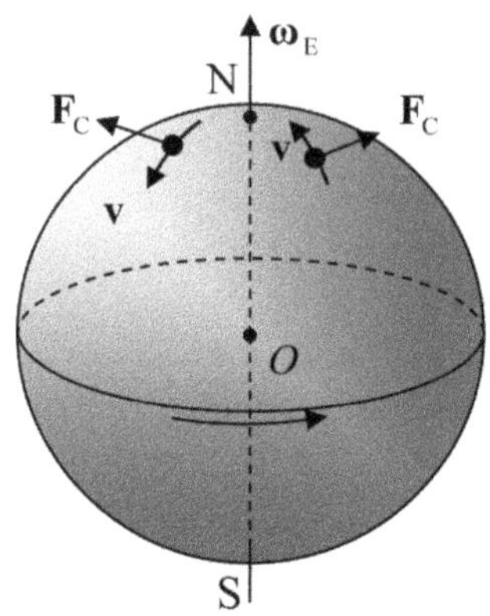

Figure 4.16. Coriolis forces due to Earth's rotation, in the Northern hemisphere.

view of an inertial-frame observer, the projectile's trajectory in the horizontal plane is a straight line. However, during the flight time t, the Earth surface slips eastward from under the trajectory by the distance $d = r\varphi = (vt)(\omega_E t) = \omega_E vt^2$, where $\varphi = \omega_E t$ is the azimuthal angle of the Earth rotation during the flight. Thus, both approaches give the same result.

Hence, the Coriolis 'force' is just a fancy (but frequently very convenient) way of description of a purely geometric effect pertinent to rotation, from the point of view of the observer participating in it. This force is responsible, in particular, for the higher right banks of rivers in the Northern hemisphere, regardless of the direction of their flow—see figure 4.16. Despite the smallness of the Coriolis force (for a typical velocity of the water in a river, $v \sim 1$ m s^{-1}, it is equivalent to acceleration $a_C \sim 10^{-2}$ cm s^{-2} $\sim 10^{-5}g$), its multi-century effects may be rather prominent[22].

Finally, the last, fourth term of Eq. (4.92), $-m\dot{\boldsymbol{\omega}} \times \mathbf{r}$, exists only when the rotation frequency changes in time, and may be interpreted as a local-position-specific addition to the first term.

The key relation (4.92), derived above from the Newton equation (4.91), may be alternatively obtained from the Lagrangian approach, which gives, as a by-product, some important insights on the momentum, as well as the relation between E and H, at rotation. Let us use Eq. (4.88) to represent the kinetic energy of the particle in an *inertial* frame in terms of $\mathbf{v}$ and $\mathbf{r}$ measured in a *rotating* frame:

$$T = \frac{m}{2}\left[\mathbf{v}_0\big|_{\text{in lab}} + (\mathbf{v} + \boldsymbol{\omega} \times r)\right]^2, \qquad (4.95)$$

and use this expression to calculate the Lagrangian function. For the relatively simple case of a point-particle motion in the field of potential forces, measured from

[22] The same force causes also the counter-clockwise circulation in the infamous 'Nor'easter' storms on the US East Coast, where the velocity $\mathbf{v}$ is directed toward its center of the cyclone, due to a lower pressure in its middle.

a reference frame that performs a pure rotation (so that $\mathbf{v}_0|_{\text{in lab}} = 0$)[23] with a constant angular velocity $\boldsymbol{\omega}$, we obtain

$$\begin{aligned} L \equiv T - U &= \frac{m}{2}v^2 + m\mathbf{v}\cdot(\boldsymbol{\omega}\times\mathbf{r}) + \frac{m}{2}(\boldsymbol{\omega}\times\mathbf{r})^2 - U \\ &\equiv \frac{m}{2}v^2 + m\mathbf{v}\cdot(\boldsymbol{\omega}\times\mathbf{r}) - U_{\text{ef}}, \end{aligned} \tag{4.96a}$$

where the effective potential energy[24],

$$U_{\text{ef}} \equiv U + U_c, \quad \text{with } U_c \equiv -\frac{m}{2}(\boldsymbol{\omega}\times\mathbf{r})^2, \tag{4.96b}$$

is just the sum of the actual potential energy U of the particle and the so-called *centrifugal potential energy*, associated with the centrifugal force Eq. (4.93):

$$\mathbf{F}_c = -\nabla U_c = -\nabla\left[-\frac{m}{2}(\boldsymbol{\omega}\times\mathbf{r})^2\right] = -m\boldsymbol{\omega}\times(\boldsymbol{\omega}\times\mathbf{r}). \tag{4.97}$$

It is straightforward to verify that the Lagrangian equations of motion derived from Eq. (4.96), considering the Cartesian components of $\mathbf{r}$ and $\mathbf{v}$ as generalized coordinates and velocities, coincide with Eq. (4.92) (with $\mathbf{a}_0|_{\text{in lab}} = \dot{\boldsymbol{\omega}} = 0$, and $\mathbf{F} = -\nabla U$). Now it is very informative to have a look at a by-product of this calculation, the generalized momentum (2.31) corresponding to particle's coordinate $\mathbf{r}$ as measured in the rotating reference frame[25],

$$\mathscr{p} \equiv \frac{\partial L}{\partial \mathbf{v}} = m(\mathbf{v} + \boldsymbol{\omega}\times\mathbf{r}). \tag{4.98}$$

According to Eq. (4.88) with $\mathbf{v}_0|_{\text{in lab}} = 0$, the expression in parentheses is just $m\mathbf{v}|_{\text{in lab}}$. However, from the point of view of the moving frame, i.e. not knowing about the simple physical sense of the vector $\mathscr{p}$, we would have a reason to speak about two different momenta of the same particle, the so-called *kinetic momentum* $\mathbf{p} = m\mathbf{v}$ and the *canonical momentum* $\mathscr{p} = \mathbf{p} + m\boldsymbol{\omega}\times\mathbf{r}$.[26]

Now let us calculate the Hamiltonian function H, defined by Eq. (2.32), and the energy E as functions of the same moving-frame variables:

[23] The similar analysis of the cases with $\mathbf{v}_0|_{\text{in lab}} \neq 0$, for example, of a mutual translational motion of the reference frames, is left for a reader's exercise.

[24] For the attentive reader who has noticed the difference between the negative sign in the expression for U_c, and the positive sign before the similar second term in Eq. (3.44): as was already discussed in chapter 3, this difference is due to the difference of assumptions. In the planetary problem, the angular momentum $\mathbf{L}$ (and hence its component L_z) is fixed, while the corresponding angular velocity $\dot{\varphi}$ is not. On the opposite, in our current discussion, the angular velocity $\boldsymbol{\omega}$ of the reference frame is assumed to be fixed, i.e. is independent on $\mathbf{r}$ and $\mathbf{v}$.

[25] Here $\partial L/\partial\mathbf{v}$ is just shorthand for a vector with Cartesian components $\partial L/\partial v_j$. In a more formal language, this is the gradient of L in the velocity space.

[26] A very similar situation arises at the motion of a particle with electric charge q in magnetic field $\mathscr{B}$. In that case the role of the additional term $\mathscr{p}-\mathbf{p} = m\boldsymbol{\omega}\times\mathbf{r}$ is played by the product $q\mathscr{A}$, where $\mathscr{A}$ is the vector-potential of the field ($\mathscr{B} = \nabla\times\mathscr{A}$)—see, e.g. *Part EM* section 9.7, and in particular Eqs. (9.183) and (9.192).

$$\begin{aligned} H &\equiv \sum_{j=1}^{3} \frac{\partial L}{\partial v_j} v_j - L = \not{p} \cdot \mathbf{v} - L = m\mathbf{v} \cdot (\mathbf{v} + \boldsymbol{\omega} \times \mathbf{r}) \\ &- \left[\frac{m}{2}v^2 + m\mathbf{v} \cdot (\boldsymbol{\omega} \times \mathbf{r}) - U_{\text{ef}}\right] = \frac{mv^2}{2} + U_{\text{ef}}, \end{aligned} \tag{4.99}$$

$$\begin{aligned} E &\equiv T + U = \frac{m}{2}v^2 + m\mathbf{v} \cdot (\boldsymbol{\omega} \times \mathbf{r}) + \frac{m}{2}(\boldsymbol{\omega} \times \mathbf{r})^2 + U \\ &= \frac{m}{2}v^2 + U_{\text{ef}} + m\mathbf{v} \cdot (\boldsymbol{\omega} \times \mathbf{r}) + m(\boldsymbol{\omega} \times \mathbf{r})^2. \end{aligned} \tag{4.100}$$

These expressions clearly show that E and H are *not* equal[27]. In hindsight, this is not surprising, because the kinetic energy (4.95), expressed in the moving-frame variables, includes a term linear in $\mathbf{v}$, and hence is not a quadratic-homogeneous function of this generalized velocity. The difference of these functions may be represented as

$$\begin{aligned} E - H &= m\mathbf{v} \cdot (\boldsymbol{\omega} \times \mathbf{r}) + m(\boldsymbol{\omega} \times \mathbf{r})^2 \\ &= m(\mathbf{v} + \boldsymbol{\omega} \times \mathbf{r}) \cdot (\boldsymbol{\omega} \times \mathbf{r}) = m\mathbf{v}\big|_{\text{in lab}} \cdot (\boldsymbol{\omega} \times \mathbf{r}). \end{aligned} \tag{4.101}$$

Now using the operand rotation rule again, we may transform this expression into an even simpler form[28]:

$$E - H = \boldsymbol{\omega} \cdot \left(\mathbf{r} \times m\mathbf{v}\big|_{\text{in lab}}\right) = \boldsymbol{\omega} \cdot (\mathbf{r} \times \not{p}) = \boldsymbol{\omega} \cdot \mathbf{L}\Big|_{\text{in lab}}. \tag{4.102}$$

As a sanity check, let us apply this general expression to the particular case of our testbed problem—see figure 2.1. In this case, the vector $\boldsymbol{\omega}$ is aligned with axis z, so that of all Cartesian components of vector $\mathbf{L}$, only the component L_z is important for the scalar product in Eq. (4.102). This component evidently equals $\omega I_z = \omega m\rho^2 = \omega mR^2 \sin^2\theta$, so that

$$E - H = m\omega^2 R^2 \sin^2\theta, \tag{4.103}$$

i.e. the same result that follows from the direct subtraction of Eqs. (2.40) and (2.41).

4.7 Problems

Problem 4.1. Calculate the principal moments of inertia for the following uniform rigid bodies:

---

[27] Please note the last form of Eq. (4.99), which shows the physical sense of the Hamiltonian function of a particle in the rotating frame very clearly, as the sum of its kinetic energy (as measured in the moving frame), and the effective potential energy (4.96b), including that of the centrifugal 'force'.

[28] Note that by definition (1.36), the angular momenta $\mathbf{L}$ of particles merely add up. As a result, equation (4.102) is valid for an arbitrary system of particles.

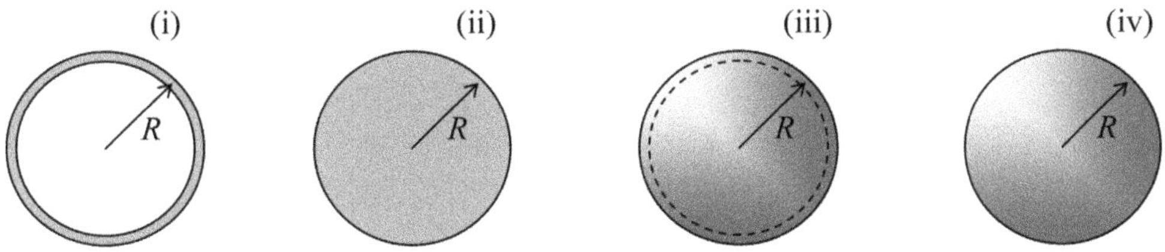

(i) a thin, plane, round hoop;
(ii) a flat round disk;
(iii) a thin spherical shell; and
(iv) a solid sphere.

Compare the results, assuming that all the bodies have the same radius R and mass M.

Problem 4.2. Calculate the principal moments of inertia for the following rigid bodies (see the figure below):

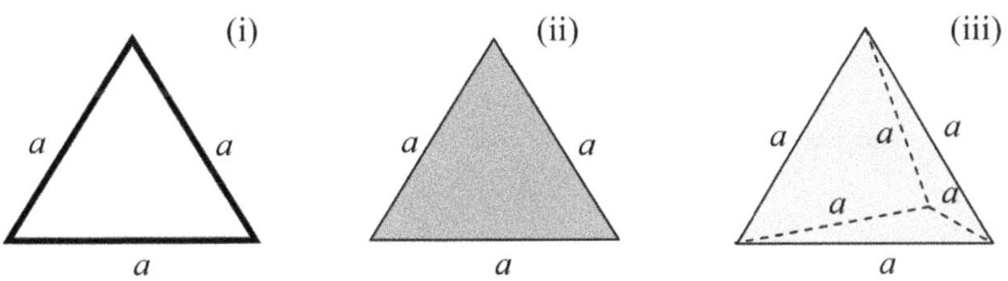

(i) an equilateral triangle made of thin rods with a uniform linear mass density μ;
(ii) a thin plate in the shape of an equilateral triangle, with a uniform areal mass density σ ; and
(iii) a tetrahedral pyramid made of a heavy material with a uniform bulk mass density ρ.

Assuming that the total mass of the three bodies is the same, compare the results and give an interpretation of their difference.

Problem 4.3. Prove that Eqs. (4.34)–(4.36) are valid for rotation of a rigid body about a fixed axis z, even if it does not pass through its center-of-mass.

Problem 4.4. The end of a uniform, thin rod of length $2l$ and mass m, initially at rest, is hit by a bullet of mass m', flying with velocity v_0 (see the figure below), which gets stuck in the rod. Use two different approaches to calculate the velocity of the opposite end of the rod right after the collision.

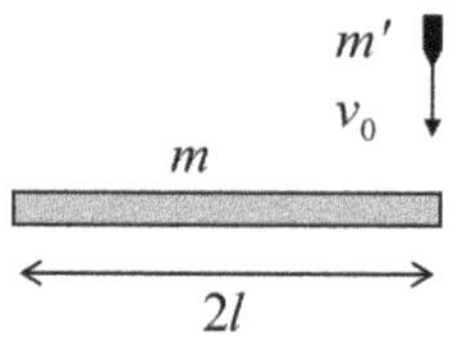

Problem 4.5. A uniform ball is placed on a horizontal plane, while rotating with an angular velocity ω_0, but having no initial linear velocity. Calculate the angular velocity after ball's slippage stops, assuming the Coulomb approximation for the kinetic friction force: $F_f = \mu N$, where N is a pressure between the surfaces, and μ is a velocity-independent coefficient.

Problem 4.6. A body may rotate about fixed horizontal axis 0′—see figure 4.5. Find the frequency of its small oscillations, in a uniform gravity field, as a function of the distance l of the axis from body's center-of-mass 0, and analyze the result.

Problem 4.7. Calculate the frequency, and sketch the mode of oscillations of a round uniform cylinder of radius R and the mass M, that may roll, without slipping, on a horizontal surface of a block of mass M'. The block, in turn, may move in the same direction, without friction, on a horizontal surface, being connected to it with an elastic spring—see the figure below.

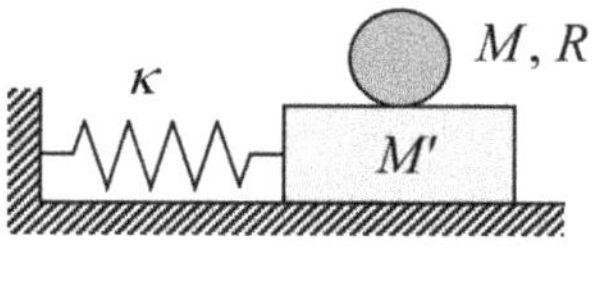

Problem 4.8. A thin uniform bar of mass M and length l is hung on a light thread of length l' (like a 'chime' bell—see the figure below). Derive the equations of motion of the system within the plane of drawing.

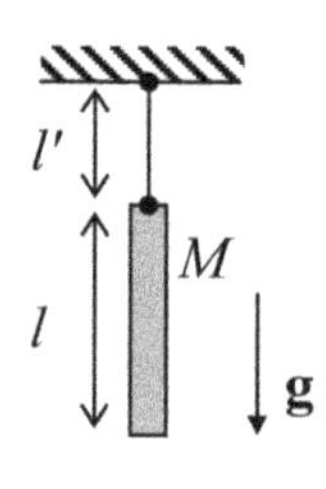

Problem 4.9. A solid, uniform, round cylinder of mass M can roll, without slipping, over a concave, round cylindrical surface of a block of mass M', in a uniform gravity field—see the figure below. The block can slide without friction on a horizontal surface. Using the Lagrangian formalism,

(i) find the frequency of small oscillations of the system near equilibrium, and
(ii) sketch the oscillation mode for the particular case $M' = M$, $R' = 2R$.

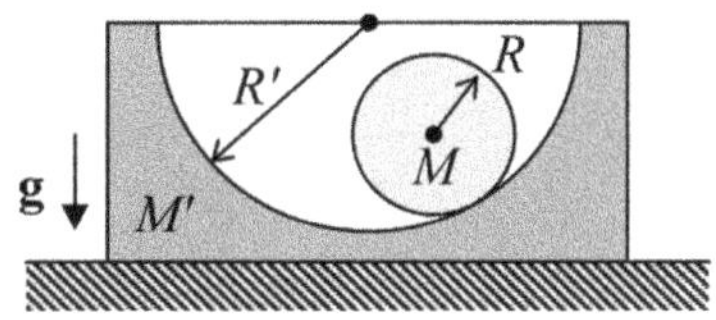

Problem 4.10. A uniform solid hemisphere of radius R is placed on a horizontal plane—see the figure below. Find the frequency of its small oscillations within a vertical plane, for two ultimate cases:

(i) there is no *friction* between the hemisphere and plane surfaces, and

(ii) the static friction is so strong that there is no *slippage* between these surfaces.

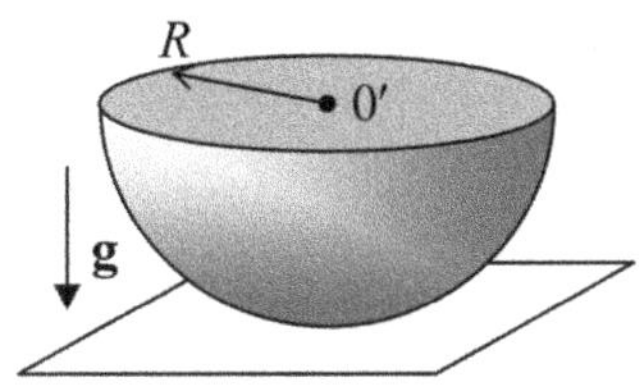

Problem 4.11. For the 'sliding ladder' problem, started in section 4.3 (see figure 4.7), find the critical value α_c of the angle α, at that the ladder loses its contact with the vertical wall, assuming that it starts sliding from the vertical position with a negligible initial velocity.

Problem 4.12. Six similar, uniform rods of length l and mass m are connected by light joints so that they may rotate, without friction, versus each other, forming a planar polygon. Initially, the polygon is at rest, and has the correct hexagon shape—see the figure below. Suddenly, an external force **F** is applied to the middle of one rod, in the direction of the hexagon's symmetry center. Calculate the accelerations: of the rod to which the force is applied (a), and of the opposite rod (a'), immediately after the application of the force.

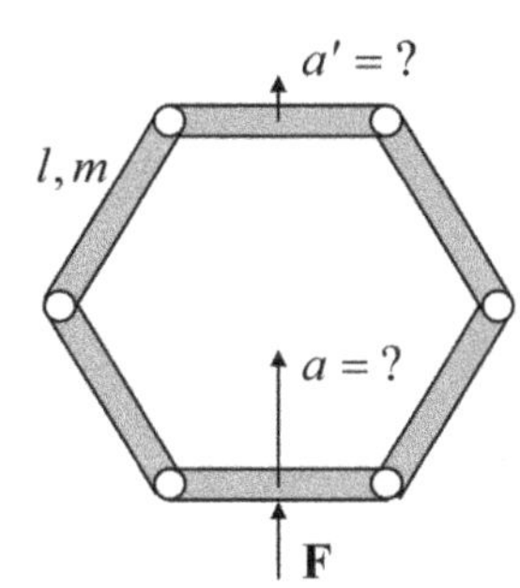

Problem 4.13. A rectangular cuboid (parallelepiped) with sides a_1, a_2, and a_3, made of a material with a constant mass density ρ, is rotated, with a constant angular velocity ω, about one of its space diagonals—see the figure below. Calculate the torque $\boldsymbol{\tau}$ necessary to sustain such rotation.

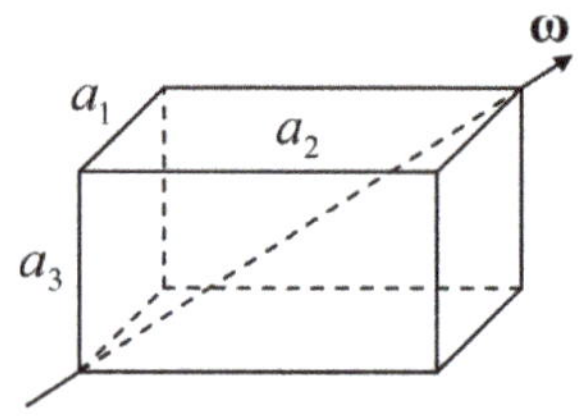

Problem 4.14. One end of a light shaft of length l is firmly attached to the center of a thin uniform solid disk of radius $R \ll l$ and mass M, whole plane is perpendicular to the shaft. Another end of the shaft is attached to a vertical axis (see the figure below) so that the shaft may rotate about the axis without friction. The disk rolls, without slippage, over a horizontal surface, so that the whole system rotates about the vertical axis with a constant angular velocity ω. Calculate the (vertical) supporting force N exerted on the disk by the surface.

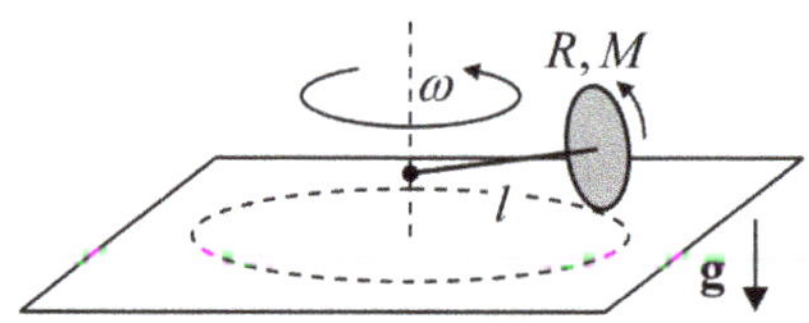

Problem 4.15. An air-filled balloon is placed inside a container filled with water that moves by inertia in free space, at negligible gravity. Suddenly, force **F** is applied to the container, pointing in a certain direction. What direction does the balloon move relative to the container?

Problem 4.16. Two planets are on the circular orbit around their common center-of-mass. Calculate the effective potential energy of a much lighter body (say, a spaceship) rotating with the same angular velocity, on the line connecting the planets. Sketch the plot of the radial dependence of U_{ef} and find out the number of so-called *Lagrange points* is which the potential energy has local maxima. Calculate their position explicitly in the limit when one of the planets is much more massive than the other one.

Problem 4.17. A small body is dropped down to the surface of Earth from height $h \ll R_E$, without initial velocity. Calculate the magnitude and direction of its deviation from the vertical, due to the Earth rotation. Estimate the effect's magnitude for a body dropped from the Empire State Building.

Problem 4.18. Calculate the height of solar tides on a large ocean, using the following simplifying assumptions: the tidal period (1/2 of Earth's day) is much longer than the period of all ocean waves, the Earth (of mass M_E) is a sphere of radius R_E, and its distance r_S from the Sun (of mass M_S) is constant and much larger than R_E.

Problem 4.19. A coin of radius r is rolled, with velocity V, on a horizontal surface without slippage. What should be coin's tilt angle θ (see the figure below) for it to roll on a circle of radius $R \gg r$? Modeling the coin as a very thin, uniform disk, and assuming that the angle θ is small, solve this problem in:

(i) an inertial ('lab') reference frame and
(ii) the non-inertial reference frame moving with the coin's center-of-mass (but not rotating with it).

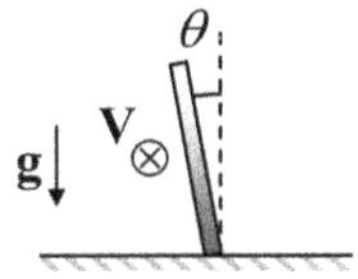

Problem 4.20.* A satellite is on a circular orbit, of radius R, around the Earth.

(i) Write the equations of motion of a small body as observed from the satellite, and simplify them for the case when the motion is limited to a close vicinity of the satellite.
(ii) Use the equations to prove that the body may be placed on an elliptical trajectory around satellite's center-of-mass, within its plane of rotation about Earth. Calculate the ellipse's orientation and eccentricity.

Problem 4.21.* A non-spherical shape of an artificial satellite may ensure its stable angular orientation relative to Earth's surface, advantageous for many practical goals. Modeling the satellite as a strongly elongated, axially symmetric body, moving around the Earth on a circular orbit of radius R, find its stable orientation, and analyze possible small oscillations of the satellite's symmetry axis about this equilibrium position.

Problem 4.22.* A rigid, straight, uniform rod of length l, with the lower end on a pivot, falls in a uniform gravity field—see the figure below. Neglecting friction, calculate the distribution of the bending torque τ along its length, and analyze the result.

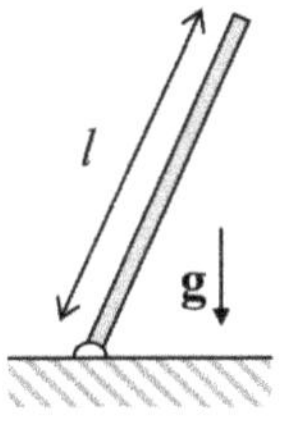

Hint: The bending torque is the net torque of the force **F** acting between two parts of the rod, mentally separated by its cross-section, about a certain 'neutral axis'[29]. As will be discussed in detail in section 7.5, at the proper definition of this axis, the

[29] Inadequate definitions of the torque are the main reason for numerous wrong solutions of this problem, posted online—readers beware!

bending torque's gradient along the rod's length is equal to $(-F)$, where F is the rod-normal ('shear') component of the force exerted by the top part of the rod on its lower part.

Problem 4.23. Let $\mathbf{r}$ be the radius-vector of a particle, as measured in a possibly non-inertial but certainly non-rotating reference frame. Taking its Cartesian components for the generalized coordinates, calculate the corresponding generalized momentum $\mathscr{p}$ of the particle, and its Hamiltonian function H. Compare $\mathscr{p}$ with $m\mathbf{v}$, and H with the particle's energy E. Derive the Lagrangian equation of motion in this approach, and compare it with Eq. (4.92) of the lecture notes.

References

[1] Kaajakari V 2009 *Practical MEMS* (Small Gear Publishing)
[2] Landau L and Lifshitz E 1976 *Mechanics* 3rd edn (Oxford: Butterworth-Heinemann)
[3] Goldstein H, Poole C and Safko J 2002 *Classical Mechanics* 3rd edn (Reading, MA: Addison-Wesley)

IOP Publishing

Classical Mechanics

Lecture notes

Konstantin K Likharev

Chapter 5

Oscillations

In this course, oscillations and waves are discussed in detail because of their key importance for fundamental and applied physics. This chapter starts with discussion of the so-called 'linear' (or 'harmonic') oscillator whose differential equation of motion is linear and hence allows the full analytical solution, and then proceeds to 'nonlinear' and parametric systems whose dynamics may only be explored by either approximate analytical or numerical methods.

5.1 Free and forced oscillations

In section 3.2 we briefly discussed oscillations in a very important Hamiltonian system—a 1D *harmonic oscillator* described by a simple 1D Lagrangian[1]

$$L \equiv T(\dot{q}) - U(q) = \frac{m}{2}\dot{q}^2 - \frac{\kappa}{2}q^2, \qquad (5.1)$$

whose Lagrange equation of motion[2],

$$m\ddot{q} + \kappa q = 0, \quad \text{i.e.} \quad \ddot{q} + \omega_0^2 q = 0, \quad \text{with } \omega_0^2 \equiv \frac{\kappa}{m} \geqslant 0, \qquad (5.2)$$

is a *linear homogeneous* differential equation. Its general solution is given by Eq. (3.16), which is frequently useful to recast into another, amplitude-phase form:

$$q(t) = u \cos \omega_0 t + v \sin \omega_0 t = A \cos(\omega_0 t - \varphi), \qquad (5.3a)$$

[1] For simplicity of notation, in this chapter I will drop indices 'ef' in the energy components T and U, and parameters such as m, κ, etc. However, the reader should still remember that T and U do not necessarily coincide with the actual kinetic and potential energies (even if those energies may be uniquely identified)—see section 3.1.

[2] ω_0 is usually called the *own frequency* of the oscillator. In quantum mechanics, the Germanized version of the same term, *eigenfrequency*, is more commonly used. In this series I will use either of the terms, depending on the context.

doi:10.1088/978-0-7503-1398-8ch5

where A is the *amplitude* and φ the *phase* of the oscillations, which are determined by the initial conditions. Mathematically, it is frequently easier to work with sinusoidal functions as complex exponents, by rewriting Eq. (5.3*a*) in one more form[3]:

$$q(t) = \text{Re}\,[Ae^{-i(\omega_0 t-\varphi)}] = \text{Re}\,[ae^{-i\omega_0 t}], \tag{5.3b}$$

where a is the *complex amplitude* of the oscillations:

$$a \equiv Ae^{i\varphi}, \quad |\,a\,| = A, \quad \text{Re}\,a = A\cos\varphi = u, \quad \text{Im}\,a = A\sin\varphi = v. \tag{5.4}$$

For an autonomous, Hamiltonian oscillator, Eq. (5.3) gives the full classical description of its dynamics. However, it is important to understand that this *free-oscillation* solution, with a constant amplitude A, is due to the conservation of energy $E = T + U = \kappa A^2/2$ of the oscillator. If its energy changes by any reason, the description needs to be generalized.

First of all, if the energy leaks out of the oscillator to its environment (the effect usually called the *energy dissipation*), the free oscillation decay with time. The simplest model of this effect is represented by an additional *linear drag* (or 'kinematic friction') *force*, proportional to the generalized velocity and directed opposite to it:

$$F_v = -\eta\dot{q}, \tag{5.5}$$

where constant η is called the *drag coefficient*[4]. The inclusion of this force modifies the equation of motion (5.2) to become

$$m\ddot{q} + \eta\dot{q} + \kappa q = 0. \tag{5.6a}$$

This equation is frequently rewritten in the form

$$\ddot{q} + 2\delta\dot{q} + \omega_0^2 q = 0, \quad \text{with } \delta \equiv \frac{\eta}{2m}, \tag{5.6b}$$

where parameter δ is called the *damping coefficient* (or just 'damping'). Note that Eq. (5.6) is still a linear homogeneous second-order differential equation, and its general solution still has the form of the sum (3.13) of two exponents of the type exp $\{\lambda t\}$, with arbitrary pre-exponential coefficients. Plugging such an exponent into Eq. (5.6), we obtain the following algebraic characteristic equation for λ:

$$\lambda^2 + 2\delta\lambda + \omega_0^2 = 0. \tag{5.7}$$

---

[3] Note that this is the so-called *physics convention*. Most engineering texts use the opposite sign in the imaginary exponent, $\exp\{-i\omega t\} \rightarrow \exp\{i\omega t\}$, with the corresponding sign implications for intermediate formulas, but (of course) similar final results for real variables.

[4] Here Eq. (5.5) is treated as a phenomenological model, but in statistical mechanics such a dissipative term may be *derived* as an average force exerted on the oscillator by its environment, at very general assumptions. As will be discussed in detail later in this series (*Part SM* chapter 5 and *Part QM* chapter 7), due to the ts numerous degrees of freedom of a typical environmental (think about the molecules of air surrounding the usual mechanical pendulum), its force also has a random component; as a result, the *dissipation* is fundamentally related to *fluctuations*. The latter effects may be neglected (as they are in this course) only if E is much higher than the energy scale of the random fluctuations—in the thermal equilibrium at temperature T, the larger of $k_B T$ and $\hbar\omega_0/2$.

Solving this quadratic equation, we get

$$\lambda_\pm = -\delta \pm i\omega_0', \quad \text{where } \omega_0' \equiv (\omega_0^2 - \delta^2)^{1/2}, \tag{5.8}$$

so that for not-very-high damping ($\delta < \omega_0$)[5] we obtain the following generalization of Eq. (5.3):

$$\begin{aligned} q_{\text{free}}(t) &= c_+ e^{\lambda_+ t} + c_- e^{\lambda_- t} \\ &= (u_0 \cos \omega_0' t + v_0 \sin \omega_0' t)\, e^{-\delta t} \\ &= A_0 e^{-\delta t} \cos(\omega_0' t - \varphi_0). \end{aligned} \tag{5.9}$$

The result shows that, besides a certain correction to the free-oscillation frequency (which is very small in the most interesting *low damping limit*, $\delta \ll \omega_0$), the energy dissipation leads to an exponential decay of oscillation amplitude with time constant $\tau = 1/\delta$:

$$A = A_0 e^{-t/\tau}, \quad \text{where } \tau \equiv \frac{1}{\delta} = \frac{2m}{\eta}. \tag{5.10}$$

A very popular, dimensionless measure of damping is the so-called *quality factor Q* (or just the *Q-factor*) that is defined as $\omega_0/2\delta$, and may be rewritten in several other useful forms:

$$Q \equiv \frac{\omega_0}{2\delta} = \frac{m\omega_0}{\eta} = \frac{(m\kappa)^{1/2}}{\eta} = \pi\frac{\tau}{\mathcal{T}} = \frac{\omega_0 \tau}{2}, \tag{5.11}$$

where $\mathcal{T} = 2\pi/\omega_0$ is the oscillation period in the absence of damping—see Eq. (3.29). Since the oscillation energy E is proportional to A^2, i.e. decays as $\exp\{-2t/\tau\}$, with the time constant $\tau/2$, the last form of Eq. (5.11) may be used to rewrite the Q-factor in one more form:

$$Q = \omega_0 \frac{E}{(-\dot{E})} \equiv \omega_0 \frac{E}{\mathscr{P}}, \tag{5.12}$$

where $\mathscr{P}$ is the dissipation power. (Two other practical ways to measure Q will be discussed below.) The range of Q-factors of important oscillators is very broad, all the way from $Q \sim 10$ for a human leg (with relaxed muscles), to $Q \sim 10^4$ for the quartz crystals used in 'electronic' clocks and watches, all the way up to $Q \sim 10^{12}$ for carefully designed microwave cavities with superconducting walls.

In contrast to the decaying free oscillations, the *forced oscillations*, induced by an external force $F(t)$, may maintain their amplitude (and hence energy) infinitely, even at non-zero damping. This process may be described using a still linear but now *inhomogeneous* differential equation

$$m\ddot{q} + \eta\,\dot{q} + \kappa q = F(t), \tag{5.13a}$$

[5] Systems with very high damping ($\delta > \omega_0$) can hardly be called oscillators, and although they are used in engineering and physics experiments (e.g. for the shock, vibration, and sound isolation), for their discussion I have to refer the interested reader to special literature—see, e.g. [1]. Let me only note that at very high damping, $\delta \gg \omega_0$, the system may be adequately described by just one parameter: the relaxation time $1/\lambda_+ \approx 2\delta/\omega_0^2 \equiv \eta/m \gg \omega_0$.

or, more conveniently for analysis, the following generalization of Eq. (5.6*b*):

$$\ddot{q} + 2\delta\dot{q} + \omega_0^2 q = f(t), \quad \text{where } f(t) \equiv F(t)/m. \tag{5.13b}$$

For a mechanical linear, dissipative 1D oscillator (5.6), under the effect of an additional external force $F(t)$, Eq. (5.13*a*) is just an expression of Newton's second law. However, according to Eq. (1.41), Eq. (5.13) is valid for any dissipative, *linear*[6] 1D system whose Gibbs potential energy (1.39) has the form $U_G(q, t) = \kappa q^2/2 - F(t)q$.

The forced-oscillation solutions may be analyzed by two mathematically equivalent methods whose relative convenience depends on the character of function $f(t)$

(i) *Frequency domain.* Let us present function $f(t)$ as a Fourier sum of sinusoidal harmonics[7]:

$$f(t) = \sum_\omega f_\omega e^{-i\omega t}. \tag{5.14}$$

Then, due to the linearity of Eq. (5.13), its general solution may be represented as a sum of the decaying free oscillations (5.9) with the frequency ω_0', independent of the function $F(t)$, and forced oscillations due to each of the Fourier components of the force[8]:

$$q(t) = q_{\text{free}}(t) + q_{\text{forced}}(t), \quad q_{\text{forced}}(t) = \sum_\omega a_\omega e^{-i\omega t}. \tag{5.15}$$

Plugging Eq. (5.15) into Eq. (5.13), and requiring that the factors before each $e^{-i\omega t}$ in both parts be equal, we obtain

$$a_\omega = f_\omega \chi(\omega), \tag{5.16}$$

where the complex function $\chi(\omega)$, which in our particular case equals

$$\chi(\omega) = \frac{1}{(\omega_0^2 - \omega^2) - 2i\omega\delta}, \tag{5.17}$$

is called either the *response function* or (particularly for non-mechanical oscillators) the *generalized susceptibility.* From here, and Eq. (5.4), the amplitude of the oscillations under the effect of a sinusoidal force is

$$A_\omega \equiv |a_\omega| = |f_\omega||\chi(\omega)|, \quad \text{with } |\chi(\omega)| = \frac{1}{\left[(\omega_0^2 - \omega^2)^2 + (2\omega\delta)^2\right]^{1/2}}. \tag{5.18}$$

[6] This is a very unfortunate, but common jargon, meaning 'the system described by linear equations of motion'.
[7] Here, in contrast to Eq. (5.3*b*), we may drop the operator Re, assuming that $f_{-\omega} = f_\omega^*$, so that the imaginary components of the sum compensate each other.
[8] In physics, this mathematical property of linear equations is frequently called the *linear superposition principle*.

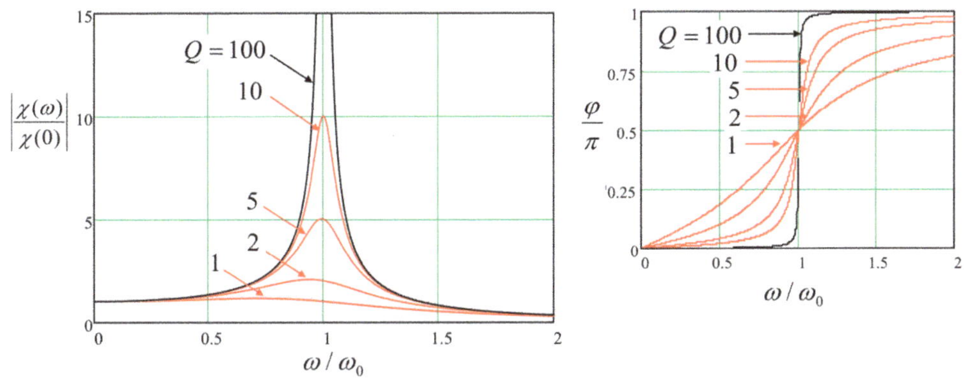

Figure 5.1. Resonance in the linear oscillator, for several values of Q.

This formula describes, in particular, an increase of the oscillation amplitude A_ω at $\omega \rightarrow \omega_0$—see the left panel in figure 5.1. At the exact resonance,

$$|\chi(\omega)|_{\omega=\omega_0} = \frac{1}{2\omega_0\delta}, \tag{5.19}$$

so that, according to Eq. (5.11), the ratio of the response magnitudes at $\omega = \omega_0$ and $\omega = 0$ ($|\chi(\omega)|_{\omega=0} = 1/\omega_0^2$) is exactly equal to the Q-factor of the oscillator. Thus, the response increase is particularly strong in the low damping limit ($\delta \ll \omega_0$, i.e. $Q \gg 1$); moreover at $Q \rightarrow \infty$ and $\omega \rightarrow \omega_0$ the response diverges. (This fact is very useful for the approximate methods to be discussed later in this chapter.) This is of course the classical description of the famous phenomenon of *resonance*, so ubiquitous in physics.

Due to the increase of the resonance peak height, its width is inversely proportional to Q.[9] Quantitatively, in the most interesting low-damping limit, i.e. at $Q \gg 1$, the reciprocal Q-factor gives the normalized value of the so-called *full-width at half-maximum* (FWHM) of the resonance curve:

$$\frac{\Delta\omega}{\omega_0} = \frac{1}{Q}. \tag{5.20}$$

Indeed, this $\Delta\omega$ is defined as the difference $(\omega_+ - \omega_-)$ between the two values of ω at which the square of the oscillator response function, $|\chi(\omega)|^2$ (proportional to the oscillation energy), equals half of its resonance value (5.19). In the low damping limit, both these points are very close to ω_0, so that in the first (linear) approximation in $(\omega - \omega_0) \ll \omega_0$, we can take $(\omega_0^2 - \omega^2) \equiv -(\omega + \omega_0)(\omega - \omega_0) \approx (-2\omega\xi) \approx (-2\omega_0\xi)$, where

$$\xi \equiv \omega - \omega_0 \tag{5.21}$$

[9] Note that the phase shift $\varphi = \arg[\chi(\omega)]$ between the oscillations and the external force (see the right panel in figure 1) makes its steepest change, by $\pi/2$, exactly within the same frequency interval $\Delta\omega$.

is a very convenient parameter called *detuning*, which will be repeatedly used later in this chapter. In this approximation, the second of Eqs. (5.18) is reduced to

$$|\chi(\omega)|^2 = \frac{1}{4\omega^2(\delta^2 + \xi^2)}. \tag{5.22}$$

As a result, the points $\omega_\pm$ correspond to $\xi^2 = \delta^2$, i.e. $\omega_\pm = \omega_0 \pm \delta = \omega_0(1 \pm 1/2Q)$, so that $\Delta\omega \equiv \omega_+ - \omega_- = \omega_0/Q$, thus proving Eq. (5.20).

(ii) *Time domain.* Returning to the general problem of linear oscillations, one may argue that Eqs. (5.9) and (5.15)–(5.17) provide a full solution of the forced oscillation problem. This is formally correct, but this solution may be very inconvenient if the external force is far from a sinusoidal function of time, in particular if it is not periodic at all. In this case, we should first calculate the complex amplitudes f_ω participating in the Fourier sum (5.14). In the case of a non-periodic $f(t)$, this is actually the Fourier integral[10],

$$f(t) = \int_{-\infty}^{+\infty} f_\omega e^{-i\omega t} dt, \tag{5.23}$$

so that f_ω should be calculated using the reciprocal Fourier transform,

$$f_\omega = \frac{1}{2\pi} \int_{-\infty}^{+\infty} f(t') e^{i\omega t'} dt. \tag{5.24}$$

Now we can use Eq. (5.16) for each Fourier component of the resulting forced oscillations, and rewrite the last of Eqs. (5.15) as

$$\begin{aligned} q_{\text{forced}}(t) &= \int_{-\infty}^{+\infty} a_\omega e^{-i\omega t} d\omega \\ &= \int_{-\infty}^{+\infty} \chi(\omega) f_\omega e^{-i\omega t} d\omega \\ &= \int_{-\infty}^{+\infty} d\omega\, \chi(\omega) \frac{1}{2\pi} \int_{-\infty}^{+\infty} dt\, f(t) e^{i\omega(t'-t)} \\ &= \int_{-\infty}^{+\infty} dt\, f(t) \left[\frac{1}{2\pi} \int_{-\infty}^{+\infty} d\omega\, \chi(\omega) e^{i\omega(t'-t)}\right], \end{aligned} \tag{5.25}$$

with the response function $\chi(\omega)$ given, in our case, by Eq. (5.17). Apart from requiring two integrations, the relation (5.25) is conceptually uncomforting: it seems to indicate that the oscillator's coordinate at time t depends not only on the external force exerted at earlier times $t' < t$, but also at future times. This would contradict one of the most fundamental principles of physics (and indeed, science as a whole), the *causality*: no effect may precede its cause.

---

[10] Let me hope that the reader knows that Eq. (5.23) may be used even for periodic functions; in such a case, f_ω is a set of equidistant delta-functions. (A reminder of the basic properties of the δ-function may be found, for example, in appendix A, section A.14.)

Fortunately, a straightforward calculation (left for a reader's exercise) shows that the response function (5.17) satisfies the following rule[11]:

$$\int_{-\infty}^{+\infty} \chi(\omega) e^{-i\omega\tau} d\omega = 0, \quad \text{for } \tau < 0. \tag{5.26}$$

This fact allows the last form of Eq. (5.25) to be rewritten in either of the following equivalent forms:

$$q_{\text{forced}}(t) = \int_{-\infty}^{t} f(t')G(t - t')dt' = \int_{0}^{\infty} f(t - \tau)\, G(\tau)d\tau, \tag{5.27}$$

where $G(\tau)$, defined as the Fourier transform of the response function,

$$G(\tau) \equiv \frac{1}{2\pi} \int_{-\infty}^{+\infty} \chi(\omega) e^{-i\omega\tau} d\omega, \tag{5.28}$$

is called the *(temporal) Green's function* of the system. According to Eq. (5.26), $G(\tau) = 0$ for all $\tau < 0$.

While the second form of Eq. (5.27) is frequently more convenient for calculations, its first form is more suitable for understanding the physical sense of the Green's function. Indeed, let us consider the particular case, when the force is a delta-function

$$f(t) = \delta(t - t'), \qquad \text{with } t' < t, \qquad \text{i.e. } \tau \equiv t - t' > 0, \tag{5.29}$$

representing an ultimately short pulse at the moment t', with a unit 'area' $\int f(t')dt'$. Substituting Eq. (5.29) into Eq. (5.27)[12], we obtain

$$q(t) = G(t - t'). \tag{5.30}$$

Thus the Green's function $G(t - t')$ is just the oscillator's response to a short pulse of force, of a unit 'area', measured at time t. Hence Eq. (5.27) expresses the linear superposition principle in the time domain: the full effect of the force $f(t)$ on an oscillator (actually, any linear system) is a sum of the effects of short pulses of duration dt' and magnitude $f(t')$, each with its own 'weight' $G(t-t')$—see figure 5.2.

This picture may be used for the calculation of the Green's function for a particular system. Indeed, Eqs. (5.29)–(5.30) mean that $G(\tau)$ is just the solution of the differential equation of motion of the system, in our case of Eq. (5.13), with the replacement $t \to \tau$ and a δ-functional right-hand side:

$$\frac{d^2G(\tau)}{d\tau^2} + 2\delta\frac{dG(\tau)}{d\tau} + \omega_0^2 G(\tau) = \delta(\tau). \tag{5.31}$$

[11] Eq. (5.26) remains true for any linear physical systems in which $f(t)$ represents a cause, and $q(t)$ its effect. Following tradition, I discuss the frequency-domain expression of this causality relation (called the *Kramers–Kronig relations*) in the *Classical Electrodynamics* part of this lecture series—see section 7.3.

[12] Technically, for this integration, t' in Eq. (5.27) should be temporarily replaced with another letter, say t''.

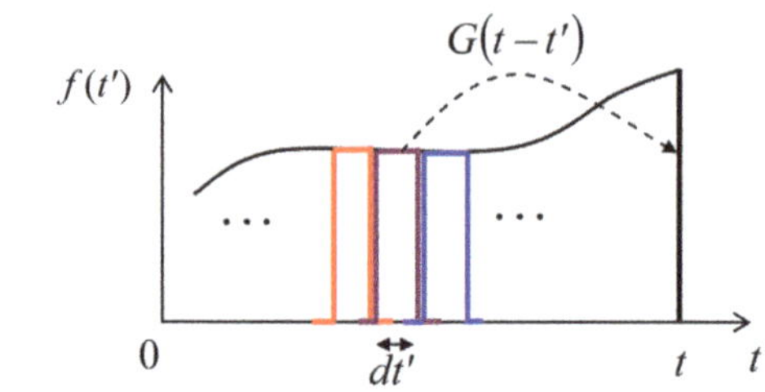

Figure 5.2. A schematic, finite-interval representation of a force $f(t)$ as a sum of short pulses, and their effects on the linear system's response at time t, given by (5.27).

Since Eq. (5.27) describes only the second term in Eq. (5.15), i.e. only the forced rather than free oscillations, we have to exclude the latter by solving Eq. (5.31) with zero initial conditions:

$$G(-0) = \frac{dG}{d\tau}(-0) = 0, \tag{5.32}$$

where $t = -0$ means the instant immediately preceding $t = 0$.

This calculation may be simplified even further. Let us integrate both sides of Eq. (5.31) over an infinitesimal interval including the origin, e.g. $[-d\tau/2, +d\tau/2]$, and then follow the limit $d\tau \to 0$. Since the Green's function has to be continuous because of its physical sense as the (generalized) coordinate, all terms on the left-hand side but the first one vanish, while the first term yields $dG/d\tau|_{+0} - dG/d\tau|_{-0}$. Due to the second of the conditions (5.32), the last of these two derivatives equals zero, while the right-hand side of Eq. (5.31) yields 1 upon the integration. Thus, $G(\tau)$ may be calculated for $\tau > 0$ (i.e. for all times when it is different from zero) by solving the *homogeneous* version of system's equation of motion for $\tau > 0$, with the following special initial conditions:

$$G(0) = 0, \quad \frac{dG}{d\tau}(0) = 1. \tag{5.33}$$

This approach gives us a convenient way for calculation of Green's functions of linear systems. In particular for the oscillator with not-very-low damping ($\delta > \omega_0$, i.e. $Q > 1/2$), imposing the boundary conditions (5.33) on the homogeneous equation's solution (5.9), we immediately obtain

$$G(\tau) = \frac{1}{\omega_0'} e^{-\delta\tau} \sin \omega_0'\tau. \tag{5.34}$$

(The same result may be obtained directly from Eq. (5.28) with the response function $\chi(\omega)$ given by Eq. (5.19). This way is, however, a little bit more cumbersome, and is left for a reader exercise.)

Relations (5.27) and (5.34) provide a very convenient recipe for solving most forced oscillation problems. As a very simple example, let us calculate the transient

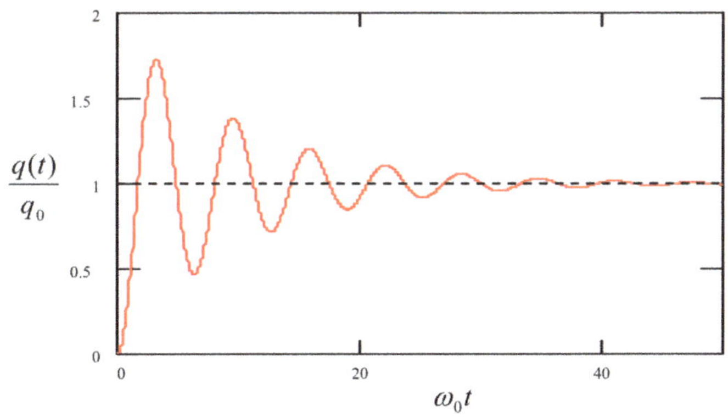

Figure 5.3. The transient process in a linear oscillator, induced by a step-like force $f(t)$, for the particular case $\delta/\omega_0 = 0.1$ (i.e. $Q = 5$).

process in an oscillator under the effect of a constant force being turned on at $t = 0$, i.e. proportional to the theta-function of time:

$$f(t) = f_0\theta(t) \equiv \begin{cases} 0, & t < 0, \\ f_0, & t > 0, \end{cases} \tag{5.35}$$

provided that at $t < 0$ the oscillator was at rest, so that in Eq. (5.15), $q_{\text{free}}(t) \equiv 0$. Then the second form of Eq. (5.27), and Eq. (5.34), yield

$$q(t) = \int_0^\infty f(t - \tau)G(\tau)d\tau = f_0\int_0^t \frac{1}{\omega_0'}e^{-\delta\tau}\sin\omega_0'\tau\, d\tau. \tag{5.36}$$

The simplest way to work out such integrals is to represent the sine function under it as the imaginary part of $\exp\{i\omega_0't\}$, and merge the two exponents, obtaining

$$q(t) = f_0\frac{1}{\omega_0'}\text{Im}\left[\frac{1}{\delta + i\omega_0'}e^{-\delta\tau - i\omega_0'\tau}\right]_0^t = \frac{F_0}{k}\left[1 - e^{-\delta t}\left(\cos\omega_0't + \frac{\delta}{\omega_0'}\sin\omega_0't\right)\right]. \tag{5.37}$$

This result, plotted in figure 5.3, is rather natural: it describes nothing more than the transient from the initial equilibrium position $q = 0$ to the new equilibrium position $q_0 = f_0/\omega_0^2 = F_0/\kappa$, accompanied by decaying oscillations. For this particular simple function $f(t)$, the same result might be also obtained by introducing a new variable $\tilde{q}(t) \equiv q(t) - q_0$ and solving the resulting *homogeneous* equation for $\tilde{q}$ (with the appropriate initial condition $\tilde{q}(0) = -q_0$). However, for more complicated functions $f(t)$ the Green's function approach is irreplaceable.

Note that for any particular linear system, its Green's function should be calculated only once, and then may be repeatedly used in Eq. (5.27) to calculate the system response to various external forces—either analytically or numerically. This property makes the Green's function approach very popular in many other fields of physics—with the corresponding generalization or re-definition of the function[13].

[13] See, e.g. *Part EM* section 2.7 and *Part QM* section 2.2.

5.2 Weakly nonlinear oscillations

In comparison with systems discussed in the last section, which are described by linear differential equations with constant coefficients and thus allow a complete and exact analytical solution, oscillations in nonlinear systems generally present a complex and, generally, analytically intractable problem. However, much insight on possible processes[14] in such systems may be gained from the discussion of the important case of *weakly nonlinear* systems that may be explored analytically. An example of such a system is given by an *anharmonic oscillator*—a 1D system whose higher terms in the potential expansion (3.10) cannot be neglected, but are small and may be accounted for approximately. If, in addition, damping is low (or negligible), and the external harmonic force exerted on the system is not too large (if any), the equation of motion is a slightly modified version of Eq. (5.13):

$$\ddot{q} + \omega^2 q = f(t, q, \dot{q}, \ldots), \tag{5.38}$$

where $\omega \approx \omega_0$ is the anticipated frequency of oscillations (whose choice is to a certain extent arbitrary—see below), and the right-hand side f is small (say, scales as some small dimensionless parameter $\varepsilon \ll 1$), and may be considered as a *perturbation*.

Since at $\varepsilon = 0$, this equation has the sinusoidal solution given by Eq. (5.3), one might naïvely think that at a non-vanishing but small ε, the approximate solution to Eq. (5.38) should be sought in the form

$$q(t) = q^{(0)} + q^{(1)} + q^{(2)} + \cdots, \quad \text{where } q^{(n)} \propto \varepsilon^n, \tag{5.39}$$

with $q^{(0)} = A\cos(\omega_0 t - \varphi) \propto \varepsilon^0$. This is a good example of an apparently impeccable mathematical reasoning that would lead to a very inefficient procedure. Indeed, let us apply it to the problem we already know the exact solution for, namely the free oscillations in a linear but damped oscillator, for this occasion assuming the damping to be very low, $\delta/\omega_0 \sim \varepsilon \ll 1$. The corresponding equation of motion (5.6) may be represented in the form (5.38) if we take $\omega = \omega_0$ and

$$f = -2\delta\dot{q}, \quad \delta \propto \varepsilon. \tag{5.40}$$

The naïve approach described above would allow us to find *small* corrections, of the order of δ, to the free, non-decaying oscillations $A\cos(\omega_0 t - \varphi)$. However, we already know from Eq. (5.9) that the main effect of damping is a gradual decrease of the free-oscillation amplitude to zero, i.e. a very *large* change of the amplitude, though at low damping, $\delta \ll \omega_0$, this decay takes large time $t \sim \tau \gg 1/\omega_0$. Hence, if we want our approximate method to be productive (i.e. to work at all time scales, in particular for forced oscillations with established, constant amplitude and phase), we need to account for the fact that the *small* right-hand side of Eq. (5.38) may eventually lead to *essential* changes of the oscillation amplitude A (and sometimes,

[14] They are frequently called *nonlinear oscillations*. This is, again, a very unfortunate but generally accepted slang term for oscillations in systems described by nonlinear equations of motion.

as we will see below, also of the oscillation phase φ) at large times, because of the *slowly accumulating* effects of the small perturbation[15].

This goal may be achieved[16] by the account of these slow changes already in the '0th approximation', i.e. the basic part of the solution in the expansion (5.39):

$$q^{(0)} = A(t)\cos[\omega t - \varphi(t)], \quad \text{with } \dot{A}, \dot{\varphi} \to 0 \quad \text{at } \varepsilon \to 0. \tag{5.41}$$

(It is evident that the solution (5.9) fits this form.) Let me discuss this approach using a particular simple but representative example of a dissipative (but high-Q) pendulum driven by a weak sinusoidal external force with a nearly resonant frequency:

$$\ddot{q} + 2\delta\dot{q} + \omega_0^2 \sin q = f_0 \cos \omega t, \tag{5.42}$$

with $|\omega - \omega_0|$, $\delta \ll \omega_0$, and the force amplitude f_0 so small that $|q| \ll 1$ at all times. From what we know about the forced oscillations from section 5.1, it is natural to identify ω in the left-hand side of Eq. (5.38) with the force frequency. Expanding $\sin q$ into the Taylor series in small q, keeping only the first two terms of this expansion, and moving all the small terms to the right-hand side, we can rewrite Eq. (5.42) in the canonical form (5.38)[17]:

$$\ddot{q} + \omega^2 q = -2\delta\dot{q} + 2\xi\omega q + \alpha q^3 + f_0 \cos \omega t \equiv f(t, q, \dot{q}). \tag{5.43}$$

Here $\alpha = \omega_0^2/6$ in the case of the pendulum (although the calculations below will be valid for any α), and the second term on the right-hand side was obtained using the approximation already employed in section 5.1: $(\omega^2 - \omega_0^2)q \approx 2\omega(\omega - \omega_0)q = 2\omega\xi q$, where $\xi \equiv \omega - \omega_0$ is the detuning parameter that was already used earlier—see Eq. (5.21).

Now, following the general recipe expressed by Eqs. (5.39) and (5.41), in the first approximation in $f \propto \varepsilon$, we may look for the solution to Eq. (5.43) in the form[18]

$$q(t) = A \cos \Psi + q^{(1)}(t), \quad \text{where } \Psi \equiv \omega t - \varphi, \quad q^{(1)} \sim \varepsilon. \tag{5.44}$$

Let us plug this solution into both parts of Eq. (5.43), keeping only the terms of the first order in ε. Thanks to our (smart:-) choice of ω on the left-hand side of that equation, the two zero-order terms in that part cancel each other. Moreover, since

[15] The same flexible approach is necessary to approximations used in quantum mechanics. The method discussed here is closer in spirit (but not identical) to the *WKB approximation* (see, e.g. *Part QM* section 2.4) rather to the *perturbation theory* varieties (*Part QM* chapter 6).

[16] The basic idea of this approach was suggested in 1920 by B van der Pol, and in some textbooks its first approximation (on which I will focus) is called the *van der Pol method.* However, in optics and quantum mechanics, it is commonly called the *rotating wave approximation* (RWA). In mathematics-oriented texts, this approach, in particular its extensions to higher approximations, is usually called either the *small parameter method* or the *asymptotic method.* The list of other scientists credited for the development of this method, its variations and extensions includes, most notably, J H Poincaré, N Krylov, N Bogolyubov, and Yu Mitroplolsky.

[17] This equation is frequently called the *Duffing equation* (or the equation of the *Duffing oscillator*), after G Duffing who was the first to carry out its (rather incomplete) analysis in 1918.

[18] For a mathematically rigorous treatment of higher approximations, see, e.g. [2]. A more layman (and somewhat verbose) discussion of various oscillatory phenomena may be found in the classical text [3].

each term on the right-hand side of Eq. (5.43) is already of the order of ε, we may drop $q^{(1)} \propto \varepsilon$ from the substitution into that side at all, because this would give us only terms $O(\varepsilon^2)$ or higher. As a result, we get the following approximate equation:

$$\ddot{q}^{(1)} + \omega^2 q^{(1)} = f^{(0)} \equiv -2\delta \frac{d}{dt}(A \cos \Psi) + 2\xi\omega A \cos \Psi + \alpha(A \cos \Psi)^3 + f_0 \cos \omega t. \quad (5.45)$$

According to Eq. (5.41), generally A and φ should be considered as (slow) functions of time. However, let us leave the analyses of the transient process and system's stability until the next section, and use Eq. (5.45) to find stationary oscillations in the system that are established after an initial transient. For that limited task, we may take A = const, φ = const, so that $q^{(0)}$ presents sinusoidal oscillations of frequency ω. Sorting the terms on the right-hand side according to their time dependence[19], we see that it has terms with frequencies ω and 3ω:

$$f^{(0)} = \left(2\xi\omega A + \frac{3}{4}\alpha A^3 + f_0 \cos \varphi\right) \cos \Psi + \left(2\delta\omega A - f_0 \sin \varphi\right) \sin \Psi + \frac{1}{4}\alpha A^3 \cos 3\Psi. \quad (5.46)$$

Now comes the main trick of the van der Pol approach: mathematically, Eq. (5.45) may be viewed as the equation of oscillations in a linear, *dissipation-free* harmonic oscillator of frequency ω (not ω_0!) under the action of an external force $f(t)$ represented by the right-hand side of the equation. In our particular case, it has three terms: two 'quadrature' components at that very frequency ω, and the third one at frequency 3ω. As we know from our analysis of this problem in section 5.1, if any of the first two components is non-vanishing, $q^{(1)}$ grows to infinity—see Eq. (5.19) with $\delta = 0$. At the same time, by the very structure of the rotating-wave approximation, $q^{(1)}$ has to be finite—moreover, small! The only way out of this contradiction is to require that the amplitudes of both quadrature components of $f^{(0)}$ with frequency ω are equal to zero:

$$2\xi\omega A + \frac{3}{4}\alpha A^3 + f_0 \cos \varphi = 0, \qquad 2\delta\omega A - f_0 \sin \varphi = 0. \quad (5.47)$$

These two *harmonic balance equations* enable us to find both parameters of the forced oscillations: their amplitude A and phase φ. In particular, the phase may be readily eliminated from this system (most easily, by expressing $\sin \varphi$ and $\cos \varphi$ from Eqs. (5.47), and then requiring the sum $\sin^2 \varphi + \cos^2 \varphi$ to equal 1), and the solution for the amplitude A recast in the following implicit but convenient form:

$$A^2 = \frac{f_0^2}{4\omega^2} \frac{1}{\xi^2(A) + \delta^2}, \quad \text{where } \xi(A) \equiv \xi + \frac{3}{8}\frac{\alpha A^2}{\omega} = \omega - \left(\omega_0 - \frac{3}{8}\frac{\alpha A^2}{\omega}\right). \quad (5.48)$$

[19] Using the second of Eqs. (5.44), $\cos\omega t$ may be rewritten as $\cos(\Psi + \varphi) \equiv \cos \Psi \cos \varphi - \sin \Psi \sin \varphi$. Then using the identity given, e.g. by Eq. (A.19): $\cos^3 \Psi = (3/4)\cos \Psi + (1/4)\cos 3\Psi$, we obtain Eq. (5.46).

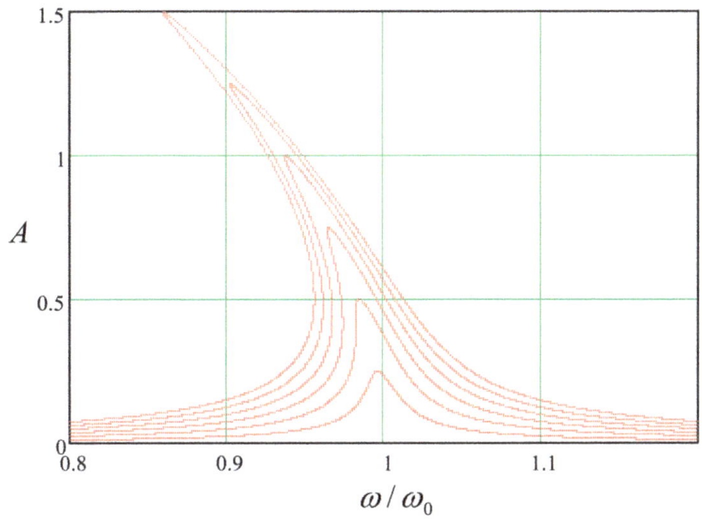

Figure 5.4. The nonlinear resonance in the Duffing oscillator, as described by Eq. (5.48), for the particular case $\alpha = \omega_0^2/6$, $\delta/\omega = 0.01$ (i.e. $Q = 50$), and several values of the parameter f_0/ω_0^2, increased by equal steps from 0.005 to 0.03.

This expression differs from Eq. (5.22) for the linear resonance in the low-damping limit only by the replacement of the detuning ξ with its effective amplitude-dependent value $\xi(A)$—or, equivalently, the replacement of the frequency ω_0 of the oscillator with its effective, amplitude-dependent value

$$\omega_0(A) = \omega_0 - \frac{3}{8}\frac{\alpha A^2}{\omega}. \tag{5.49}$$

The physical meaning of $\omega_0(A)$ is simple: this is just the frequency of free oscillations of amplitude A in a similar nonlinear system, but with zero damping[20]. Indeed, for $\delta = 0$ and $f_0 = 0$ we could repeat our calculations, assuming that ω is an amplitude-dependent eigenfrequency $\omega_0(A)$. Then the second of Eqs. (5.47) is trivially satisfied, while the second gives Eq. (5.49). Eq. (5.48) enables us to draw the curves of this *nonlinear resonance* by just bending the linear resonance plots (figure 5.1) according to the so-called *skeleton curve* expressed by Eq. (5.49). Figure 5.4 shows the result of this procedure. Note that at small amplitude, $\omega(A) \to \omega_0$, and we return to the usual, 'linear' resonance Eq. (5.22).

To bring our solution to its logical completion, we should still find the first perturbation $q^{(1)}(t)$ from what is left of Eq. (5.45). Since the structure of this equation

[20] This effect of the pendulum's frequency dependence on its oscillation amplitude was described as early as 1673 by C Huygens, who had earlier invented the pendulum clock, increasing time-keeping accuracy by ~3 orders of magnitude.

is similar to Eq. (5.13) with the force of frequency 3ω and zero damping, we may use Eqs. (5.16) and (5.17) to obtain

$$q^{(1)}(t) = -\frac{1}{32\omega^2}\alpha A^3 \cos 3(\omega t - \varphi). \tag{5.50}$$

Adding this perturbation (note the negative sign!) to the sinusoidal oscillation (5.41), we see that as the amplitude A of oscillations in a system with $\alpha > 0$ (e.g. a pendulum) grows, their waveform becomes a bit more 'blunt' near the maximum deviations from the equilibrium.

The same Eq. (5.50) also allows an estimate of the range of validity of our first approximation: since it has been based on the assumption $|q^{(1)}| \ll |q^{(0)}| \leqslant A$, for this particular problem we have to require $\alpha A^2/32\omega^2 \ll 1$. For a pendulum (with $\alpha = \omega_0^2/6$), this condition becomes $A^2 \ll 192$. Although numerical coefficients in such strong inequalities should be taken with a grain of salt, the very large magnitude of this particular coefficient gives a good hint that the method should give very good results even for relatively large oscillations with $A \sim 1$. In section 5.7 below, we will see that this is indeed the case.

From the mathematical viewpoint, the next step would be to write the next approximation as

$$q(t) = A\cos\Psi + q^{(1)}(t) + q^{(2)}(t), \qquad q^{(2)} \sim \varepsilon^2, \tag{5.51}$$

and plug it into the Duffing equation (5.43), which (thanks to our special choice of $q^{(0)}$ and $q^{(1)}$) would retain only $\ddot{q}^{(2)} + \omega^2 q^{(2)}$ on its left-hand side. Again, requiring that amplitudes of two quadrature components of frequency ω on the right-hand side to be zero, we may obtain the second-order corrections to A and φ. Then we may use the remaining part of the equation to calculate $q^{(2)}$, and then go after the third-order terms, etc. However, for most purposes the sum $q^{(0)} + q^{(1)}$, and sometimes even just the crudest approximation $q^{(0)}$ alone, are completely sufficient. For example, according to Eq. (5.50), for a simple pendulum ($\alpha = \omega_0^2/6$) swinging as much as between the opposite horizontal positions ($A = \pi/2$), the first-order correction $q^{(1)}$ is of the order of 0.5%. (Soon beyond this value, completely new dynamic phenomena begin—see section 5.7 below, but they cannot be described by these successive approximations at all.) For such reasons, higher approximations are rarely pursued for particular systems.

5.3 Reduced equations

A much more important issue is the stability of the solutions described by Eq. (5.48). Indeed, figure 5.4 shows that within a certain range of parameters, these equations give three different values for the oscillation amplitude (and phase), and it is important to understand which of these solutions are stable. Since these solutions are not the fixed points in the sense discussed in section 3.2 (each point in figure 5.4 represents a nearly sinusoidal oscillation), their stability analysis needs a more general approach that would be valid for oscillations with amplitude and phase slowly evolving in time. This approach will also enable the analysis of non-stationary (in particular the initial transient) processes, which are of key importance for some dynamic systems.

First of all, let us formalize the way the harmonic balance equations, such as Eq. (5.47), are obtained for the general case (5.38)—rather than for the particular Eq. (5.43) considered in the last section. After plugging in the 0th approximation Eq. (5.41) into the right-hand side of Eq. (5.38) we have to require the amplitudes of its both quadrature components of frequency ω to be zero. From the standard Fourier analysis we know that these requirements may be represented as

$$\overline{f^{(0)} \sin \Psi} = 0, \quad \overline{f^{(0)} \cos \Psi} = 0, \tag{5.52}$$

where the top bar means time averaging—in our current case, over the period $2\pi/\omega$ of the right-hand side of Eq. (5.52), with the arguments calculated in the 0th approximation:

$$\begin{aligned} &f^{(0)} \equiv f(t, q^{(0)}, \dot{q}^{(0)}, \cdots) \equiv f(t, A \cos \Psi, -A\omega \sin \Psi, \ldots), \\ &\text{with} \quad \Psi = \omega t - \varphi. \end{aligned} \tag{5.53}$$

Now, for a transient process the contribution of $q^{(0)}$ on the left-hand side of Eq. (5.38) is not zero any longer, because both amplitude and phase may be slow functions of time—see Eq. (5.41). Let us calculate this contribution. The exact result would be

$$\begin{aligned} \ddot{q}^{(0)} + \omega^2 q^{(0)} &\equiv \left(\frac{d^2}{dt^2} + \omega^2\right) A \cos(\omega t - \varphi) \\ &= (\ddot{A} + 2\dot{\varphi}\omega A - \dot{\varphi}^2 A) \cos(\omega t - \varphi) - 2\dot{A}(\omega - \dot{\varphi}) \sin(\omega t - \varphi). \end{aligned} \tag{5.54}$$

However, in the first approximation in ε, we may neglect the second derivative of A, and also the squares and products of the first derivatives of A and φ (which are all of the second order in ε), so that Eq. (5.54) is reduced to

$$\ddot{q}^{(0)} + \omega^2 q^{(0)} \approx 2A\dot{\varphi}\omega \cos(\omega t - \varphi) - 2\dot{A}\omega \sin(\omega t - \varphi). \tag{5.55}$$

On the right-hand side of Eq. (5.53), we can neglect the time derivatives of the amplitude and phase, because this part is already proportional to the small parameter. Hence, in the first order in ε, Eq. (5.38) becomes

$$\ddot{q}^{(1)} + \omega^2 q^{(1)} = f_{\text{ef}}^{(0)} \equiv f^{(0)} - (2A\dot{\varphi}\omega \cos \Psi - 2\dot{A}\omega \sin \Psi). \tag{5.56}$$

Now, applying Eq. (5.52) to the function $f_{\text{ef}}^{(0)}$, and taking into account that the time averages of $\sin^2 \Psi$ and $\cos^2 \Psi$ are both equal to 1/2, while the time average of the product $\sin\Psi \cos\Psi$ vanishes, we get a pair of so-called *reduced equations* (alternatively called the 'RWA equations' or the 'van der Pol equations') for the time evolution of the amplitude and the phase:

$$\dot{A} = -\frac{1}{\omega} \overline{f^{(0)} \sin \Psi}, \qquad \dot{\varphi} = \frac{1}{\omega A} \overline{f^{(0)} \cos \Psi}. \tag{5.57a}$$

Extending the definition Eq. (5.4) of the complex amplitude of oscillations to their slow evolution in time, $a(t) \equiv A(t) \exp\{i\varphi(t)\}$, and differentiating this relation, the two equations (5.57*a*) may be also rewritten in the form of either one equation for a:

$$\dot{a} = \frac{i}{\omega}\overline{f^{(0)}e^{i(\Psi+\varphi)}} \equiv \frac{i}{\omega}\overline{f^{(0)}e^{i\omega t}}, \tag{5.57b}$$

or two equations for the real and imaginary parts of $a(t) = u(t) + iv(t)$:

$$\dot{u} = -\frac{1}{\omega}\overline{f^{(0)}\sin\omega t}, \quad \dot{v} = \frac{1}{\omega}\overline{f^{(0)}\cos\omega t}. \tag{5.57c}$$

The first-order harmonic balance Eq. (5.52) are evidently just the particular case of the reduced equations (5.57) for stationary oscillations ($\dot{A} = \dot{\varphi} = 0$)[21].

Superficially, the system (5.57*a*) of two coupled, first-order differential equations may look more complex than the initial, second-order differential Eq. (5.38), but actually it is usually much simpler. For example, let us spell them out for the easy case of free oscillations a linear oscillator with damping. For that, we may reuse the ready Eq. (5.46) by taking $\alpha = f_0 = 0$, and thus turning Eq. (5.57*a*) into

$$\dot{A} = -\frac{1}{\omega}\overline{f^{(0)}\sin\Psi} \equiv -\frac{1}{\omega}\overline{(2\xi\omega A\cos\Psi + 2\delta\omega A\sin\Psi)\sin\Psi} \equiv -\delta A, \tag{5.58a}$$

$$\dot{\varphi} = \frac{1}{\omega A}\overline{f^{(0)}\cos\Psi} \equiv \frac{1}{\omega A}\overline{(2\xi\omega A\cos\Psi + 2\delta\omega A\sin\Psi)\cos\Psi} \equiv \xi. \tag{5.58b}$$

The solution of Eq. (5.58*a*) gives us the same 'envelope' law $A(t) = A(0)e^{-\delta t}$ as the exact solution (5.10) of the initial differential equation, while the elementary integration of Eq. (5.58*b*) yields $\varphi(t) = \xi\, t + \varphi(0) = \omega t - \omega_0 t + \varphi(0)$. This means that our approximate solution,

$$q^{(0)}(t) = A(t)\cos[\omega t - \varphi(t)] = A(0)e^{-\delta t}\cos[\omega_0 t - \varphi(0)], \tag{5.59}$$

agrees with the exact Eq. (5.9), and misses only the correction (5.8) of the oscillation frequency. (This correction is of the second order in δ, i.e. of the order of ε^2, and hence beyond the accuracy of our first approximation.) It is remarkable how nicely the reduced equations recover the proper frequency of free oscillations in this autonomous system—in which the very notion of ω is ambiguous.

The result is different at forced oscillations. For example, for the (generally nonlinear) Duffing oscillator described by Eq. (5.43) with $f_0 \neq 0$, Eqs. (5.57*a*) yield the reduced equations

$$\dot{A} = -\delta A + \frac{f_0}{2\omega}\sin\varphi, \quad A\dot{\varphi} = \xi(A)\,A + \frac{f_0}{2\omega}\cos\varphi, \tag{5.60}$$

which are valid for an arbitrary function $\xi(A)$, provided that the nonlinear detuning remains much smaller than the oscillation frequency. Here (after a transient), the

[21] One may ask why we cannot stick to just the one, most compact, complex-amplitude form (5.57*b*) of the reduced equations. The main reason is that when the function $f(q, \dot{q}, t)$ is nonlinear, we cannot replace its real arguments, such as $q = A\cos(\omega t - \varphi)$, with their complex-function representations such as $a\exp\{-i\omega t\}$ (as could be done in the linear problems considered in section 4.1), and we need to use real variables, such as either $\{A, \varphi\}$ or $\{u, v\}$, anyway.

amplitude and phase tend to the stationary states described by Eq. (5.47). This means that φ becomes a constant, so that $q^{(0)} \to A\cos(\omega t - \text{const})$, i.e. the reduced equations again automatically recover the correct frequency of the solution, in this case equal to that of the external force.

Note that each stationary oscillation regime, with a certain amplitude and phase, corresponds to a fixed point of the reduced equations, so that the stability of those fixed points determines that of the oscillations. In what follows, we will carry out such an analysis for several simple systems of key importance for physics and engineering.

5.4 Self-oscillations and phase locking

The motivation for B van der Pol to develop his method was the analysis of one more type of oscillatory motion: *self-oscillations*. Several systems, e.g. electronic radio-frequency (RF) amplifiers with positive feedback, and optical media with quantum level population inversion, provide convenient means for the compensation, and even over-compensation, of the intrinsic energy losses in oscillators. Phenomenologically, this effect may be described as the change of sign of the damping coefficient δ from positive to negative. Since for small oscillations the equation of motion is still linear, we may use Eq. (5.9) to describe its general solution. This equation shows that at $\delta < 0$, even infinitesimal deviations from equilibrium (say, due to unavoidable fluctuations) lead to oscillations with exponentially growing amplitude. Of course, in any real system such a growth cannot persist infinitely, and shall be limited by this or that effect—e.g. in the above examples, respectively, by amplifier's saturation and quantum level population's exhaustion.

In many cases, the amplitude limitation may be described reasonably well by making the following replacement:

$$2\delta\dot{q} \to 2\delta\dot{q} + \beta\dot{q}^3, \tag{5.61}$$

with $\beta > 0$. Let us analyze the effects of such *nonlinear damping*, applying the van der Pol's approach to the corresponding homogeneous differential equation (which also carries his name):

$$\ddot{q} + 2\delta\dot{q} + \beta\dot{q}^3 + \omega_0^2 q = 0. \tag{5.62}$$

Carrying out the dissipative and detuning terms to the right-hand side, and taking them for f in the canonical Eq. (5.38), we can easily calculate the right-hand sides of the reduced equations (5.57a), obtaining[22]

$$\dot{A} = -\delta(A)\,A, \qquad \text{where } \delta(A) \equiv \delta + \frac{3}{8}\beta\omega^2 A^2, \tag{5.63a}$$

$$A\dot{\varphi} = \xi\,A. \tag{5.63b}$$

[22] For that, one needs to use the trigonometric identity $\sin^3\Psi = (3/4)\sin\Psi - (1/4)\sin 3\Psi$—see, e.g. Eq. (A.19).

The second of these equations has exactly the same form as Eq. (5.58*b*) for the case of decaying oscillations and hence shows that the self-oscillations (if they happen, i.e. if $A \neq 0$) have the own frequency ω_0 of the oscillator—see Eq. (5.59). Equation (5.63*a*) is more substantive. If the initial damping δ is positive, it has only the trivial fixed point, $A_0 = 0$ (that describes the oscillator at rest), but if δ is negative, there is also another fixed point,

$$A_1 = \left(\frac{8\,|\,\delta\,|}{3\beta\omega^2}\right)^{1/2}, \qquad \text{for } \delta < 0, \tag{5.64}$$

which describes steady self-oscillations with a non-zero amplitude A_1.

Let us apply the general approach discussed in section 3.2, the linearization of equations of motion, to this reduced equation. For the trivial fixed point $A_0 = 0$, the linearization of Eq. (5.63*a*) is reduced to discarding the nonlinear term in the definition of the amplitude-dependent damping $\delta(A)$. The resulting linear equation evidently shows that the system's equilibrium point, $A = A_0 = 0$, is stable at $\delta > 0$ and unstable at $\delta < 0$. (We have already discussed this *self-excitation condition* above.) The linearization near the non-trivial fixed point A_1 requires a bit more math: in the first order in $\tilde{A} \equiv A - A_1 \to 0$, we obtain

$$\begin{aligned}\dot{\tilde{A}} \equiv \dot{A} &= -\delta(A_1 + \tilde{A}) - \frac{3}{8}\beta\omega^2(A_1 + \tilde{A})^3 \approx -\delta\,\tilde{A} - \frac{3}{8}\beta\omega^2 3A_1^{\,2}\tilde{A} \\ &= (-\delta + 3\delta)\,\tilde{A} = 2\delta\tilde{A},\end{aligned} \tag{5.65}$$

where Eq. (5.64) has been used to eliminate A_1. We see that the fixed point A_1 (and hence the whole process) is stable as soon as it exists ($\delta < 0$)—similar to the situation in our 'testbed problem' (figure 2.1), except that in our current, dissipative system, the stability is 'actual' rather than 'orbital'—see section 5.6 for more.

Now let us consider another important problem: the effect of an external sinusoidal force on a self-excited oscillator. If the force is sufficiently small, its effects on the self-excitation condition and the oscillation amplitude are negligible. However, if the frequency ω of such a weak force is close to the own frequency ω_0 of the oscillator, it may lead to a very important effect of *phase locking*[23]—also called the 'synchronization', although the latter term also has a much broader meaning. At this effect, the oscillator's frequency deviates from ω_0, and becomes exactly equal to the external force's frequency ω, within a certain range

$$-\Delta \leqslant \omega - \omega_0 < +\Delta. \tag{5.66}$$

In order to prove this fact, and also to calculate the phase-locking range width 2Δ, we may repeat the calculation of the right-hand sides of the reduced equations (5.57*a*), adding term $f_0 \cos \omega t$ to the right-hand side of Eq. (5.62)—see Eqs. (5.42)–(5.43). This addition modifies Eqs. (5.63) as follows[24]:

[23] Apparently, the mutual phase-locking of two pendulum clocks was first noticed by the same C Huygens.

[24] Actually, this result should be evident, even without calculations, from the comparison of Eqs. (5.60) and (5.63).

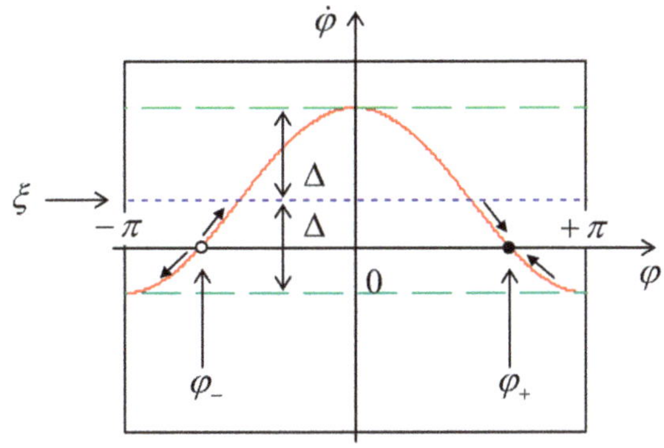

Figure 5.5. The phase plane of a phase-locked oscillator, for the particular case $\xi = \Delta/2$, $f_0 > 0$.

$$\dot{A} = -\delta(A)\, A + \frac{f_0}{2\omega} \sin \varphi, \tag{5.67a}$$

$$A\dot{\varphi} = \xi\, A + \frac{f_0}{2\omega} \cos \varphi. \tag{5.67b}$$

If the system is self-excited, and the external force is weak, its effect on the oscillation amplitude is small, and in the first approximation in f_0 we can take A to be constant and equal to the value A_1 given by Eq. (5.64). Plugging this approximation into Eq. (5.67*b*), we obtain a very simple equation[25]

$$\dot{\varphi} = \xi + \Delta \cos \varphi, \tag{5.68}$$

where in our current case

$$\Delta \equiv \frac{f_0}{2\omega A_1}. \tag{5.69}$$

Within the range $-|\Delta| < \xi < +|\Delta|$, Eq. (5.68) has two fixed points on each 2π-segment of the variable φ:

$$\varphi_\pm = \pm\cos^{-1}\left(-\frac{\xi}{\Delta}\right) + 2\pi n. \tag{5.70}$$

It is easy to linearize Eq. (5.68) near each point to analyze their stability in our usual way; however, let me use this case to demonstrate another convenient way to achieve this in 1D systems, using the so-called *phase plane*—the plot of the right-hand side of Eq. (5.68) as a function of φ (see figure 5.5).

Since the positive values of this function correspond to the growth of φ in time, and vice versa, we may draw the arrows showing the direction of phase evolution. From these graphics, it is clear that one of these fixed points (for $f_0 > 0$, φ_+) is stable,

[25] This equation is ubiquitous in phase-locking systems, including even some digital electronic circuits used for that purpose.

while its counterpart (in this case, φ_-) is unstable. Hence the magnitude of Δ given by Eq. (5.69) is indeed the phase-locking range (or rather it half) that we wanted to find. Note that the range is proportional to the amplitude of the phase-locking signal—perhaps the most important feature of this effect.

In order to complete our simple analysis, based on the assumption of fixed oscillation amplitude, we need to find the condition of its validity. For that, we may linearize Eq. (5.67*a*), for the stationary case, near the value A_1, just as we have done in Eq. (5.65) for the transient process. The stationary result,

$$\tilde{A} \equiv A - A_1 = \frac{1}{2\,|\,\delta\,|}\frac{f_0}{2\omega}\sin\varphi_\pm \approx A_1\left|\frac{\Delta}{2\delta}\right|\sin\varphi_\pm, \tag{5.71}$$

shows that our assumption, $|\tilde{A}| \ll A_1$, and hence the final result (5.69), are valid if the phase-locking range, 2Δ, is much smaller than $4\,|\delta\,|$.

5.5 Parametric excitation

In both problems solved in the last section, the stability analysis was easy because it could be carried out for just one slow variable, *either* amplitude *or* phase. More generally, such analysis of the reduced equations involves both these variables. The classical example of such a situation is provided by one important physical phenomenon—the *parametric excitation* of oscillations. An elementary example of such excitation is given by a pendulum with a variable parameter, for example the suspension length $l(t)$—see figure 5.6. Experiments (including those with playground swings) and numerical simulations show that if the length is changed (*modulated*) periodically, with a frequency 2ω that is close to $2\omega_0$ and a sufficiently large swing Δl, the equilibrium position of the pendulum becomes unstable and it starts swinging with frequency ω equal to *exactly* half of the modulation frequency (and hence only *approximately* equal to the average eigenfrequency ω_0 of the oscillator).

For an elementary analysis of this effect we may consider the simplest case when the oscillations are small. At the lowest point ($\theta = 0$), where the pendulum moves with the highest velocity v_{max}, the string's tension $\mathscr{T}$ is *higher* than mg by the centripetal force: $\mathscr{T}_{max} = mg + mv_{max}^2/l$. In contrast, at the maximum deviation of the pendulum from the equilibrium, the force is *lower* than mg, because of

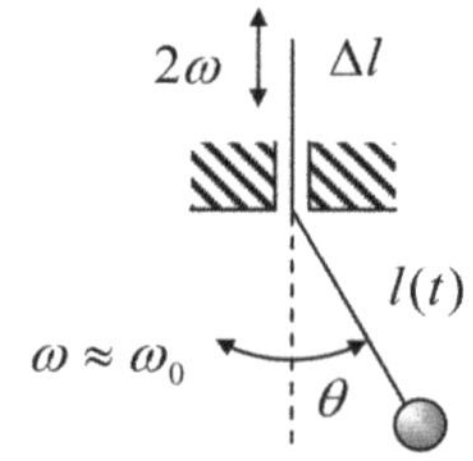

Figure 5.6. Parametric excitation of a pendulum.

the suspension line's tilt: $\mathscr{T}_{\min} = mg\cos\theta_{\max}$. Using the energy conservation, $E = mv_{\max}^2/2 = mgl(1 - \cos\theta_{\max})$, we may express these values as $\mathscr{T}_{\max} = mg + 2E/l$ and $\mathscr{T}_{\min} = mg - E/l$. Now, if during each oscillation period the string is pulled up slightly by Δl (with $|\Delta l| \ll l$) at each of its two passages through the lowest point, and is allowed to go down by the same amount at each of two points of the maximum deviation, the net work of the external force per period is positive:

$$\mathscr{W} \approx 2(\mathscr{T}_{\max} - \mathscr{T}_{\min})\Delta l \approx 6\frac{\Delta l}{l}E, \tag{5.72}$$

and hence results in an increase of the oscillator's energy. If the so-called *modulation depth* $\Delta l/2l$ is sufficient, this increase may overcompensate the energy drained out by damping. Quantitatively, Eq. (5.10) shows that low damping ($\delta \ll \omega_0$) leads to the following energy decrease,

$$\Delta E \approx -4\pi\frac{\delta}{\omega_0}E, \tag{5.73}$$

per oscillation period. Comparing Eqs. (5.72) and (5.73), we see that the net energy flow into the oscillations is positive, $\mathscr{W} + \Delta E > 0$, i.e. oscillation amplitude has to grow, if[26]

$$\frac{\Delta l}{l} > \frac{2\pi\delta}{3\omega_0} \equiv \frac{\pi}{3Q}. \tag{5.74}$$

Since this result is independent of the oscillation energy E, the growth of energy and amplitude is exponential (until E becomes so large that some of our assumptions fails), so that Eq. (5.74) is the condition of parametric excitation—in this simple model.

However, this result does not account for a possible difference between the oscillation frequency ω and the eigenfrequency ω_0, and also does not clarify whether the best phase shift between the oscillations and parameter modulation, assumed in the above calculation, may be sustained automatically. In order to address these issues, we may apply the van der Pol approach to a simple but reasonable model:

$$\ddot{q} + 2\delta\dot{q} + \omega_0^2(1 + \mu\cos 2\omega t)q = 0, \tag{5.75}$$

describing the parametric excitation in a linear oscillator with a sinusoidal modulation of the parameter $\omega_0^2(t)$. Rewriting this equation in the canonical form (5.38),

$$\ddot{q} + \omega^2 q = f(t, q, \dot{q}) \equiv -2\delta\dot{q} + 2\xi\omega q - \mu\omega_0^2 q\cos 2\omega t, \tag{5.76}$$

[26] A modulation of a pendulum's mass (say, by periodic pumping water in and out of a suspended bottle) gives a qualitatively similar result. Note, however, that parametric oscillations cannot be excited by modulating *any* oscillator's parameter—for example, oscillator's damping coefficient (at least if it stays positive at all times), because it does not change the system's energy, just the energy drain rate.

and assuming that the dimensionless ratios δ/ω and $|\xi|/\omega$, and the modulation depth μ are all much less than 1, we may use the general Eqs. (5.57*a*) to obtain the following reduced equations:

$$\begin{aligned} \dot{A} &= -\delta\, A - \frac{\mu\omega}{4} A \sin 2\varphi, \\ A\dot{\varphi} &= A\xi - \frac{\mu\omega}{4} A \cos 2\varphi. \end{aligned} \tag{5.77}$$

These equations evidently have a fixed point, with $A_0 = 0$, but its stability analysis (though possible) is not absolutely straightforward, because the phase φ of oscillations is undetermined at that point. In order to avoid this (technical rather than conceptual) difficulty, we may use, instead of the real amplitude and phase of oscillations, either their complex amplitude $a = A \exp\{i\varphi\}$, or its Cartesian components u and v—see Eq. (5.4). Indeed, for our function f, Eq. (5.57*b*) gives

$$\dot{a} = (-\delta + i\xi)a - i\frac{\mu\omega}{4}a^*, \tag{5.78}$$

while Eqs. (5.57*c*) yield

$$\begin{aligned} \dot{u} &= -\delta u - \xi v - \frac{\mu\omega}{4}v, \\ \dot{v} &= -\delta v + \xi u - \frac{\mu\omega}{4}u. \end{aligned} \tag{5.79}$$

We see that in contrast to Eqs. (5.77), in the Cartesian coordinates $\{u, v\}$ the trivial fixed point $a_0 = 0$ (i.e. $u_0 = v_0 = 0$) is absolutely regular. Moreover, Eqs. (5.78) and (5.79) are already linear, so they do not require any additional linearization. Thus we may use the same approach as was already used in sections 3.2 and 5.1, i.e. look for the solution of Eq. (5.79) in the exponential form $\exp\{\lambda t\}$. However, now we are dealing with two variables, and should allow them to have, for each value of λ, a certain ratio u/v. For that, we may take the partial solution in the form

$$u = c_u e^{\lambda t}, \qquad v = c_v e^{\lambda t}, \tag{5.80}$$

where constants c_u and c_v are frequently called the *distribution coefficients.* Plugging this solution into Eqs. (5.79), we obtain for them the following system of two linear algebraic equations:

$$\begin{aligned} (-\delta - \lambda)c_u + \left(-\xi - \frac{\mu\omega}{4}\right)c_v &= 0, \\ \left(\xi - \frac{\mu\omega}{4}\right)c_u + (-\delta - \lambda)c_v &= 0. \end{aligned} \tag{5.81}$$

The characteristic equation of this system, i.e. the condition of compatibility of Eqs. (5.81),

$$\begin{vmatrix} -\delta-\lambda & -\xi-\dfrac{\mu\omega}{4} \\ \xi-\dfrac{\mu\omega}{4} & -\delta-\lambda \end{vmatrix} \equiv \lambda^2+2\delta\lambda+\delta^2+\xi^2-\left(\frac{\mu\omega}{4}\right)^2=0, \tag{5.82}$$

has two roots:

$$\lambda_\pm = -\delta \pm \left[\left(\frac{\mu\omega}{4}\right)^2-\xi^2\right]^{1/2}. \tag{5.83}$$

Requiring the fixed point to be unstable, Re $\lambda_+>0$, we get the parametric excitation condition

$$\frac{\mu\omega}{4} > (\delta^2+\xi^2)^{1/2}. \tag{5.84}$$

Thus the parametric excitation may indeed happen without any artificial phase control: the arising oscillations self-adjust their phase to pick up energy from the external source responsible for the parameter variation.

Our key result (5.84) may be compared with two other calculations. First, in the case of negligible damping ($\delta = 0$), Eq. (5.84) turns into condition $\mu\omega/4 > |\xi|$. This result may be compared with the well-developed theory of the so-called *Mathieu equation*, whose canonical form is

$$\frac{d^2y}{dv^2} + (a - 2b\cos 2v)y = 0. \tag{5.85}$$

With the substitutions $y \to q$, $v \to \omega t$, $a \to (\omega_0/\omega)^2$, $b \to -\mu/2$, this equation is just a particular case of Eq. (5.75) for $\delta = 0$. In terms of Eq. (5.85), the result of our analysis may be re-written just as $b > |a - 1|$, and is supposed to be valid for $b \ll 1$. This condition is shown in figure 5.7 together with the numerically calculated[27] stability boundaries of the Mathieu equation.

One can see that the van der Pol approximation works just fine within its applicability limit (and a bit beyond :-), although it fails to predict some other features of the Mathieu equation, such as the existence of higher, more narrow regions of parametric excitation (at $a \approx n^2$, i.e. $\omega_0 \approx \omega/n$, for all integer n), and some spill-over of the stability region into the lower half-plane $a < 0$.[28] The reason for these failures is the fact that, as can be seen in figure 5.7, these phenomena do not appear in the first approximation in the parameter modulation amplitude $\mu \propto \varepsilon$, that is the realm of applicability of the reduced equation (5.79).

[27] Such calculations may be substantially simplified by the use of the so-called *Floquet theorem*, which is also the mathematical basis for the discussion of wave propagation in periodic media—see the next chapter.

[28] This region describes, for example, the counter-intuitive stability of an inverted pendulum with the periodically modulated length, within a limited range of the modulation depth μ.

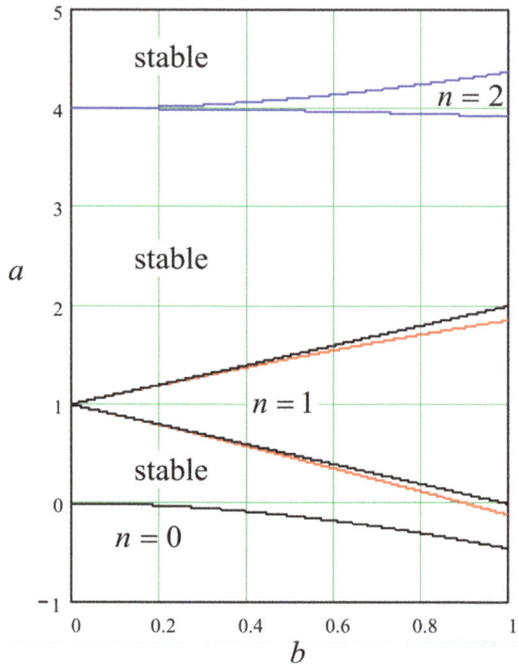

Figure 5.7. Stability boundaries of the Mathieu equation (5.85), as calculated: numerically (curves) and using the reduced equations (5.79) (dashed straight lines). In the regions numbered by various n, the trivial solution $y = 0$ of the equation is unstable, i.e. its general solution $y(v)$ includes an exponentially growing term.

In the opposite case of non-zero damping but exact tuning ($\xi = 0$, $\omega \approx \omega_0$), Eq. (5.84) gives

$$\mu > \frac{4\delta}{\omega_0} \equiv \frac{2}{Q}. \tag{5.86}$$

This condition may be compared with Eq. (5.74), taking $\Delta l/l = 2\mu$. The comparison shows that while the structure of these conditions is similar, the numerical coefficients are different by a factor close to 2. The first reason for this difference is that the instant parameter change at optimal moments of time is more efficient then the smooth, sinusoidal variation described by Eq. (5.75). Even more significantly, the change of the pendulum's length modulates not only its frequency $\omega_0 \equiv (g/l)^{1/2}$ as Eq. (5.75) implies, but also its *mechanical impedance* $Z \equiv (gl)^{1/2}$—the notion to be discussed in detail in the next chapter. (The analysis of the general case of the simultaneous modulation of ω_0 and Z is left for a reader exercise.)

Before moving on, let me summarize the most important differences between the parametric and forced oscillations:

(i) Parametric oscillations completely disappear outside of their excitation range, while the forced oscillations have a non-zero amplitude for any frequency and amplitude of the external force—see Eq. (5.18).

(ii) Parametric excitation may be described by a linear *homogeneous* equation—e.g. Eq. (5.75)—which cannot predict any finite oscillation amplitude within the excitation range, even at finite damping. In order to describe stationary parametric oscillations, some nonlinear effect has to be taken into account. (Again, I am leaving analyses of such effects for a reader exercise.)

One more important feature of the parametric oscillations will be discussed at the end of the next section.

5.6 Fixed-point classification

The reduced equations (5.79) give us a good pretext for a brief discussion of an important general topic of dynamics: fixed points of a system described by two time-independent, first-order differential equations with time-independent coefficients[29]. After their linearization near a fixed point, the equations for deviations can always be expressed in the form similar to Eq. (5.79):

$$\begin{aligned}\dot{\tilde{q}}_1 &= M_{11}\tilde{q}_1 + M_{12}\tilde{q}_2,\\ \dot{\tilde{q}}_2 &= M_{21}\tilde{q}_1 + M_{22}\tilde{q}_2,\end{aligned} \tag{5.87}$$

where $M_{jj'}$ (with $j, j' = 1, 2$) are some real scalars, which may be viewed upon as the elements of a 2×2 matrix M. Looking for an exponential solution of the type (5.80),

$$\tilde{q}_1 = c_1 e^{\lambda t}, \quad \tilde{q}_2 = c_2 e^{\lambda t}, \tag{5.88}$$

we obtain a more general system of two linear equations for the distribution coefficients $c_{1,2}$:

$$\begin{aligned}(M_{11} - \lambda)c_1 + M_{12}c_2 &= 0,\\ M_{21}c_1 + (M_{22} - \lambda)c_2 &= 0.\end{aligned} \tag{5.89}$$

These equations are consistent if

$$\begin{vmatrix} M_{11} - \lambda & M_{12} \\ M_{21} & M_{22} - \lambda \end{vmatrix} = 0, \tag{5.90}$$

giving us a quadratic characteristic equation

$$\lambda^2 - \lambda(M_{11} + M_{22}) + (M_{11}M_{22} - M_{12}M_{21}) = 0. \tag{5.91}$$

Its solution[30],

[29] Autonomous systems described by a single, second-order homogeneous differential equation, say $F(q, \dot{q}, \ddot{q}) = 0$, also belong to this class, because we may always treat the generalized velocity $\dot{q} \equiv v$ as a new variable, and use this definition as one first-order differential equation, and the initial equation, in the form $F(q, v, \dot{v}) = 0$, as the second first-order equation.

[30] In the terms of linear algebra, $\lambda_\pm$ are the *eigenvalues*, and the corresponding sets of the distribution coefficients $[c_1, c_2]_\pm$, the *eigenvectors* of the matrix M with elements $M_{jj'}$.

$$\lambda_\pm = \frac{1}{2}(M_{11} + M_{22}) \pm \frac{1}{2}[(M_{11} - M_{22})^2 + 4M_{12}M_{21}]^{1/2}, \tag{5.92}$$

shows that the following situations are possible:

A. The expression under the square root, $(M_{11} - M_{22})^2 + 4M_{12}M_{21}$, is positive. In this case, both characteristic exponents $\lambda_\pm$ are real, and we can distinguish three sub-cases:

(i) Both λ_+ and λ_- are negative. As Eq. (5.88) shows, in this case the deviations $\tilde{q}$ tend to zero at $t \to \infty$, i.e. the fixed point is stable. Because of generally different magnitudes of exponents $\lambda_\pm$, the process represented on the phase plane $[\tilde{q}_1, \tilde{q}_2]$ (see figure 5.8a, with the solid arrows, for an example) may be

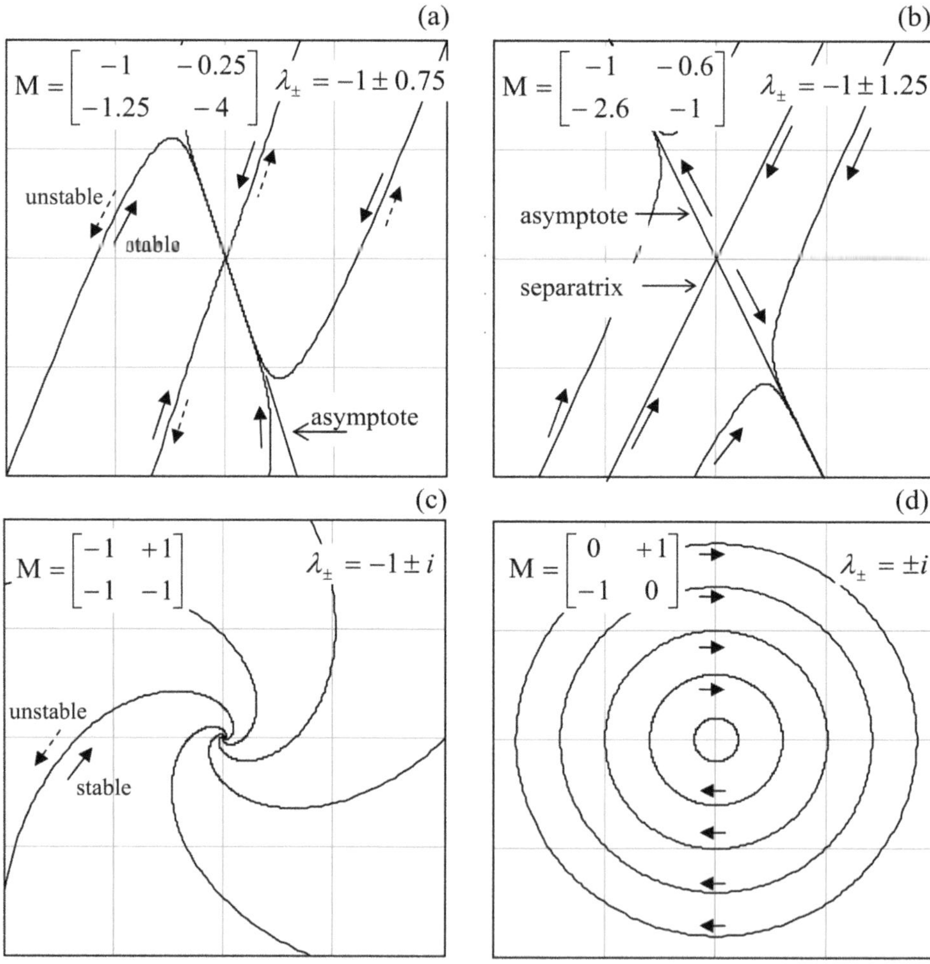

Figure 5.8. Typical trajectories on the phase-plane $[\tilde{q}_1, \tilde{q}_2]$ near fixed points of different types: (a) node, (b) saddle, (c) focus, and (d) center. The particular values of the matrix M, used in the first three panels, correspond to Eq. (5.81) for the parametric excitation, with $\xi = \delta$, and three different values of the parameter $\mu\omega/4\delta$: (a) 1.25, (b) 1.6, and (c) 0.

seen as consisting of two stages: first, a faster (with the rate $|\lambda_-| > |\lambda_+|$) relaxation to a linear *asymptote*[31], and then a slower decline, with the rate $|\lambda_+|$, along this line, i.e. at the virtually fixed ratio of the variables. Such a fixed point is called the *stable node*.

(ii) Both λ_+ and λ_- are positive. This case of an *unstable node* differs from the previous one only by the direction of motion along the phase-plane trajectories—see the dashed arrows in figure 5.8a. Here the variable ratio is also approaching a constant soon, now the one corresponding to $\lambda+ > \lambda_-$.

(iii) Finally, in the case of a *saddle* ($\lambda_+ > 0$, $\lambda_- < 0$) the system dynamics is different (figure 5.8b): after the rate $|\lambda_-|$ relaxation to the asymptote, the perturbation starts to grow, with the rate λ_+, along one of two opposite directions. (The direction is determined on which side of another straight line, called the *separatrix*, the system has been initially.) So the saddle[32] is an unstable fixed point.

B. The expression under the square root in Eq. (5.92), $(M_{11} - M_{22})^2 + 4M_{12}M_{21}$, is negative. In this case the square root is imaginary, making the real parts of both roots equal, Re $\lambda_\pm = (M_{11} + M_{22})/2$, and their imaginary parts equal but opposite. As a result, here there can be just two types of fixed points:

(i) *Stable focus*, at $(M_{11} + M_{22}) < 0$. The phase-plane trajectories are spirals going to the origin (i.e. toward the fixed point)—see figure 5.8c with the solid arrow.

(ii) *Unstable focus*, taking place at $(M_{11} + M_{22}) > 0$, differs from the stable focus only by the direction of motion along the phase trajectories—see the dashed arrow in the same figure 5.8c.

C. Frequently, the border case, $M_{11} + M_{22} = 0$, corresponding to the orbital ('indifferent') stability already discussed in section 3.2, is also distinguished, and the corresponding fixed point is referred to as the *center*—see figure 5.8d. Considering centers as a separate category makes sense because such fixed points are typical for Hamiltonian systems, whose first integral of motion may be frequently represented as the distance of the phase point from a fixed point. For example, introducing new variables $\tilde{q}_1 \equiv \tilde{q}, \quad \tilde{q}_2 \equiv m\dot{\tilde{q}}_1$, we may rewrite Eq. (3.12) of a harmonic oscillator without dissipation (again, with indices 'ef' dropped for brevity), as a system of two first-order differential equations:

$$\dot{\tilde{q}}_1 = \frac{1}{m}\tilde{q}_2, \quad \dot{\tilde{q}}_2 = -\kappa\tilde{q}_1, \tag{5.93}$$

i.e. as a particular case of Eq. (5.87), with $M_{11} = M_{22} = 0$, and $M_{12}M_{21} = -\kappa/m \equiv -\omega_0^2 < 0$, and hence $(M_{11} - M_{22})^2 + 4M_{12}M_{21} = -4\omega_0^2 < 0$, and $M_{11} + M_{22} = 0$. On

[31] The asymptote direction may be found by plugging the value λ_+ back into Eq. (5.89) and finding the corresponding ratio c_1/c_2. Note that the separation of the system's evolution into the two stages is conditional, being most vivid in the case of a large difference between the exponents λ_+ and λ_-.

[32] The term 'saddle' is due to the fact that system's dynamics in this case is qualitatively similar to those of a viscous motion in the 2D potential $U(\tilde{q}_1, \tilde{q}_2)$ having the shape of a horse saddle (or a mountain pass).

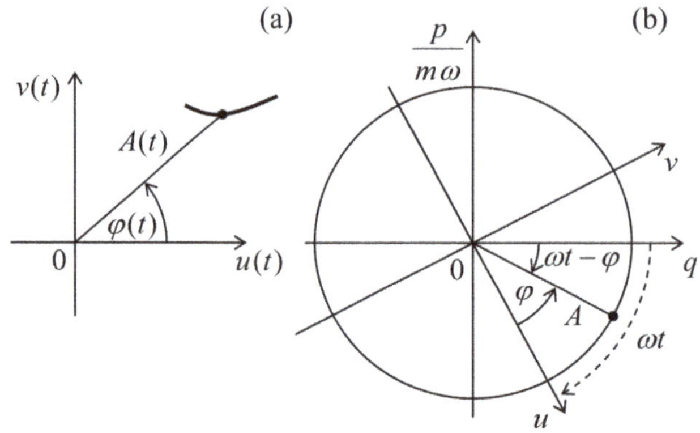

Figure 5.9. (a) Representation of a sinusoidal oscillation (point) and a slow transient process (line) on the Poincaré plane and (b) the relation between 'fast' phase plane to the 'slow' (Poincaré) plane.

the symmetrized phase plane $[\tilde{q}_1, \tilde{q}_2/Z]$, where the parameter $Z \equiv (\kappa m)^{1/2} \equiv m\omega_0$ is the oscillator's impedance, the sinusoidal oscillations of amplitude A are represented by a circle of radius A about the center-type fixed point $A = 0$. In the case when $\tilde{q}_1 \equiv \tilde{q}$ is the linear coordinate q of an actual mechanical oscillator, so that $\tilde{q}_2 \equiv m\dot{\tilde{q}}_1$ is its linear momentum $p = m\dot{q}$, such a circular trajectory corresponds to the conservation of the oscillator's energy

$$E \equiv T + U \equiv \frac{p^2}{2m} + \frac{\kappa q^2}{2} \equiv \frac{\kappa}{2}\left[\tilde{q}_1^{\,2} + \left(\frac{\tilde{q}_2}{Z}\right)^2\right] = \frac{\kappa A^2}{2} = \text{const}. \tag{5.94}$$

This is a convenient moment for a brief discussion of the so-called *Poincaré* (or 'slow-variable', or 'stroboscopic') *plane*[33]. From the point of view of the basic Eq. (5.41), the sinusoidal oscillations $q(t) = A\cos(\omega t - \varphi)$, described by a circular trajectory on the actual (symmetrized) phase plane, correspond to a fixed point $\{A, \varphi\}$, which may conveniently be represented by a steady geometric point on a plane with these polar coordinates—see figure 5.9a. (As follows from Eq. (5.4), the Cartesian coordinates of the point on that plane are just the variables $u \equiv A\cos\varphi$ and $v \equiv A\cos\varphi$ used, in particular, in the last section.) The quasi-sinusoidal process (5.41), with slowly changing A and φ, may be represented by a slow motion of that point on this Poincaré plane.

Figure 5.9b shows a convenient way to visualize the relation between the actual phase plane of an oscillator, with the 'fast' symmetrized coordinates q and p/Z, and the Poincaré plane with 'slow' coordinates u and v: the latter plane rotates relative to the former one, about the origin, clockwise, with the angular velocity ω[34]. Another, 'stroboscopic' way to generate the Poincaré plane pattern is to have a fast glance at the 'real' phase-plane just once during the oscillation period $\mathcal{T} = 2\pi/\omega$.

---

[33] Named after J H Poincaré (1854–1912), who is credited, among many other achievements, with his contributions to special relativity (see, e.g. *Part EM* chapter 9), and the basic idea of unstable trajectories, responsible for the deterministic chaos—to be discussed in chapter 9 of this volume.

[34] This notion of phase-plane rotation is the origin of the term 'rotating wave approximation', RTA. (The word 'wave' is an artifact of this method's wide application in classical and quantum optics.)

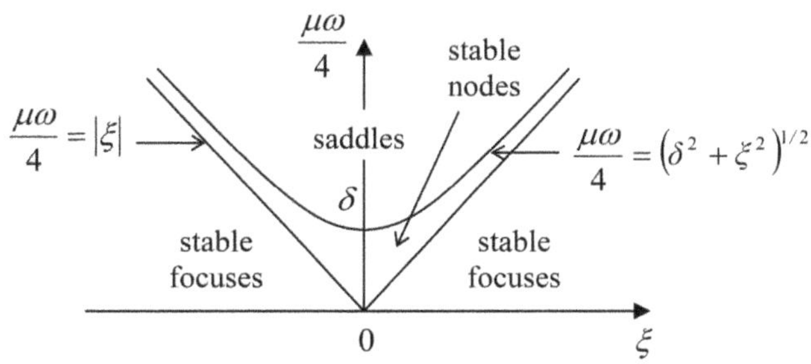

Figure 5.10. The types of the trivial fixed point of a parametric oscillator.

In many cases, the representation on the Poincaré plane is more convenient than that on the 'real' phase-plane. In particular, we have already seen that the reduced equations for such important phenomena as the phase-locking and the parametric oscillations, whose original differential equations include time explicitly, are time-independent—see, e.g. Eqs. (5.75) and (5.79) describing the latter effect. This simplification brings the equations into the category considered in this section, and enables the classification of their fixed points, which may shed additional light on their dynamic properties.

In particular, figure 5.10 shows the classification of the fixed points of a parametric oscillator, which follows from equation (5.83). As the parameter modulation depth μ is increased, the type of the trivial fixed point $A_1 = 0$ on the Poincaré plane changes from a stable focus (pertinent to a simple oscillator with damping) to a stable node and then to a saddle describing the parametric excitation. In the last case, the two directions of the perturbation growth, so prominently featured in figure 5.8b, correspond to the two possible values of the oscillation phase φ, with the phase choice determined by initial conditions.

This double degeneracy of the parametric oscillation's phase could already be noticed from Eq. (5.77), because they are evidently invariant with respect to the replacement $\varphi \to \varphi + \pi$. Moreover, the degeneracy is not an artifact of the reduced equations, because the initial equation (5.75) is already invariant with respect to the corresponding replacement $q(t) \to q(t - \pi/\omega)$. This invariance means that all other characteristics (e.g. the amplitude) of the parametric oscillations excited with either of two phases are *absolutely* similar. At the dawn of the computer age (in the late 1950s and early 1960s), there were substantial attempts, particularly in Japan, to use this property for storage and processing digital information coded in the phase-binary form. Though these attempts have not survived competition with simpler approaches, based on voltage-binary coding, some current trends in the development of prospective reversible and quantum computers may be traced back to this idea.

5.7 Numerical approaches

If the amplitude of oscillations, for whatever reason, becomes so large that the nonlinear terms in the equation describing the oscillator are comparable to its linear terms, numerical methods are virtually the only avenue available for their theoretical study. In Hamiltonian 1D systems, such methods may be applied directly to the integral (3.26), but dissipative and/or parametric systems typically lack first integrals of motion similar to Eq. (3.24), so that the initial differential equation has to be solved.

Let us discuss the general idea of such methods on the example of what mathematicians call the *Cauchy problem* (finding the solution for all moments of time, starting from the known initial conditions) for first-order differential equation

$$\dot{q} = f(t, q). \tag{5.95}$$

(The generalization to a system of several such equations is straightforward.) Breaking the time axis into small, equal steps h (figure 5.11) we can reduce the equation integration problem to finding the function value at the next time point, $q_{n+1} \equiv q(t_{n+1}) \equiv q(t_n + h)$ from the previously found value $q_n = q(t_n)$—and, if necessary, the values of q at other previous time steps.

In the simplest approach (called the *Euler method*), q_{n+1} is found using the following formula:

$$\begin{aligned} q_{n+1} &= q_n + k, \\ k &\equiv h f(t_n, q_n). \end{aligned} \tag{5.96}$$

This approximation is equivalent to the replacement of the genuine function $q(t)$, on the segment $[t_n, t_{n+1}]$, with the two first terms of its Taylor expansion in point t_n:

$$q(t_n + h) \approx q(t_n) + \dot{q}(t_n)h \equiv q(t_n) + hf(t_n, q_n). \tag{5.97}$$

This approximation has an error proportional to h^2. One could argue that making the step h sufficiently small, the Euler method's error might be made arbitrary small, but even with the number-crunching power of modern computers, the computation time necessary to reach sufficient accuracy may be too high for large problems[35]. In addition, the increase of the number of time steps, which is necessary at $h \to 0$, increases the total rounding errors, and eventually may cause an increase, rather than the reduction of the overall error of the computed result.

A more efficient way is to modify Eq. (5.96) to include the terms of the second order in h. There are several ways to do this, for example using the *second-order Runge–Kutta* method:

$$\begin{aligned} q_{n+1} &= q_n + k_2, \\ k_2 &\equiv hf\left(t_n + \frac{h}{2}, q_n + \frac{k_1}{2}\right), \quad k_1 \equiv hf(t_n, q_n). \end{aligned} \tag{5.98}$$

One can readily check that this method gives the exact result if the function $q(t)$ is a quadratic polynomial, and hence in the general case its errors are of the order of h^3.

[35] In addition, the Euler method is not time-reversible—the handicap which may be essential for integration of Hamiltonian systems described by systems of second-order differential equations. However, this drawback may be readily overcome by the so-called *leapfrogging*—the overlap of time steps h for a generalized coordinate and the corresponding generalized velocity.

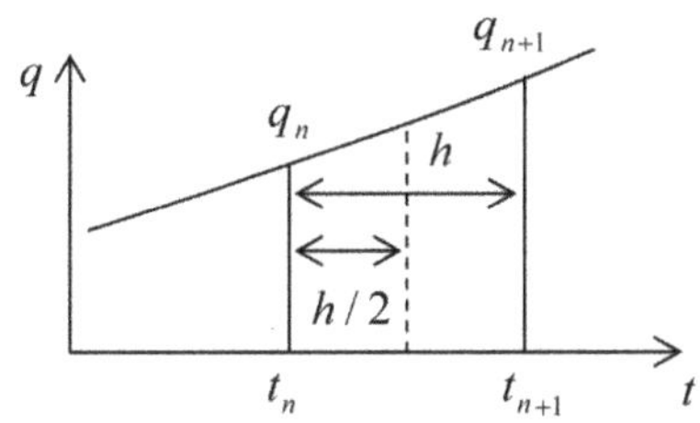

Figure 5.11. The basic notions used at numerical integration of ordinary differential equations.

We see that the main idea here is to first break the segment $[t_n, t_{n+1}]$ in half (see figure 5.11), then evaluate the right-hand side of the differential equation (5.95) at the point intermediate (in both t and q) between the points number n and $(n + 1)$, and then use this information to predict q_{n+1}.

The advantage of the Runge–Kutta approach is that it may be extended to the fourth order, without an additional breaking of the interval $[t_n, t_{n+1}]$:

$$
\begin{aligned}
&q_{n+1} = q_n + \frac{1}{6}(k_1 + 2k_2 + 2k_3 + k_4),\\
&k_4 \equiv hf\left(t_n + h,\, q_n + k_3\right), \quad k_3 \equiv hf\left(t_n + \frac{h}{2},\, q_n + \frac{k_2}{2}\right),\\
&k_2 \equiv hf\left(t_n + \frac{h}{2},\, q_n + \frac{k_1}{2}\right), \quad k_1 \equiv hf(t_n, q_n).
\end{aligned}
\tag{5.99}
$$

This method reaches much lower error, $O(h^5)$, without being not too cumbersome. These features have made the fourth-order Runge–Kutta the default method in most numerical libraries. Its extension to higher orders is possible, but requires more complex formulas, and is justified only for some special cases, e.g. very abrupt functions $q(t)$[36]. The most frequent enhancement of the method is the automatic adjustment of the step h to reach the specified accuracy, but not make more calculations than necessary.

Figure 5.12 shows a typical example of an application of that method to the very simple problem of a damped linear oscillator, for two values of fixed time step h (expressed in terms of the number N of such steps per oscillation period). The black lines connect the points obtained by the fourth-order Runge–Kutta method, while the points connected with the green lines represent the exact analytical solution (5.22). Few-percent errors start to appear only at as few as $\sim$10 time steps per period, so that the method is indeed very efficient.

Let me hope that the discussion in the next section will make the conveniences and the handicaps of the numerical approach to the solution of problems of nonlinear dynamics very clear.

[36] The most popular approaches in such cases are the *Richardson extrapolation*, the *Bulirsch–Stoer algorithm*, and a set of *prediction–correction techniques*, e.g. the *Adams–Bashforth–Moulton method*—see the literature recommended in appendix A, section A.16(iii).

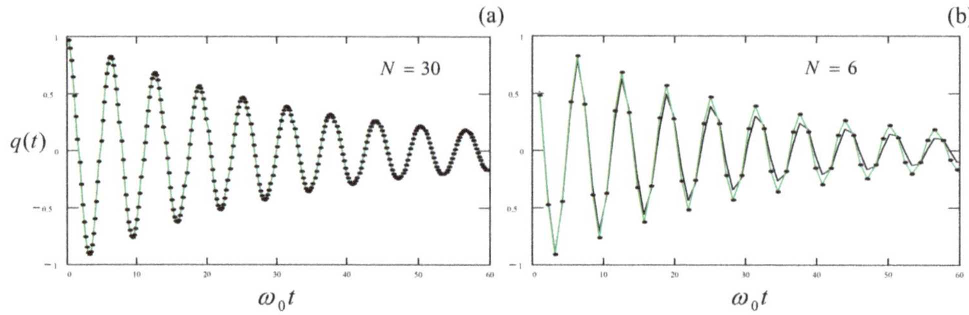

Figure 5.12. Results of the Runge–Kutta solution of Eq. (5.6) (with $\delta/\omega_0 = 0.03$) for: (a) 30 and (b) 6 points per oscillation period. The results are shown by points; the black and green lines are only the guides for the eye.

5.8 Higher harmonic and subharmonic oscillations

Figure 5.13 shows the numerically calculated[37] transient process and stationary oscillations in a linear oscillator and in a very representative nonlinear system, the pendulum described by Eq. (5.42), both with the same resonance frequency ω_0 for small oscillations. Both systems are driven by a sinusoidal external force of the same amplitude and frequency—in this illustration, equal to the small-oscillation own frequency ω_0 of both systems. The plots show that despite a very substantial amplitude of the pendulum oscillations (an angle amplitude of about one radian), their waveform remains almost exactly sinusoidal[38]. On the other hand, the non-linearity affects the oscillation amplitude very substantially. These results imply that the corresponding reduced equation (5.60), which is based on the assumption (5.41), may work very well far beyond its formal restriction $|q| \ll 1$.

Still, the waveform of oscillations in a nonlinear system always differs from that of the applied force—in our case, from the sine function of frequency ω. This fact is frequently formulated as the generation, by the system, of *higher harmonics*. Indeed, the Fourier theorem tells us that any non-sinusoidal periodic function of time may be represented as a sum of its basic harmonic of frequency ω and higher harmonics with frequencies $n\omega$, with integer $n > 1$.

Note that an effective generation of higher harmonics is only possible with adequate nonlinearity of the system. For example, consider the nonlinear term αq^3 used in the equations explored in sections 5.2 and 5.3. If the waveform $q(t)$ is approximately sinusoidal, such term will have only the basic (first) and the third harmonics—see, e.g. Eq. (5.50). As another example, the 'pendulum nonlinearity' $\sin q$ cannot produce, without a constant component ('bias') in $q(t)$, any even harmonic, including the second one. The most efficient generation of harmonics may be achieved using systems with the sharpest nonlinearities—e.g. semiconductor

[37] All numerical results shown in this section have been obtained by the fourth-order Runge–Kutta method with the automatic step adjustment which guarantees the relative error of the order of 10^{-4}—much smaller than the pixel size in the plots.

[38] In this particular case, the higher harmonic content is about 0.5%, dominated by the third harmonic whose amplitude and phase are in a very good agreement with Eq. (5.50).

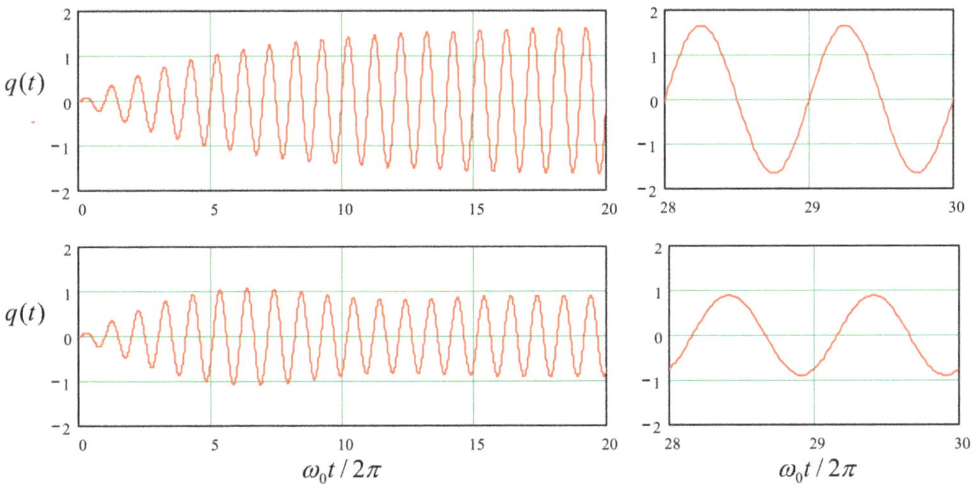

Figure 5.13. The oscillations induced by a similar sinusoidal external force (turned on at $t = 0$) in two systems with the same small-oscillation frequency ω_0 and low damping: a linear oscillator (two top panels) and a pendulum (two bottom panels). In all cases, $\delta/\omega_0 = 0.03$, $f_0 = 0.1$, and $\omega = \omega_0$.

diodes whose current may follow an exponential dependence on the applied voltage through several orders of magnitude[39].

Another way to increase the contents of an nth higher harmonic in a nonlinear oscillator is to reduce the excitation frequency ω to $\sim\omega_0/n$, so that the oscillator resonated at the frequency $n\omega \approx \omega_0$ of the desirable harmonic. For example, figure 5.14a shows the oscillations in a pendulum described by the same Eq. (5.42), but driven at frequency $\omega = \omega_0/3$. One can see that the third harmonic amplitude may be comparable with that of the basic harmonic, in particular if the external frequency is additionally lowered (figure 5.14b) to accommodate for the deviation of the effective frequency $\omega_0(A)$ of own oscillations from its small-oscillation value ω_0—see Eq. (5.49), figure 5.4 and their discussion in section 5.2 above.

However, numerical modeling of nonlinear oscillators, as well as experiments with their physical implementations, bring more surprises. For example, the bottom panel of figure 5.15 shows the oscillations in a pendulum under the effect of a strong sinusoidal force with a frequency ω close to $3\omega_0$. One can see that at some parameter values and initial conditions, the system's oscillation spectrum is heavily contributed (almost dominated) by the third *sub*harmonic, i.e. the Fourier component of frequency $\omega/3 \approx \omega_0$.

This counter-intuitive phenomenon of such *subharmonic generation* may be explained as follows. Let us assume that subharmonic oscillations of frequency $\omega/3 \approx \omega_0$ have somehow appeared, and coexist with the forced oscillations of frequency 3ω:

[39] This method is used in practice, for example, for the generation of electromagnetic waves with frequencies in the terahertz range (10^{12}–10^{13} Hz), which still lacks efficient electronic self-oscillators.

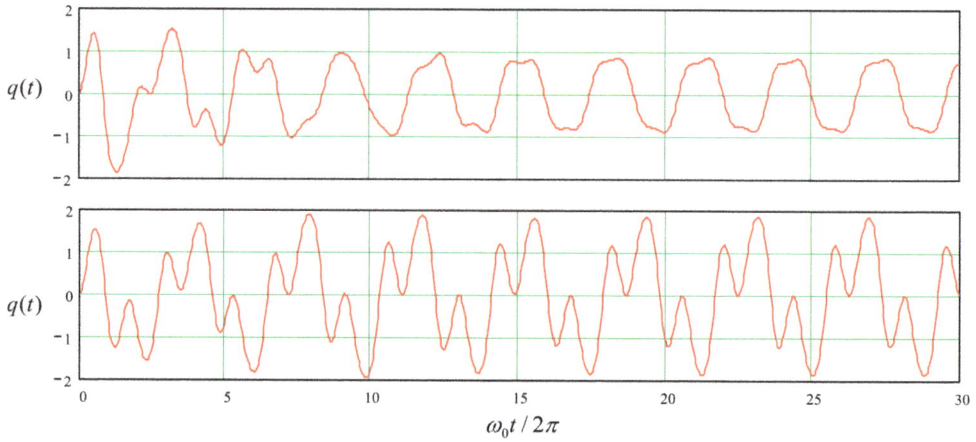

Figure 5.14. The oscillations induced in a pendulum, with damping $\delta/\omega_0 = 0.03$, driven by a sinusoidal external force of amplitude $f_0 = 0.75$, and frequency $\omega_0/3$ (top panel) and $0.8\omega_0/3$ (bottom panel).

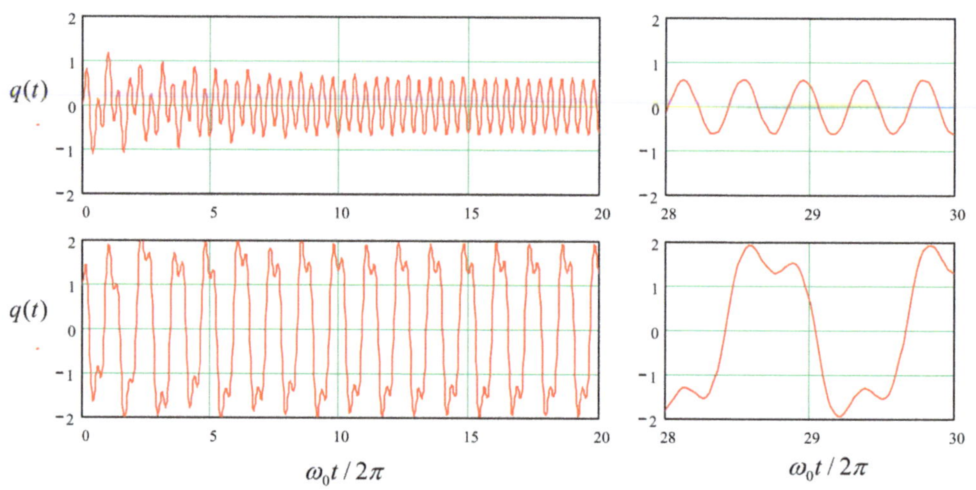

Figure 5.15. The oscillations sustained in a pendulum with $\delta/\omega_0 = 0.03$, by a sinusoidal external force of amplitude $f_0 = 3$ and frequency $3\omega_0 \times 0.8$, at initial conditions $q(0) = 0$ (the top row) and $q(0) = 1$ (the bottom row).

$$q(t) \approx A\cos\Psi + A_{\text{sub}}\cos\Psi_{\text{sub}}, \quad \text{where } \Psi \equiv \omega t - \varphi, \quad \Psi_{\text{sub}} \equiv \frac{\omega t}{3} - \varphi_{\text{sub}}. \quad (5.100)$$

Then the leading nonlinear term, αq^3, of the Taylor expansion of the pendulum's nonlinearity $\sin q$, is proportional to

$$\begin{aligned} q^3 &= (A\cos\Psi + A_{\text{sub}}\cos\Psi_{\text{sub}})^3 \\ &= A^3\cos^3\Psi + 3A^2A_{\text{sub}}\cos^2\Psi\cos\Psi_{\text{sub}} \\ &\quad + 3AA_{\text{sub}}^2\cos\Psi\cos^2\Psi_{\text{sub}} + A_{\text{sub}}^3\cos^3\Psi_{\text{sub}}. \end{aligned} \quad (5.101)$$

While the first and the last terms of this expression depend only of amplitudes of the individual components of oscillations, the two middle terms are more interesting because they produce so-called *combinational frequencies* of the two components. For our case, the third term,

$$3AA_{\text{sub}}^2 \cos\Psi \cos^2\Psi_{\text{sub}} = \frac{3}{4}AA_{\text{sub}}^2 \cos(\Psi - 2\Psi_{\text{sub}}) + \dots, \quad (5.102)$$

is of special importance, because it produces, in addition to other combinational frequencies, the subharmonic component with the total phase

$$\Psi - 2\Psi_{\text{sub}} = \frac{\omega t}{3} - \varphi + 2\varphi_{\text{sub}}. \quad (5.103)$$

Thus, within a certain range of the mutual phase shift between the Fourier components, this nonlinear contribution is synchronous with the subharmonic oscillations, and describes the interaction that can deliver to it the energy from the external force, so that the oscillations may be sustained. Note, however, that the amplitude of the term (5.102) describing this energy exchange is proportional to the square of A_{sub}, and vanishes at the linearization of the equations of motion near the trivial fixed point. This means that the point is always stable, i.e. the third subharmonic cannot be self-excited and always needs an initial 'kick-off'—compare the two panels of figure 5.15. The same is evidently true for higher subharmonics.

Only the second subharmonic is a special case. Indeed, let us make a calculation similar to Eq. (5.101), by replacing Eq. (5.100) with

$$q(t) \approx A\cos\Psi + A_{\text{sub}}\cos\Psi_{\text{sub}}, \quad \text{where } \Psi \equiv \omega t - \varphi, \quad \Psi_{\text{sub}} \equiv \frac{\omega t}{2} - \varphi_{\text{sub}}, \quad (5.104)$$

for a nonlinear term proportional to q^2:

$$\begin{aligned} q^2 = (A\cos\Psi + A_{\text{sub}}\cos\Psi_{\text{sub}})^2 &= A^2\cos^2\Psi \\ &+ 2AA_{\text{sub}}\cos\Psi\cos\Psi_{\text{sub}} + A_{\text{sub}}^2\cos^2\Psi_{\text{sub}}. \end{aligned} \quad (5.105)$$

Here the combinational-frequency term capable of supporting the second subharmonic,

$$\begin{aligned} 2AA_{\text{sub}}\cos\Psi\cos\Psi_{\text{sub}} &= AA_{\text{sub}}\cos(\Psi - \Psi_{\text{sub}}) \\ &= AA_{\text{sub}}\cos(\omega t - \varphi + \varphi_{\text{sub}}) + \dots, \end{aligned} \quad (5.106)$$

is linear in the subharmonic amplitude, i.e. survives the equation linearization near the trivial fixed point. This mean that the second subharmonic may arise spontaneously, from infinitesimal fluctuations.

Moreover, such excitation of the second subharmonic is very similar to the parametric excitation that was discussed in detail in section 5.5, and this similarity is not coincidental. Indeed, let us redo the expansion (5.105) at a somewhat different assumption—that the oscillations are a sum of the forced oscillations at the external force frequency ω, and an *arbitrary but weak* perturbation:

$$q(t) = A\cos(\omega t - \varphi) + \tilde{q}(t), \qquad \text{with } |\tilde{q}| \ll A. \quad (5.107)$$

Then, neglecting the small term proportional to $\tilde{q}^2$, we obtain

$$q^2 \approx A^2 \cos^2(\omega t - \varphi) + 2\tilde{q}(t)A\cos(\omega t - \varphi). \tag{5.108}$$

Besides the inconsequential phase φ, the second term in the last formula is *exactly* similar to the term describing the parametric effects in Eq. (5.75). This fact means that for a weak perturbation, a system with a quadratic nonlinearity in the presence of a strong 'pumping' signal of frequency ω is equivalent to a system with parameters changing in time with frequency ω. This fact is broadly used for the parametric excitation at high (e.g. optical) frequencies where the mechanical means of parameter modulation (see, e.g. figure 5.5) are not practicable. The necessary quadratic nonlinearity at optical frequencies may be provided by a *non-centrosymmetric nonlinear crystal*, e.g. the β-phase barium borate (BaB_2O_4).

Before finishing this chapter, let me elaborate a bit on a general topic: the relation between the numerical and analytical approaches to problems of dynamics (and physics as a whole). We have just seen that sometimes numerical solutions, such as those shown in figure 5.15b, may give vital clues for previously unanticipated phenomena such as the excitation of subharmonics. (The phenomenon of deterministic chaos, which will be discussed in chapter 9, presents another example of such 'numerical discoveries'.) One might also argue that in the absence of exact analytical solutions, numerical simulations may be the main theoretical tool for the study of such phenomena. These hopes are, however, muted by the general problem that is frequently called the *curse of dimensionality*[40], in which the last word refers to the number of input parameters of the problem to be solved[41].

Indeed, let us have another look at figure 5.15. We have been lucky to find a new phenomenon, the third subharmonic generation, for a particular set of parameters—in this case, five of them: $\delta/\omega_0 = 0.03$, $\omega/\omega_0 = 2.4$, $f_0 = 3$, $q(0) = 1$, and $dq/dt\,(0) = 0$. Could we tell anything about how common this effect is? Are subharmonics with different n possible in the system? The only way to address these questions computationally is to carry out similar numerical simulations at many points of the d-dimensional (in this case, $d = 5$) space of parameters. Say we have decided that breaking the reasonable range of each parameter to $N = 100$ points is sufficient. (For many problems, even more points are necessary—see, e.g. section 9.1.) Then the total number of numerical experiments to carry out is $N^d = (10^2)^5 = 10^{10}$—not a simple task even for powerful modern computing facilities. (In addition to the pure number of required CPU cycles, consider the storage and analysis of the results.) For many important problems of nonlinear dynamics, e.g. turbulence, the parameter dimensionality d is substantially larger, and the computer resources necessary for even one numerical experiment are much greater.

[40] This term was coined in 1957 by R Bellman in the context of optimal control theory (where the dimensionality means the number of parameters affecting the system under control), but gradually has spread throughout quantitative sciences using numerical methods.

[41] In *Part EM* section 1.2, I discuss the curse's implications for a different case, when both analytical and numerical solutions to the same problem are possible.

In the view of the curse of dimensionality, approximate analytical considerations, such as those outlined above for the subharmonic excitation, are invaluable. More generally, physics used to stand on two legs, experiment and (analytical) theory. The enormous progress of computer performance during the last few decades has provided it with one more point of support (a tail?)—numerical simulation. This does not mean we can afford to cut and throw away any of the legs we are standing on.

5.9 Problems

Problem 5.1. For a system with the response function given by Eq. (5.17), prove Eq. (5.26), and use an approach different to the one used in section 5.1 to derive Eq. (5.34).

Hint: You may like to use the *Cauchy integral theorem* and the Cauchy integral formula for analytical functions of complex variables[42].

Problem 5.2. A square-wave pulse of force (see the figure below) is exerted on a linear oscillator with eigenfrequency ω_0 (with no damping), initially at rest. Calculate the law of motion $q(t)$, sketch it, and interpret the result.

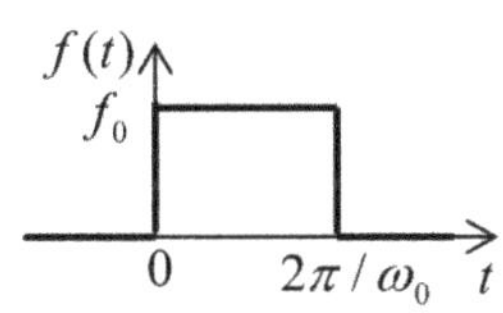

Problem 5.3. At $t = 0$, a sinusoidal external force $F(t) = F_0 \cos \omega t$, is applied to a linear oscillator with eigenfrequency ω_0 and damping δ, which was at rest at $t \leqslant 0$.

(i) Derive the general expression for the time evolution of the oscillator's coordinate, and interpret the result.

(ii) Spell out the result for the case of resonance ($\omega = \omega_0$) in a system with low damping ($\delta \ll \omega_0$), and, in particular, explore the limit $\delta \to 0$.

Problem 5.4. A pulse of external force $F(t)$, with a finite duration $\mathcal{T}$, is exerted on a linear oscillator, initially at rest in its equilibrium position. Neglecting dissipation, calculate the change of oscillator's energy, using two different methods, and compare the results.

Problem 5.5. For a system with the following Lagrangian function:

$$L = \frac{m}{2}\dot{q}^2 - \frac{\kappa}{2}q^2 + \frac{\varepsilon}{2}\dot{q}^2 q^2,$$

calculate the frequency of free oscillations as a function of their amplitude A, at $A \to 0$, using two different approaches.

[42] See, e.g. Eq. (A.91).

Problem 5.6. For a system with the Lagrangian function

$$L = \frac{m}{2}\dot{q}^2 - \frac{\kappa}{2}q^2 + \varepsilon\dot{q}^4,$$

with small parameter ε, use the van der Pol method to find the frequency of free oscillations as a function of their amplitude.

Problem 5.7. On the plane $[a_1, a_2]$ of two real, time-independent parameters a_1 and a_2, find the regions in which the fixed point of the following system of equations,

$$\dot{q}_1 = a_1(q_2 - q_1),$$
$$\dot{q}_2 = a_2q_1 - q_2,$$

is unstable, and sketch the regions of each fixed-point type—stable and unstable nodes, focuses, etc.

Problem 5.8. Solve problem 5.3(ii) using the reduced equations, and compare the result with the exact solution.

Problem 5.9. Use the reduced equations to analyze the forced oscillations in an oscillator with weak nonlinear damping, described by equation

$$\ddot{q} + 2\delta\dot{q} + \omega_0^2 q + \beta\dot{q}^3 = f_0 \cos \omega t,$$

with $\omega \approx \omega_0$; β, $\delta > 0$; and $\beta\omega A^2 \ll 1$. In particular, find the stationary amplitude of forced oscillations and analyze their stability. Discuss the effect(s) of the nonlinear term on the resonance.

Problem 5.10. Within the approach discussed in section 5.4, calculate the average frequency of a self-oscillator outside the range of its phase-locking by an external sinusoidal force.

Problem 5.11.* Use the reduced equations to analyze stability of the forced nonlinear oscillations described by the Duffing equation (5.43). Relate the result to the slope of resonance curves (figure 5.4).

Problem 5.12. Use the van der Pol method to find the condition of parametric excitation of the oscillator described by the following equation:

$$\ddot{q} + 2\delta\dot{q} + \omega_0^2(t)q = 0,$$

where $\omega_0^2(t)$ is the square-wave function shown in the figure to the right, with $\omega \approx \omega_0$.

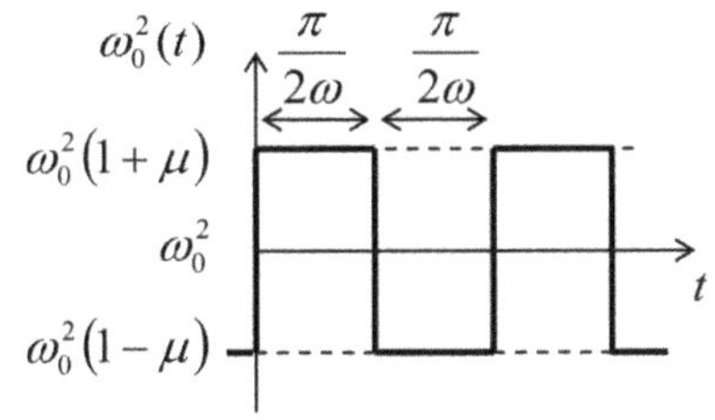

Problem 5.13. Use the van der Pol method to analyze parametric excitation of an oscillator with a weak nonlinear damping, described by equation

$$\ddot{q} + 2\delta\dot{q} + \beta\dot{q}^3 + \omega_0^2(1 + \mu\cos 2\omega t)q = 0,$$

with $\omega \approx \omega_0$; $\beta, \delta > 0$; and μ, $\beta\omega A^2 \ll 1$. In particular, find the amplitude of stationary oscillations and analyze their stability.

Problem 5.14.* Adding nonlinear term αq^3 to the left-hand side of Eq. (5.76),

(i) find the corresponding addition to the reduced equations,
(ii) calculate the stationary amplitude A of the parametric oscillations,
(iii) find the type and stability of each fixed point of the reduced equations, and
(iv) sketch the Poincaré phase-plane of the system in major parameter regions.

Problem 5.15. Use the van der Pol method to find the condition of parametric excitation of an oscillator with weak modulation of both the effective mass $m(t) = m_0(1 + \mu_m \cos 2\omega t)$ and the spring constant $\kappa(t) = \kappa_0[1 + \mu_\kappa \cos(2\omega t - \psi)]$, with the same frequency $2\omega \approx 2\omega_0$, at arbitrary modulation depths ratio μ_m/μ_k and phase shift ψ. Interpret the result in terms of modulation of the instantaneous frequency $\omega(t) \equiv [\kappa(t)/m(t)]^{1/2}$ and the mechanical impedance $Z(t) \equiv [\kappa(t)m(t)]^{1/2}$ of the oscillator.

Problem 5.16.* Find the condition of parametric excitation of a nonlinear oscillator described by equation

$$\ddot{q} + 2\delta\dot{q} + \omega_0^2 q + \gamma q^2 = f_0 \cos 2\omega t,$$

with sufficiently small δ, γ, f_0, and $\xi \equiv \omega - \omega_0$.

References

[1] Harris C and Piersol A 2002 *Shock and Vibration Handbook* 5th edn (New York: McGraw-Hill)
[2] Mitropolsky Yu and Dao N 2004 *Applied Asymptotic Methods in Nonlinear Oscillations* (Berlin: Springer)
[3] Andronov A, Vitt A and Khaikin S 2011 *Theory of Oscillators* (New York: Dover)

IOP Publishing

Chapter 6

From oscillations to waves

In this chapter, the discussion of oscillations is extended to systems with two and more degrees of freedom. This extension naturally leads to another key notion of physics—the waves, *in particular those in uniform 1D systems. (A more general discussion of elastic waves in continua is deferred until the next chapter.) However, even the limited scope of the models analyzed in this chapter will still enable us to discuss such important general aspects of waves as their dispersion, phase and group velocities, impedance, reflection, and attenuation.*

6.1 Two coupled oscillators

Let us discuss oscillations in systems with several degrees of freedom, starting from the simplest case of two linear (harmonic), dissipation-free, 1D oscillators. If the oscillators are independent of each other, the Lagrangian function of their system may be represented as a sum of two independent terms of the type (5.1):

$$L = L_1 + L_2, \quad L_{1,2} = T_{1,2} - U_{1,2} = \frac{m_{1,2}}{2}\dot{q}_{1,2}^2 - \frac{\kappa_{1,2}}{2}q_{1,2}^2. \tag{6.1}$$

Correspondingly, Eqs. (2.19) for $q_j = q_{1,2}$ yield two independent equations of motion of the oscillators, each one being similar to Eq. (5.2):

$$m_{1,2}\ddot{q}_{1,2} + \Omega_{1,2}^2 q_{1,2} = 0, \quad \text{where } \Omega_{1,2}^2 = \frac{\kappa_{1,2}}{m_{1,2}}. \tag{6.2}$$

(In the context of what follows, $\Omega_{1,2}$ are sometimes called the *partial frequencies*.) This means that in this simplest case, an arbitrary motion of the system is just a sum of independent sinusoidal oscillations at two frequencies equal to the partial frequencies (6.2).

doi:10.1088/978-0-7503-1398-8ch6

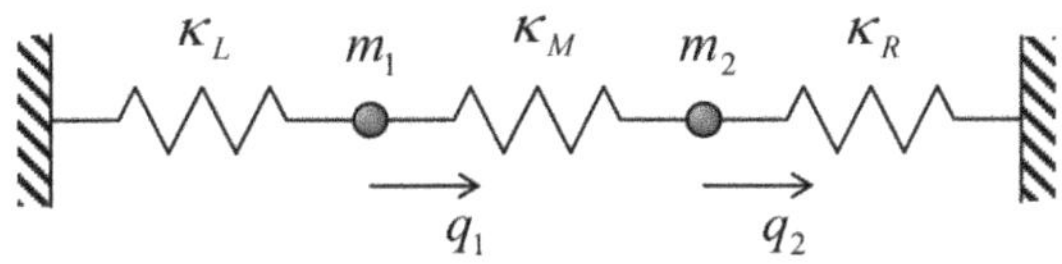

Figure 6.1. A simple system of two coupled linear oscillators.

However, as soon as the oscillators are *coupled* (i.e. interact), the full Lagrangian L contains an additional *mixed* term L_{int} depending on both generalized coordinates q_1 and q_2 and/or generalized velocities. As a simple example, consider the simplest system shown in figure 6.1, where two small masses $m_{1,2}$ are constrained to move in only one dimension (shown horizontal), and are kept between two stiff walls with three springs.

In this case the kinetic energy is still separable, $T = T_1 + T_2$, but the total potential energy, consisting of the elastic energies of three springs, is not:

$$U = \frac{\kappa_L}{2} q_1^2 + \frac{\kappa_M}{2}(q_1 - q_2)^2 + \frac{\kappa_R}{2} q_2^2, \tag{6.3a}$$

where $q_{1,2}$ are the horizontal displacements of particles from their equilibrium positions. It is convenient to rewrite this expression as

$$U = \frac{\kappa_1}{2} q_1^2 + \frac{\kappa_2}{2} q_2^2 - \kappa q_1 q_2,$$
$$\text{where } \kappa_1 \equiv \kappa_L + \kappa_M, \quad \kappa_2 \equiv \kappa_R + \kappa_M, \quad \kappa \equiv \kappa_M, \tag{6.3b}$$

showing that the Lagrangian $L = T - U$ of this system contains, besides the partial terms (6.1), a *bilinear* interaction term:

$$L = L_1 + L_2 + L_{\text{int}}, \quad L_{\text{int}} = \kappa q_1 q_2. \tag{6.4}$$

The resulting Lagrange equations of motion are

$$\begin{aligned} m_1 \ddot{q}_1 + m_1 \Omega_1^2 q_1 &= \kappa q_2, \\ m_2 \ddot{q}_2 + m_2 \Omega_2^2 q_2 &= \kappa q_1. \end{aligned} \tag{6.5}$$

Thus the interaction leads to an effective generalized force κq_2 exerted on subsystem 1 by subsystem 2, and the reciprocal effective force κq_1.

Please note two important aspects of this (otherwise rather simple) system of equations. First, in contrast to the actual physical interaction forces (such as $F_{12} = -F_{21} = \kappa_M(q_2 - q_1)$ for our system[1]) the effective forces on the right-hand side of Eq. (6.5) do not obey Newton's third law. Second, the forces are proportional to the same coefficient κ; this feature is a result of the general bilinear structure (6.4) of the interaction energy, rather than of any special symmetry.

[1] Using these expressions, Eqs. (5.5) may be readily obtained from Newton's laws, but the Lagrangian approach used above will make their generalization, in the next section, more straightforward.

We already know how to solve Eq. (6.5), because it is still a system of linear and homogeneous differential equations, so that its general solution is a sum of particular solutions of the form similar to Eq. (5.88),

$$q_1 = c_1 e^{\lambda t}, \quad q_2 = c_2 e^{\lambda t}, \tag{6.6}$$

for all possible values of λ. These values may be found by plugging Eqs. (6.6) into Eqs. (6.5), and requiring the resulting system of two linear, homogeneous algebraic equations for the distribution coefficients $c_{1,2}$,

$$\begin{aligned} m_1\lambda^2 c_1 + m_1\Omega_1^2 c_1 &= \kappa c_2, \\ m_2\lambda^2 c_2 + m_2\Omega_2^2 c_2 &= \kappa c_1, \end{aligned} \tag{6.7}$$

to be self-consistent. In our particular case, we obtain a characteristic equation,

$$\begin{vmatrix} m_1(\lambda^2 + \Omega_1^2) & -\kappa \\ -\kappa & m_2(\lambda^2 + \Omega_2^2) \end{vmatrix} = 0, \tag{6.8}$$

which is quadratic in λ^2, and thus allows a simple solution:

$$\begin{aligned} (\lambda^2)_\pm &= -\frac{1}{2}(\Omega_1^2 + \Omega_2^2) \mp \left[\frac{1}{4}(\Omega_1^2 + \Omega_2^2)^2 - \Omega_1^2\Omega_2^2 + \frac{\kappa^2}{m_1 m_2}\right]^{1/2} \\ &\equiv -\frac{1}{2}(\Omega_1^2 + \Omega_2^2) \mp \left[\frac{1}{4}(\Omega_1^2 - \Omega_2^2)^2 + \frac{\kappa^2}{m_1 m_2}\right]^{1/2}. \end{aligned} \tag{6.9}$$

According to Eqs. (6.2) and (6.3*b*), for any positive values of spring constants, the product $\Omega_1\Omega_2 = (\kappa_L + \kappa_M)(\kappa_R + \kappa_M)/(m_1m_2)^{1/2}$ is always larger than $\kappa/(m_1m_2)^{1/2} = \kappa_M/(m_1m_2)^{1/2}$, so that the square root in Eq. (6.9) is always smaller than $(\Omega_1^2+\Omega_2^2)/2$. As a result, both values of λ^2 are negative, i.e. the general solution to Eq. (6.5) is a sum of four terms, each proportional to $\exp\{\pm i\omega_\pm t\}$, where both *own frequencies* ('eigen-frequencies') $\omega_\pm \equiv i\lambda_\pm$ are real:

$$\omega_\pm^2 \equiv -\lambda_\pm^2 = \frac{1}{2}(\Omega_1^2 + \Omega_2^2) \pm \left[\frac{1}{4}(\Omega_1^2 - \Omega_2^2)^2 + \frac{\kappa^2}{m_1 m_2}\right]^{1/2}. \tag{6.10}$$

A plot of these eigenfrequencies as a function of one of the partial frequencies (say, Ω_1), with the other partial frequency fixed, gives us the famous *anti-crossing* (also called the 'avoided crossing' or 'non-crossing') *diagram*—see figure 6.2. One can see that at weak coupling, frequencies $\omega_\pm$ are close to the partial frequencies $\Omega_{1,2}$ everywhere apart from a narrow range near the anti-crossing point $\Omega_1 = \Omega_2$. Most remarkably, when passing through this region, ω_+ smoothly 'switches' from following Ω_2 to following Ω_1 and vice versa.

The reason for this counter-intuitive behavior may be found by examining the distribution coefficients $c_{1,2}$ corresponding to each branch of the diagram, which may be obtained by plugging the corresponding value of $\lambda_\pm = -i\omega_\pm$ back into

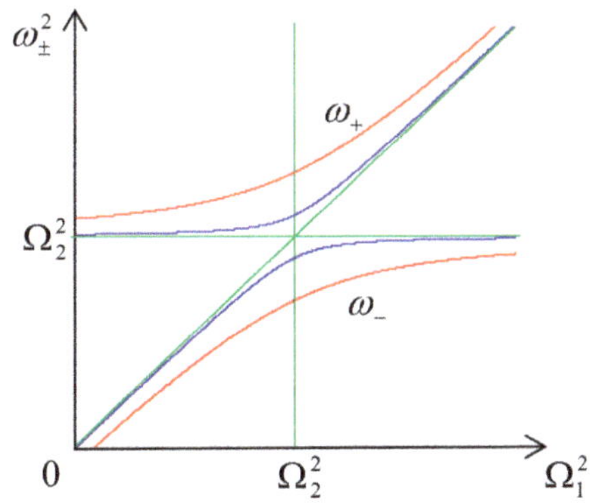

Figure 6.2. The anti-crossing diagram for two values of the normalized coupling strength $\kappa/(m_1m_2)^{1/2}\Omega_2^2$: 0.3 (red lines) and 0.1 (blue lines). In this plot, Ω_1 is assumed to be changed by varying κ_1 rather than m_1, but in the opposite case the diagram is qualitatively similar.

Eq. (6.7). For example, at the anti-crossing point $\Omega_1 = \Omega_2 \equiv \Omega$, Eq. (6.10) is reduced to

$$\omega_\pm^2 = \Omega^2 \pm \frac{\kappa}{(m_1m_2)^{1/2}} = \Omega^2\left(1 \pm \frac{\kappa}{(\kappa_1\kappa_2)^{1/2}}\right). \tag{6.11}$$

Plugging this expression back into any of Eqs. (6.7), we see that for the two branches of the anti-crossing diagram, the distribution coefficient ratio is the same in magnitude but opposite in sign:

$$\left(\frac{c_1}{c_2}\right)_\pm = \mp\left(\frac{m_2}{m_1}\right)^{1/2}, \quad \text{at } \Omega_1 = \Omega_2. \tag{6.12}$$

In particular, if the system is symmetric ($m_1 = m_2$, $\kappa_L = \kappa_R$), then at the upper branch, corresponding to $\omega_+ > \omega_-$, we obtain $c_1 = -c_2$. This means that in this *hard mode*[2], masses oscillate in anti-phase: $q_1(t) \equiv -q_2(t)$. The resulting substantial extension/compression of the middle spring yields an additional returning force which increases the oscillation frequency. In contrast, at the lower branch, corresponding to ω_-, the particle oscillations are in-phase: $c_1 = c_2$, i.e., $q_1(t) \equiv q_2(t)$, so that the middle spring is never stretched at all. As a result, the *soft mode* oscillation frequency ω_- is lower than ω_+, and does not depend on κ:

$$\omega_-^2 = \Omega^2 - \frac{\kappa}{m} = \frac{\kappa_L}{m} = \frac{\kappa_R}{m}. \tag{6.13}$$

Note that for both modes, the oscillations equally engage both particles.

[2] In physics, the term 'mode' is typically used to describe the distribution of a variable in space, pertaining to the oscillations at a single frequency. In our current case, when the notion of space is reduced to two oscillator numbers, the 'mode' means just a set of two distribution coefficients $c_{1,2}$ for a particular eigenfrequency.

Far from the anti-crossing point, the situation is completely different. Indeed, an absolutely similar calculation of $c_{1,2}$ shows that on each branch of the diagram, the magnitude of one of the distribution coefficients is much larger than that of its counterpart. Hence, in this limit any particular mode of oscillations involves virtually only one particle. A slow change of system parameters, bringing it through the anti-crossing, results, first, in a maximal delocalization of each mode at $\Omega_1 = \Omega_2$, and then in the restoration of the localization, but in a different partial degree of freedom.

We could readily carry out similar calculations for the case when the systems are coupled via their velocities, $L_{\text{int}} = m\dot{q}_1\dot{q}_2$, where m is a coupling coefficient—not necessarily a certain physical mass[3]. The results are generally similar to those discussed above, again with the maximum level splitting at $\Omega_1 = \Omega_2 \equiv \Omega$:

$$\omega_\pm^2 = \frac{\Omega^2}{1 \mp | m |/(m_1 m_2)^{1/2}} \approx \Omega^2\left(1 \pm \frac{| m |}{(m_1 m_2)^{1/2}}\right), \tag{6.14}$$

the last relation being valid for weak coupling. The generalization to the case of both coordinate and velocity coupling is also straightforward—see the next section.

Note that the anti-crossing diagram shown in figure 6.2 is even more ubiquitous in quantum mechanics because, due to the time-oscillatory character of the Schrödinger equation solutions, a weak coupling of any two quantum states leads to a qualitatively similar behavior of the eigenfrequencies $\omega_\pm$ of the system, and hence of the *eigenenergies* (‘*energy levels*’) $E_\pm = \hbar\omega_\pm$ of the system.

One more property of weakly coupled oscillators, a periodic slow transfer of energy from one oscillator to the other and back, particularly well pronounced at the anti-crossing point $\Omega_1 = \Omega_2$, is also more important for quantum than for classical mechanics. This is why I refer the reader to *Part QM* of this series for a detailed discussion of this phenomenon.

6.2 *N* coupled oscillators

The calculations of the previous section may be readily generalized to the case of arbitrary number (say, N) coupled harmonic oscillators, with an arbitrary type of coupling. It is evident that in this case Eq. (6.4) should be replaced with

$$L = \sum_{j=1}^{N} L_j + \sum_{j,j'=1}^{N} L_{jj'}. \tag{6.15}$$

Moreover, we can generalize the above expression for the mixed terms $L_{jj'}$, taking into account their possible dependence not only on the generalized coordinates, but

[3] In mechanics, with $q_{1,2}$ standing for the actual linear displacements of particles, such coupling is hard to implement, but there are many dynamic systems of non-mechanical nature in which such coupling is the most natural one. The simplest example is the system of two LC (‘tank’) circuits, with either capacitive or inductive coupling. Indeed, as was discussed in section 2.2, for such a system, the very notions of the potential and kinetic energies are conditional and interchangeable.

on the generalized velocities, in a bilinear form similar to Eq. (6.4). The resulting Lagrangian may be represented in a compact form,

$$L = \sum_{j,j'=1}^{N} \left(\frac{m_{jj'}}{2} \dot{q}_j \dot{q}_{j'} - \frac{\kappa_{jj'}}{2} q_j q_{j'} \right), \tag{6.16}$$

where the off-diagonal terms are index-symmetric: $m_{jj'} = m_{j'j}$, $\kappa_{jj'} = \kappa_{j'j}$, and the factors 1/2 compensate the double counting of each term with $j \neq j'$, taking place at the summation over two independently running indices. One may argue that Eq. (6.16) is quite general if we still want to keep the equations of motion linear—as they always are if the oscillations are small enough.

Plugging Eq. (6.16) into the general form (2.19) of the Lagrange equation, we obtain N equations of motion of the system, one for each value of index $j' = 1, 2,\ldots, N$:

$$\sum_{j=1}^{N} \left(m_{jj'} \ddot{q}_j + \kappa_{jj'} q_j \right) = 0. \tag{6.17}$$

Just as in the previous section, let us look for a particular solution to this system in the form

$$q_j = c_j e^{\lambda t}. \tag{6.18}$$

As a result, we are getting a system of N linear, homogeneous algebraic equations,

$$\sum_{j=1}^{N} \left(m_{jj'} \lambda^2 + \kappa_{jj'} \right) c_j = 0, \tag{6.19}$$

for the set of N distribution coefficients c_j. The condition that this system is self-consistent is that the determinant of its matrix equals zero:

$$\mathrm{Det}\left(m_{jj'} \lambda^2 + \kappa_{jj'} \right) = 0. \tag{6.20}$$

This characteristic equation is an algebraic equation of degree N for λ^2, and so has N roots $(\lambda^2)_n$. For any Hamiltonian system with stable equilibrium, matrices $m_{jj'}$ and $\kappa_{jj'}$ ensure that all these roots are real and negative. As a result, the general solution to Eq. (6.17) is the sum of $2N$ terms proportional to $\exp\{\pm i\omega_n t\}$, $n = 1, 2,\ldots, N$, where all N eigenfrequencies ω_n are real.

Plugging each of these $2N$ values of $\lambda = \pm i\omega_n$ back into a particular set of linear Eqs. (6.17), one can find the corresponding set of distribution coefficients $c_{j\pm}$. Generally, the coefficients are complex, but in order to keep $q_j(t)$ real, the coefficients c_{j+} corresponding to $\lambda = +i\omega_n$ and c_{j-} corresponding to $\lambda = -i\omega_n$ have to be complex conjugate of each other. Since the sets of the distribution coefficients may be different for each λ_n, they should be marked with two indices, j and n. Thus, at general initial conditions, the time evolution of jth coordinate may be represented as

$$q_j = \frac{1}{2} \sum_{n=1}^{N} \left(c_{jn} \exp\{+i\omega_n t\} + c_{jn}^* \exp\{-i\omega_n t\} \right) \equiv \mathrm{Re} \sum_{n=1}^{N} c_{jn} \exp\{i\omega_n t\}. \tag{6.21}$$

This formula shows very clearly again the physical sense of the distribution coefficients c_{jn}: a set of these coefficients, with different values of index j but the same n, gives the complex amplitudes of oscillations of the coordinates for the special initial conditions that ensure purely sinusoidal motion of the system, with frequency ω_n.

The calculation of the eigenfrequencies and distribution coefficients of a particular coupled system with many degrees of freedom from Eqs. (6.19)–(6.20) is a task that frequently may only be done numerically[4]. Let us discuss just two particular but very important cases. First, let all the coupling coefficients be small in the following sense, $|m_{jj'}| \ll m_j \equiv m_{jj}$ and $|\kappa_{jj'}| \ll \kappa_j \equiv \kappa_{jj}$, for all $j \neq j$, and all partial frequencies $\Omega_j \equiv (\kappa_j/m_j)^{1/2}$ be not too close to each other:

$$\frac{\left|\Omega_j^2 - \Omega_{j'}^2\right|}{\Omega_j^2} \gg \frac{\left|\kappa_{jj'}\right|}{\kappa_j}, \frac{\left|m_{jj'}\right|}{m_j}, \quad \text{for all } j \neq j'. \tag{6.22}$$

(Such a situation frequently happens if parameters of the system are 'random' in the sense that they do not follow any special, simple rule—for example, resulting from some simple symmetry of the system.) The results of the previous section imply that in this case the coupling does not produce a noticeable change of oscillation frequencies: $\{\omega_n\} \approx \{\Omega_j\}$. In this situation, oscillations at each eigenfrequency are heavily concentrated in one degree of freedom, i.e. in each set of the distribution coefficients c_{jn} (for a given n), one coefficient's magnitude is much larger than all others.

Now let the conditions (6.22) be valid for all but one pair of partial frequencies, say Ω_1 and Ω_2, while these two frequencies are so close that coupling of the corresponding partial oscillators becomes essential. In this case the approximation $\{\omega_n\} \approx \{\Omega_j\}$ is still valid for all other degrees of freedom, and the corresponding terms may be neglected in Eqs. (6.19) for $j = 1$ and 2. As a result, we return to Eq. (6.7) (perhaps generalized for velocity coupling) and hence to the anti-crossing diagram (figure 6.2) discussed in the previous section. As a result, an extended change of only one partial frequency (say, Ω_1) of a weakly coupled system produces a sequence of eigenfrequency anti-crossings—see figure 6.3.

6.3 1D waves

The second case when the general results of the last section may be simplified are coupled systems with a considerable degree of symmetry. Perhaps the most important of them are the uniform systems that may sustain traveling and standing *waves*. Figure 6.4a shows a simple example of such a system—a long uniform chain of particles[5], of mass m, connected with light, elastic springs, pre-stretched with the tension force $\mathscr{T}$ to have equal lengths d.

[4] Fortunately, very effective algorithms have been developed for this *matrix diagonalization* task—see, e.g. references in appendix A, section A.16(iii)–(iv). For example, the popular MATLAB package was initially created exactly for this purpose. (The 'MAT' in its name stands for 'matrix' rather than 'mathematics'.)

[5] To some extent, this is a generalization of the two-particle system considered in section 6.1—see figure 6.1.

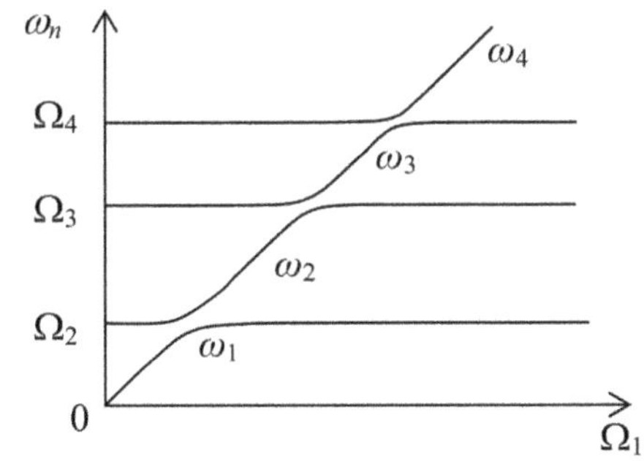

Figure 6.3. The level anti-crossing in a system of N weakly coupled oscillators—schematically.

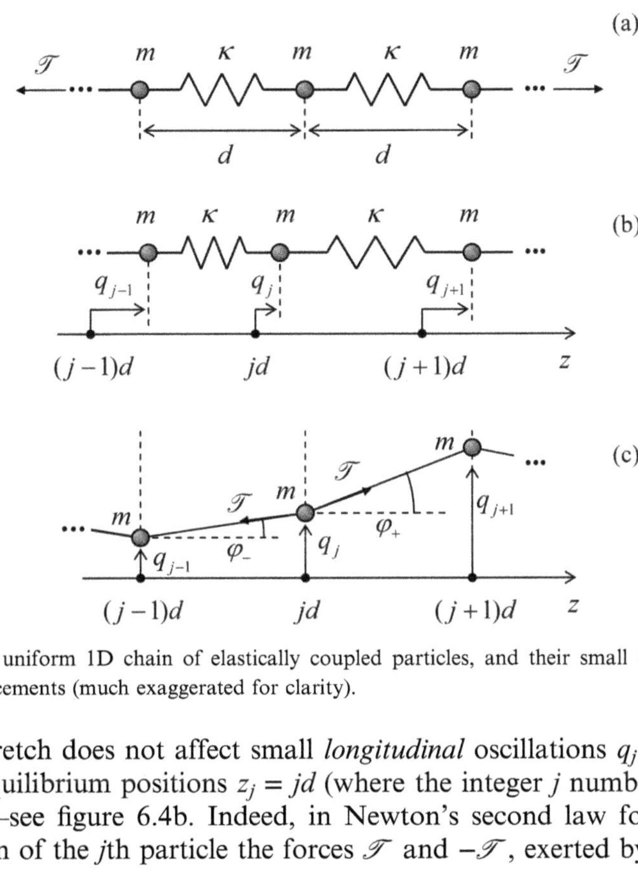

Figure 6.4. (a) A uniform 1D chain of elastically coupled particles, and their small (b) longitudinal and (c) traverse displacements (much exaggerated for clarity).

This pre-stretch does not affect small *longitudinal* oscillations q_j of the particles about their equilibrium positions $z_j = jd$ (where the integer j numbers the particles sequentially)—see figure 6.4b. Indeed, in Newton's second law for such a longitudinal motion of the jth particle the forces $\mathscr{T}$ and $-\mathscr{T}$, exerted by the springs on the right and on the left of it, cancel. However, the elastic additions, $\kappa\Delta q$, to these forces are generally different:

$$m\ddot{q}_j = \kappa(q_{j+1} - q_j) - \kappa(q_j - q_{j-1}). \tag{6.23}$$

In contrast, for *transverse* oscillations within one plane (figure 6.4c), the net transverse component of the pre-stretch force exerted on the jth particle, $\mathscr{T}_t = \mathscr{T}(\sin\varphi_+ - \sin\varphi_-)$, where $\varphi_\pm$ are the force direction angles, does not vanish. As a result, the contributions into this force from small transverse oscillations, with $|q_j| \ll d, \mathscr{T}/\kappa$, are negligible. Also, due to the first of these strong conditions, the angles $\varphi_\pm$ are small, and may be approximated, respectively, as $\varphi_+ \approx (q_{j+1} - q_j)/d$ and $\varphi_- \approx (q_j - q_{j-1})/d$. Plugging these expressions into a similar approximation, $\mathscr{T}_t \approx \mathscr{T}(\varphi_+ - \varphi_-)$ for the transverse force, we see that it may be expressed as $\mathscr{T}(q_{j+1} - q_j)/d - \mathscr{T}(q_j - q_{j-1})/d$, i.e. is absolutely similar to that in the longitudinal case, just with the replacement $\kappa \to \mathscr{T}/d$. As a result, we may write the equation of motion of the jth particle for these two cases in the same form:

$$m\ddot{q}_j = \kappa_{\rm ef}(q_{j+1} - q_j) - \kappa_{\rm ef}(q_j - q_{j-1}), \tag{6.24}$$

where $\kappa_{\rm ef}$ is the 'effective spring constant', equal to κ for the longitudinal oscillations, and to $\mathscr{T}/d$ for the transverse oscillations[6].

Apart from the (formally) infinite size of the system, Eq. (6.24) is evidently just a particular case of Eq. (6.17), and thus its particular solution may be looked at in the form (6.18), where in the light of our previous experience we may immediately take $\lambda^2 \equiv -\omega^2$. With this substitution, Eq. (6.24) gives the following simple form of the general system (6.17) for the distribution coefficients c_j:

$$(-m\omega^2 + 2\kappa_{\rm ef})\, c_j - \kappa_{\rm ef} c_{j+1} - \kappa_{\rm ef} c_{j-1} = 0. \tag{6.25}$$

Now comes the most important conceptual step toward the wave theory. The *translational symmetry* of Eq. (6.25), i.e. its invariance to the replacement $j \to j + 1$, allows it to have particular solutions, of the following form:

$$c_j = a e^{i\alpha j}, \tag{6.26}$$

where the coefficient α may depend on ω (and the system's parameters), but not on the particle number j. Indeed, plugging Eq. (6.26) into Eq. (6.25) and cancelling the common factor $e^{i\alpha j}$, we see that it is indeed identically satisfied, provided that α obeys the following algebraic equation:

$$(-m\omega^2 + 2\kappa_{\rm ef}) - \kappa_{\rm ef} e^{+i\alpha} - \kappa_{\rm ef} e^{-i\alpha} = 0. \tag{6.27}$$

The physical sense of the solution (6.26) becomes clear if we use it and Eq. (6.18), with $\lambda = \mp i\omega$, to write

$$q_j(t) = \mathrm{Re}\left[a \exp\left\{i\left(kz_j \mp \omega t\right)\right\}\right] = \mathrm{Re}\left[a \exp\left\{ik\left(z_j \mp v_{\rm ph} t\right)\right\}\right], \tag{6.28}$$

where the *wave number* k is defined as $k \equiv \alpha/d$. Eq. (6.28) describes a *sinusoidal*[7] *traveling wave* of particle displacements, which propagates, depending on the sign

[6] The re-derivation of Eqs. (6.23) from the Lagrangian formalism, with the simultaneous strict proof that the small oscillations in the longitudinal direction and the two mutually perpendicular transverse directions are all independent of each other, is a very good exercise, left for the reader.

[7] In optics, and quantum mechanics, such waves are usually called *monochromatic*; I will not use this term until the corresponding parts (*Part EM and Part QM*) of this series.

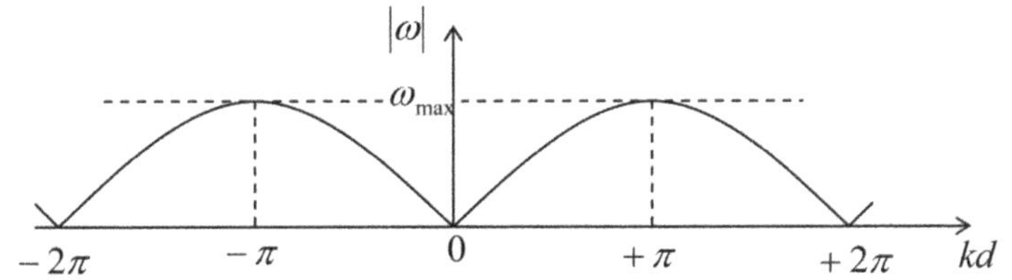

Figure 6.5. The dispersion relation (6.30).

before v_{ph}, to the right or to the left along the particle chain, with the so-called *phase velocity*

$$v_{ph} \equiv \frac{\omega}{k}. \tag{6.29}$$

Perhaps the most important characteristic of a wave system is the so-called *dispersion relation*, i.e. the relation between the wave's frequency ω and its wave number k—one may say, between the temporal and spatial frequencies of the wave. For our current system, this relation is given by Eq. (6.27) with $\alpha \equiv kd$. Taking into account that $(2 - e^{+i\alpha} - e^{-i\alpha}) \equiv 2(1 - \cos\alpha) \equiv 4\sin^2(\alpha/2)$, the dispersion relation may be rewritten in a simpler form:

$$\omega = \pm\omega_{max} \sin\frac{\alpha}{2} = \pm\omega_{max} \sin\frac{kd}{2}, \qquad \text{where } \omega_{max} \equiv 2\left(\frac{\kappa_{ef}}{m}\right)^{1/2}. \tag{6.30}$$

This result, sketched in figure 6.5, is rather remarkable in several aspects. I will discuss them in detail, because most of these features are typical for waves of any type (even including the 'de Broglie waves', i.e. wavefunctions in quantum mechanics), in particular those in periodic structures.

First, at low frequencies, $\omega \ll \omega_{max}$, the dispersion relation (6.30) is linear:

$$\omega = \pm vk, \qquad \text{where } v \equiv \left|\frac{d\omega}{dk}\right|_{k=0} = \frac{\omega_{max}d}{2} = \left(\frac{\kappa_{ef}}{m}\right)^{1/2} d. \tag{6.31}$$

Plugging Eq. (6.31) into Eq. (6.29), we see that the constant v plays, in the low-frequency limit, the role of the same phase velocity for waves of any frequency. Due to its importance, this *acoustic wave*[8] limit will with the subject of the next section.

Second, when the wave frequency *is* comparable with ω_{max}, the dispersion relation is *not* linear, and the system is *dispersive*. This means that as a wave, whose Fourier spectrum has several essential components with frequencies of the order of ω_{max}, travels along the structure, its *waveform* (which may be defined as the shape of the line connecting all points $q_j(z)$, at the same time) changes[9]. This effect may be analyzed by

[8] This term is purely historical. Although the usual sound waves in air belong to this class, the waves we are discussing may have frequencies both well below and well above the human ear's sensitivity range.

[9] The waveform deformation due to *dispersion* (which we are considering now) should be clearly distinguished from its possible change due to *attenuation*, i.e. energy loss—which is *not* taken into account is our current energy-conserving model; see section 6.6 below.

representing the general solution of Eq. (6.24) as the sum (more generally, an integral) of components (6.28) with different complex amplitudes a:

$$q_j(t) = \mathrm{Re}\int_{-\infty}^{+\infty} a_k \exp\left\{i\left[kz_j - \omega(k)t\right]\right\}dk. \qquad (6.32)$$

This notation emphasizes the dependence of the component wave amplitudes a_k and frequencies ω on the wave number k. While the latter dependence is given by the dispersion relation, in our current case by Eq. (6.30), the function a_k is determined by the initial conditions. For applications, the case when a_k is substantially different from zero only in a narrow interval, of a width $\Delta k \ll k_0$ around some central value k_0, is of special importance. (The Fourier transform reciprocal to Eq. (6.32) shows that this is true, in particular, for the so-called *wave packet*—a sinusoidal wave modulated by a spatial envelope function of a large width $\Delta z \sim 1/\Delta k \gg 1/k_0$—see, e.g. figure 6.6.)

Using the strong inequality $\Delta k \ll k_0$, the wave packet's propagation may be analyzed by expending the dispersion relation $\omega(k)$ into the Taylor series at point k_0, and, in the first approximation in $\Delta k/k_0$, restricting the expansion by its first two terms:

$$\omega(k) \approx \omega_0 + \frac{d\omega}{dk}\bigg|_{k=k_0}\tilde{k}, \text{ where } \omega_0 \equiv \omega(k_0), \qquad \text{and } \tilde{k} \equiv k - k_0. \qquad (6.33)$$

In this approximation, Eq. (6.32) yields

$$\begin{aligned} q_j(t) &\approx \mathrm{Re}\int_{-\infty}^{+\infty} a_k \exp\left\{i\left[\left(k_0 + \tilde{k}\right)z_j - \left(\omega_0 + \frac{d\omega}{dk}\bigg|_{k=k_0}\tilde{k}\right)t\right]\right\}dk \\ &= \mathrm{Re}\left[\exp\left\{i\left(k_0 z_j - \omega_0 t\right)\right\}\right. \\ &\quad \left.\times \int_{-\infty}^{+\infty} a_k \exp\left\{i\tilde{k}\left(z_j - \frac{d\omega}{dk}\bigg|_{k=k_0} t\right)\right\}dk\right]. \end{aligned} \qquad (6.34)$$

Comparing this expression with the initial form of the wave packet,

$$q_j(0) = \mathrm{Re}\int_{-\infty}^{+\infty} a_k e^{ikz_j}dk \equiv \mathrm{Re}\left[\exp\left\{ik_0 z_j\right\}\int_{-\infty}^{+\infty} a_k \exp\left\{i\tilde{k}z_j\right\}dk\right], \qquad (6.35)$$

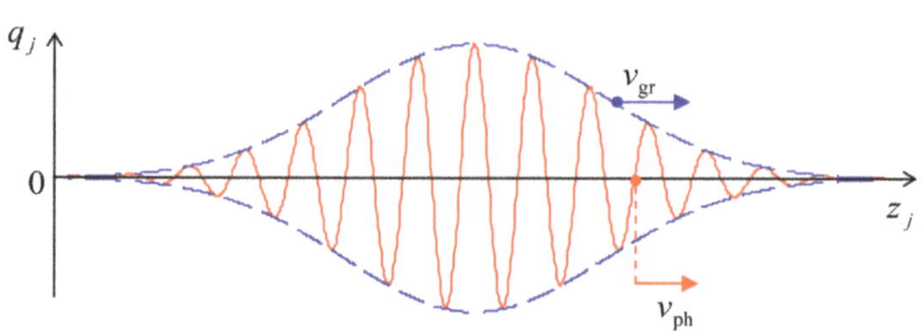

Figure 6.6. The phase and group velocities of a wave packet.

and taking into account that the phase factors before the integrals in the last forms of Eqs. (6.34) and (6.37) do not affect its envelope, we see that in this approximation the envelope sustains its initial form and propagates along the system with the so-called *group velocity*

$$v_{gr} \equiv \left.\frac{d\omega}{dk}\right|_{k=k_0}. \tag{6.36}$$

With the exception of the acoustic wave limit (6.31), this velocity, which characterizes the propagation of the waveform's envelope, is different from the phase velocity (6.29), which describes the propagation of the 'carrier' sinusoidal wave, e.g. the position of one of its zeros—see the red and blue arrows in figure 6.6[10].

Next, for our particular dispersion relation (6.30), the difference between v_{ph} and v_{gr} increases as ω approaches ω_{max}, with the group velocity (6.36) tending to zero, while the phase velocity staying virtually constant. The physics of such a maximum frequency available for the wave propagation may be readily understood by noticing that according to Eq. (6.30), at $\omega = \omega_{max}$, the wave number k equals $n\pi/d$, where n is an odd integer, and hence the phase shift $\alpha \equiv kd$ is an odd multiple of π. Plugging this value into Eq. (6.28), we see that at $\omega = \omega_{max}$, the oscillations of two adjacent particles are in anti-phase, for example:

$$\begin{aligned} q_0(t) &= a\exp\{-i\omega t\}, \\ q_1(t) &= a\exp\{i(\pi - \omega t)\} = -a\exp\{-i\omega t\} = -q_0(t). \end{aligned} \tag{6.37}$$

It is clear, in particular from figure 6.4b for longitudinal oscillations, that at such a phase shift, all the springs are maximally stretched/compressed (just as in the hard mode of the two coupled oscillators analyzed in section 6.1), so that it is natural that this mode has the highest frequency.

This fact invites a natural question: what happens with the system if it is being excited at a frequency $\omega > \omega_{max}$, say by an external force applied at the system's boundary? Reviewing the calculations that have led to the dispersion relation (6.30), we see that they are valid not only for real but also any complex values of k. In particular, at $\omega > \omega_{max}$ it gives

$$k = \frac{(2n-1)\pi}{d} \pm \frac{i}{\Lambda}, \quad \text{where } n = 1, 2, 3, \ldots, \ \Lambda \equiv \frac{d}{2\cosh^{-1}(\omega/\omega_{max})}. \tag{6.38}$$

Plugging this relation into Eq. (6.28), we see that the wave's amplitude becomes an exponential function of the particle's position:

$$\left| q_j \right| = |\,a\,|\, e^{\pm j\,\mathrm{Im}\,kd} \propto \exp\left\{\pm z_j/\Lambda\right\}. \tag{6.39}$$

[10] Taking into account the next term in the Taylor expansion of function $\omega(q)$, proportional to $d^2\omega/dq^2$, we would find that the dispersion leads to a gradual change of the envelope form. Such changes play an important role in quantum mechanics, so they will be discussed in *Part QM* of these lecture notes.

Physically this means that penetrating into the structure, the wave decays exponentially (from the excitation point), dropping by a factor of $e \approx 3$ at the so-called *penetration depth* Λ. (According to Eq. (6.38), this depth decreases as the frequency is increased beyond $\omega_{\max}$, but rather slowly, at $\omega \sim \omega_{\max}$ remaining of the order of the distance d between the adjacent particles.) Such a limited penetration is a very common property of waves, including the electromagnetic waves penetrating into various plasmas and superconductors, and the quantum-mechanical de Broglie waves penetrating into classically forbidden regions of space. Note that this effect of 'wave expulsion' from the media does not require any energy dissipation.

Finally, one more fascinating feature of the dispersion relation (6.30) is its periodicity: if the relation is satisfied by some wave number $k_0(\omega)$, it is also satisfied at any $k_n(\omega) = k_0(\omega) + 2\pi n/d$, where n is an integer. This property is independent of the particular dynamics of the system, and is a common property of all systems that are d-periodic in the usual ('direct') space. It has particularly important implications for the quantum de Broglie waves in periodic structures, e.g. crystals[11].

6.4 Acoustic waves

Now let us return to the limit of low-frequency, dispersion-free *acoustic waves*, with $|\omega| \ll \omega_0$, propagating with the frequency-independent velocity (6.31). Such waves are the general property of *any* elastic continuous medium and obey a simple (and very important) partial differential equation. In order to derive it, let us note that in the acoustic-wave limit, $|kd| \ll 1$,[12] the phase shift $\alpha \equiv kd$ is very small. This means that the differences $q_{j+1}(t) - q_j(t)$ and $q_j(t) - q_{j-1}(t)$, participating in (6.24), are relatively small and may be approximated with $\partial q/\partial j \equiv \partial q/\partial(z/d) \equiv d(\partial q/\partial z)$, with the derivatives taken at middle points between the particles: respectively, $z_+ \equiv (z_{j+1} - z_j)/2$ and $z_- \equiv (z_j - z_{j-1})/2$. Here z is now considered as a continuous argument (and hence the system, as a *1D continuum*), and the particle displacement $q(z, t)$, as a continuous function of space and time. In this approximation, the sum of the last two terms of Eq. (6.24) is equal to $-\kappa d\,[\partial q/\partial z(z_+) - \partial q/\partial z(z_-)]$, and may be similarly approximated with $-\kappa d^{\,2}(\partial^{\,2}q/\partial z^{\,2})$, with the second derivative taken at point $(z_+ - z_-)/2 = z_j$, i.e. exactly at the same point as the time derivative. As the result, the *set* of *ordinary* differential equations (6.24), for different j, is reduced to just *one partial* differential equation

$$m\frac{\partial^2 q}{\partial t^2} - \kappa_{\text{ef}} d^2 \frac{\partial^2 q}{\partial z^2} = 0. \tag{6.40a}$$

Using Eq. (6.31), we may rewrite this *1D wave equation* in a more general form

$$\left(\frac{1}{v^2}\frac{\partial^2}{\partial t^2} - \frac{\partial^2}{\partial z^2}\right) q(z, t) = 0. \tag{6.40b}$$

---

[11] For more detail see, e.g. *Part QM* section 2.5.

[12] Strictly speaking, in accordance with the discussion in the end of the previous section, in this reasoning, k means the distance of the wave number from the closest point $2\pi n/d$—see figure 6.5 again.

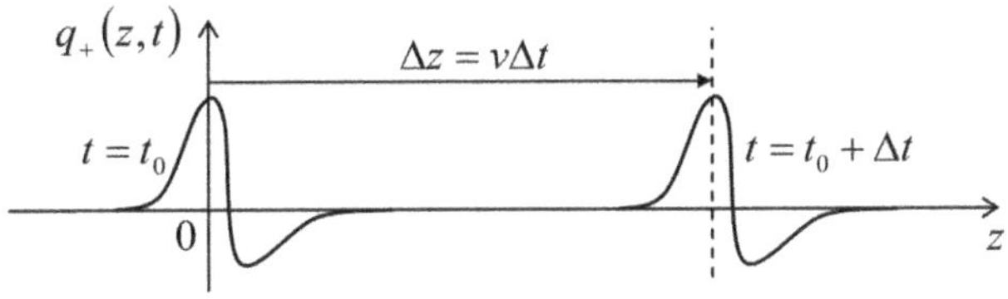

Figure 6.7. Propagation of a traveling wave in a dispersion-free 1D system.

The most important property of the wave equation (6.41), which may be verified by an elementary substitution, is that it is satisfied by either of two *traveling wave* solutions (or their linear superposition):

$$q_+(z,\ t) = f_+(t - z/v), \qquad q_-(z,\ t) = f_-(t + z/v), \tag{6.41}$$

where $f_\pm$ are *any* smooth functions of one argument. The physical sense of these solutions may be revealed by noticing that the displacements $q_\pm$ do not change at the addition of an arbitrary change Δt to their time argument, provided that it is accompanied by an addition of the proportional addition of $\mp v\Delta t$ to their space argument. This means that with the time, the waveforms just move (respectively, to the left or to the right), with the constant speed v, exactly retaining their form—see figure 6.7[13].

Returning to the simple model shown in figure 6.4, let me emphasize that the velocity v is different for the waves of two types: for the longitudinal waves (with $\kappa_{\text{ef}} = \kappa$, see figure 6.4b),

$$v = v_l \equiv \left(\frac{\kappa}{m}\right)^{1/2} d, \tag{6.42}$$

while for the transverse waves (with $\kappa_{\text{ef}} = \mathcal{T}/d$, see figure 6.4c):

$$v = v_t = \left(\frac{\mathcal{T}}{md}\right)^{1/2} d \equiv \left(\frac{\mathcal{T}d}{m}\right)^{1/2} \equiv \left(\frac{\mathcal{T}}{\mu}\right)^{1/2}, \tag{6.43}$$

where the constant $\mu \equiv m/d$ has a simple physical sense of the particle chain's mass per unit length. Evidently, these velocities, in the same system, may be rather different.

The wave equation (6.40), with its only parameter v, may conceal the fact that any wave-supporting system is characterized by one more key parameter. In our current model (figure 6.4), this parameter may be revealed by calculating the forces $F_\pm(z,\ t)$ accompanying any traveling waves (6.41). For example, in the acoustic wave limit

[13] From the point of view of Eq. (6.40), the only requirement for the 'smoothness' of the functions $f_\pm$ is to be doubly differentiable. However, we should not forget that in our case the wave equation is only the approximation of the discrete Eq. (6.24), so that according to Eq. (6.30), the traveling waveform conservation is limited by the acoustic wave limit condition $\omega \ll \omega_{\text{max}}$, which should be fulfilled for any Fourier component of these functions.

($|kd| \ll 1$) we are considering now, the force exerted by the jth particle on its right neighbor, may be approximated as

$$F(z_j, t) \equiv \kappa_{\text{ef}}\left[q_j(t) - q_{j+1}(t)\right] \approx -\kappa_{\text{ef}}\frac{\partial q}{\partial z}\bigg|_{z=z_j} d, \tag{6.44}$$

where, as was discussed above, κ_{ef} equals κ for the longitudinal waves and $\mathscr{T}/d$ for the transverse waves. But for the traveling waves (6.41), the partial derivatives $\partial q_\pm/\partial z$ are equal to $\mp(df_\pm/dt)/v$, so that the corresponding forces are equal to

$$F_\pm = \mp\frac{\kappa_{\text{ef}}d}{v}\frac{df_\pm}{dt}, \tag{6.45}$$

i.e. are proportional to the particle's velocities $u = \partial q/\partial t$ in these waves[14], $u_\pm = df_\pm/dt$, for the same z and t. This means that the ratio

$$\frac{F_\pm(z, t)}{u_\pm(z, t)} = -\kappa_{\text{ef}}d\frac{\partial q_\pm/\partial z}{\partial q_\pm/\partial t} = -\kappa_{\text{ef}}d\frac{\left(\mp df_\pm/dt\right)/v}{df_\pm/dt} = \pm\frac{\kappa_{\text{ef}}d}{v}, \tag{6.46}$$

depends only on the wave propagation direction, but is independent of z and t, and also of the propagating waveform. Its magnitude,

$$Z \equiv \frac{\kappa_{\text{ef}}d}{v} = (\kappa_{\text{ef}}m)^{1/2}, \tag{6.47}$$

characterizing the dynamic 'stiffness' of the system for the propagating waves, is called the *wave impedance*[15]. Note that the impedance is determined by the *product* of the system's generic parameters κ_{ef} and m, while the wave velocity (6.31) is proportional to their *ratio*, so that these two parameters are completely independent, and both are important. According to Eq. (6.47), the wave impedance, just as the wave velocity, is also different for the longitudinal and transverse waves [16]:

$$Z_l = \frac{\kappa d}{v_l} \equiv (\kappa m)^{1/2}, \qquad Z_t = \frac{\mathscr{T}}{v_t} \equiv (\mathscr{T}\mu)^{1/2}. \tag{6.48}$$

One of the wave impedance's key functions is to scale the power carried by a traveling wave:

$$\mathscr{P}_\pm \equiv F_\pm(z, t)u_\pm(z, t) = -\kappa_{\text{ef}}d\frac{\partial q_\pm}{\partial z}\frac{\partial q_\pm}{\partial t} = \pm\frac{\kappa_{\text{ef}}d}{v}\left(\frac{df_\pm}{dt}\right)^2 \equiv \pm Z\left(\frac{df_\pm}{dt}\right)^2. \tag{6.49}$$

Two remarks about this important result. First, the sign of $\mathscr{P}$ depends only on the direction of the wave propagation, but not on the waveform. Second, the instant value

[14] Of course, the *particle's* velocity u (which is proportional to the wave amplitude) should not be confused with the *wave's* velocity v (which is independent of this amplitude).

[15] This notion is strangely missing from many physics (though not engineering!) textbooks.

[16] Note that the first of these expressions for Z coincides with the one used for a single oscillator in section 5.6. In that case, Z may be also recast in a form similar to Eq. (6.45), namely, as the ratio of the force and velocity amplitudes at free oscillations.

of the power does not change if we move with the wave in question, i.e. measure $\mathscr{P}$ at points with $z \pm vt = \text{const}$. This is natural, because in the Hamiltonian system we are considering, the wave energy is conserved. Hence, the wave impedance Z characterizes the energy *transfer* along the system rather than its *dissipation*.

Another important function of the wave impedance notion becomes clear when we consider waves in non-uniform systems. Indeed, our previous analysis assumed that the 1D system supporting the waves (figure 6.4) is exactly periodic, i.e. macroscopically uniform, and extends all the way from $-\infty$ to $+\infty$. Now let us examine what happens when this is not true. The simplest, and very important example of such non-uniform systems is a sharp *interface*, i.e. a point (say, $z = 0$) at which system parameters, otherwise constant, experience a jump—see figure 6.8.

Since the parameters are constant on each side of the interface, the wave equation (6.40) and its partial solutions (6.41) are is still valid for $z < 0$ and $z > 0$ (in the former case, with primed parameters). However, the jump of parameters at the interface leads to a partial *reflection* of the incident wave from the interface, so that at least on the side of the incidence (in the case shown in figure 6.8, for $z \geqslant 0$), we need to use two such terms, one describing the incident wave and another the reflected wave:

$$q(z, t) = \begin{cases} f_-'(t + z/v'), & \text{for } z \leqslant 0, \\ f_-(t + z/v) + f_+(t - z/v), & \text{for } z \geqslant 0\,. \end{cases} \tag{6.50}$$

In order to find the relations between the functions f_-, f_+, and f_-' (of which the first one, describing the incident wave, may be considered known), we may use two boundary conditions at $z = 0$. First, the displacement $q_0(t)$ of the particle at the interface has to be the same whether it is considered a part of the left or right subsystem, and hence participates in Eq. (6.50) for both $z \leqslant 0$ and $z \geqslant 0$. This gives us the first boundary condition:

$$f_-'(t) = f_-(t) + f_+(t). \tag{6.51}$$

On the other hand, the forces exerted on the interface from the left and from the right should also have an equal magnitude, because the interface may be considered as an object with a vanishing mass, and any non-vanishing net force would give it an infinite (i.e. unphysical) acceleration. This gives us the second boundary condition:

$$Z'\frac{df_-'(t)}{dt} = Z\left[\frac{df_-(t)}{dt} - \frac{df_+(t)}{dt}\right]. \tag{6.52}$$

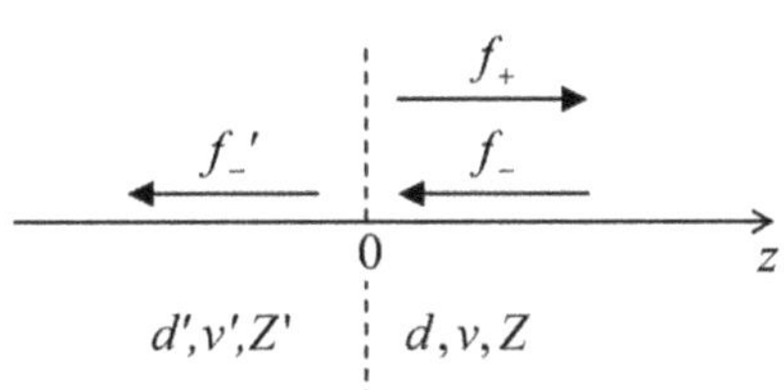

Figure 6.8. Partial reflection of a wave from a sharp interface.

Integrating both parts of this equation over time, and neglecting the integration constant (which describes a common displacement of all particles rather than their oscillations), we obtain

$$Z'f_-'(t) = Z\left[f_-(t) - f_+(t)\right]. \tag{6.53}$$

Now solving the system of two linear equations for $f_+(t)$ and $f_+'(t)$, we see that both these functions are proportional to the incident waveform:

$$f_+(t) = \mathcal{R}\, f_-(t), \qquad f_-'(t) = \mathcal{T}\, f_-(t), \tag{6.54}$$

with the following *reflection* ($\mathcal{R}$) and *transmission* ($\mathcal{T}$) *coefficients*:

$$\mathcal{R} = \frac{Z - Z'}{Z + Z'}, \qquad \mathcal{T} = \frac{2Z}{Z + Z'}. \tag{6.55}$$

Later in this series, we will see that with the appropriate re-definition of the impedance, these relations are also valid for waves of other physical nature (including the de Broglie waves in quantum mechanics) propagating in 1D continuous structures, and also in continua of higher dimensions, at the normal wave incidence on the interface[17]. Note that the coefficients $\mathcal{R}$ and $\mathcal{T}$ give the ratios of wave *amplitudes*, rather than their *powers*. Combining Eqs. (6.49) and (6.55), we get the following relations for the powers (either at the interface, or at the corresponding points of the reflected and transmitted waves):

$$\frac{\mathscr{P}_+}{\mathscr{P}_-} = \left(\frac{Z - Z'}{Z + Z'}\right)^2, \qquad \frac{\mathscr{P}_-'}{\mathscr{P}_-} = \frac{4ZZ'}{(Z + Z')^2}. \tag{6.56}$$

Note that $\mathscr{P}_- + \mathscr{P}_+ = \mathscr{P}_-'$, again reflecting the wave energy conservation.

Perhaps the most important corollary of Eqs. (6.55)–(6.56) is that the reflected wave completely vanishes, i.e. the incident wave is completely transmitted through the interface ($\mathscr{P}_+' = \mathscr{P}_+$), if the so-called *impedance matching condition* $Z' = Z$ is satisfied, even if the wave velocities v (6.31) are different on the left and right sides. In contrast, the equality of the acoustic velocities in two continua does *not* guarantee the full transmission of their interface. Again, this is a very general result.

Finally, let us note that for the important particular case of a sinusoidal incident wave[18],

$$f_-(t) = \mathrm{Re}\,[a\, e^{-i\omega t}], \qquad \text{so that } f_+(t) = \mathrm{Re}\,[\mathcal{R} a\, e^{-i\omega t}], \tag{6.57}$$

---

[17] See the corresponding parts of the lecture notes *Part QM* section 2.3 and *Part EM* section 7.3.

[18] In the acoustic wave limit, when the impedances Z and Z', and hence the reflection coefficient $\mathcal{R}$, are real, $\mathcal{R}$ and Z may be taken from under the Re operator. However, in their current, more general form, Eqs. (6.55)–(6.57) are also valid for the case of arbitrary frequencies, $\omega \sim \omega_{\max}$, when $\mathcal{R}$ and Z may be complex.

where a is the complex amplitude of the wave, the total wave (6.50) on the right of the interface is

$$\begin{aligned} q(z, t) &= \mathrm{Re}\,[a\, e^{-i\omega(t+z/v)} + \mathcal{R}\, a\, e^{-i\omega(t-z/v)}] \\ &\equiv \mathrm{Re}\,[\, a\,(e^{-ikz} + \mathcal{R}\, e^{+ikz})\, e^{-i\omega t}], \quad \text{for } z \geqslant 0, \end{aligned} \tag{6.58}$$

while, according to Eq. (6.45), the corresponding force distribution is

$$\begin{aligned} F(z, t) &= F_{-}(z, t) + F_{+}(z, t) = -Z\frac{\partial f_{-}(t - z/v)}{\partial t} + Z\frac{\partial f_{-}(t - z/v)}{\partial t} \\ &= \mathrm{Re}\,[i\omega Z a\,(e^{-ikz} - \mathcal{R}\, e^{+ikz})\, e^{-i\omega t}]. \end{aligned} \tag{6.59}$$

These expressions will be used in the next section.

6.5 Standing waves

Now let us consider the two limits in which Eq. (6.55) predicts a *total wave reflection* ($\mathcal{T} = 0$): $Z'/Z \to \infty$ (when $\mathcal{R} = -1$) and $Z'/Z \to 0$ (when $\mathcal{R} = +1$). According to Eq. (6.53), the former limit corresponds to $f_{-}(t) + f_{+}(t) \equiv q(0, t) = 0$, i.e. to vanishing oscillations at the interface. This means that this particular limit describes a perfectly *rigid boundary*, not allowing the system's end to oscillate at all. In this case, Eqs. (6.58)–(6.59) yield

$$q(z, t) = \mathrm{Re}\,[a(e^{-ikz} - e^{+ikz})\, e^{-i\omega t}] \equiv -2\,\mathrm{Re}\,[ae^{-i\omega t}]\sin kz, \tag{6.60}$$

$$F(z, t) = \mathrm{Re}\,[i\omega Z a(e^{-ikz} + e^{+ikz})\, e^{-i\omega t}] \equiv 2\omega Z\,\mathrm{Re}\,[ae^{-i(\omega t - \pi/2)}]\cos kz. \tag{6.61}$$

This means that we may interpret the process on the right of the interface using two mathematically equivalent, but physically quite different, languages: either as the sum of two *traveling waves* (the incident one and the reflected one, traveling in opposite directions), or a single *standing wave*. Note that in contrast to the traveling wave (figure 6.9a, cf figure 6.7), in the standing sinusoidal wave (figure 6.9b) all particles oscillate in time with one phase.

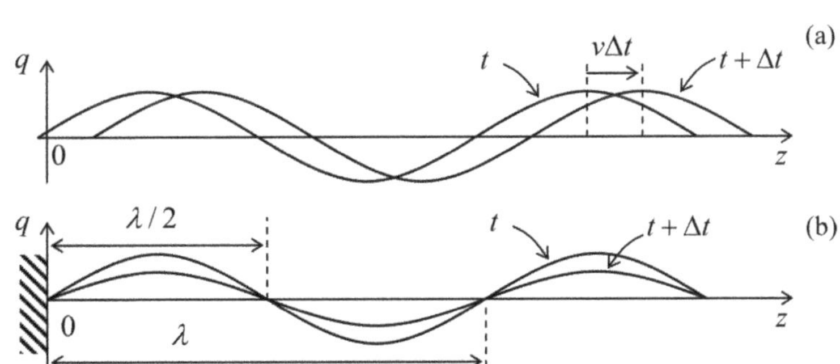

Figure 6.9. The time evolution of (a) a traveling sinusoidal wave and (b) a standing sinusoidal wave at a rigid boundary.

Note also that the phase of the force oscillations (6.61) is shifted, both in space and in time, by $\pi/2$ relative to the particle displacement oscillations. (In particular, at the rigid boundary the force amplitude reaches its maximum.) As a result, the average power flow vanishes, so that the average energy of the standing wave does not change, although its instantaneous energy still oscillates at each spatial point between its kinetic and potential components—just as in usual harmonic oscillations.

A similar standing wave, but with a maximum of displacement q, and with a zero ('node') of force F, is formed near the open boundary, with $Z'/Z \to 0$, and hence $\mathcal{R} = +1$.

Now I have to explain why I have used the sinusoidal waveform for the wave reflection analysis. Let us consider a 1D wave system, which obeys Eq. (6.40), of a finite length l, limited by two rigid walls (located, say, at $z = 0$ and $z = l$), which impose the following boundary conditions,

$$q(0, t) = q(l, t) = 0, \tag{6.62}$$

on its oscillations. Naturally, a sinusoidal traveling wave, induced in the system, will be reflected from both ends, forming the standing wave patterns of the type (6.60) near each of them. These two patterns are compatible if l is exactly equal to an integer number (say, n) of $\lambda/2$, where $\lambda \equiv 2\pi/k$ is the wavelength:

$$l = n\frac{\lambda}{2} \equiv n\frac{\pi}{k}. \tag{6.63}$$

This requirement yields the following spectrum of possible wave numbers of the standing waves,

$$k_n = n\frac{\pi}{l}, \tag{6.64}$$

where the list of possible integers n may be limited to non-negative values: $n = 1, 2, 3, \ldots$ (Indeed, negative values give absolutely similar waves (6.60), while $n = 0$ yields $k_n = 0$, and the corresponding wave vanishes at all points: $\sin(0z) \equiv 0$.) In the acoustic wave limit we are discussing, described by Eq. (6.31), $\omega = \pm vk$, enabling us to translate the wave number spectrum into that of possible standing-wave oscillations of the system[19]:

$$\omega_n = vk_n = n\frac{\pi v}{l}, \qquad \text{with } n = 1, 2, 3, \ldots \tag{6.65}$$

Now let us note that this spectrum, and the corresponding standing wave patterns[20],

$$q^{(n)}(z, t) = 2\,\mathrm{Re}\,[a_n \exp\{-i\omega_n t\}] \sin k_n z, \qquad \text{for } 0 \leqslant z \leqslant l, \tag{6.66}$$

may be obtained in a different way, by a direct solution of the wave equation (6.40) with the boundary conditions (6.62). Indeed, let us look for the general

[19] Negative values of ω may be dropped, because they give similar real functions $q(z, t)$.

[20] They describe, in particular, the well-known transverse standing waves on a guitar string.

solution of this partial differential equation in the so-called *variable-separated* form[21]

$$q(z, t) = \sum_n Z_n(z)T_n(t), \tag{6.67}$$

where each partial product $Z_n(z)T_n(t)$ is supposed to satisfy the equation on its own. Plugging this partial solution into Eq. (6.40), and then dividing all its terms by the same product, Z_nT_n, we may rewrite the result as

$$\frac{1}{v^2}\frac{1}{T_n}\frac{d^2T_n}{dt^2} = \frac{1}{Z_n}\frac{d^2Z_n}{dz^2}. \tag{6.68}$$

Here comes the main punch of the variable separation method: since the left-hand side of the equation may depend only on t, and its right-hand side only on z, Eq. (6.68) may be valid only if the both sides are constant. Denoting this constant as $-k_n^2$, we get two similar ordinary differential equations,

$$\frac{d^2Z_n}{dz^2} + k_n^2Z_n = 0, \qquad \frac{d^2T_n}{dt^2} + \omega_n^2T_n = 0, \qquad \text{where } \omega_n^2 \equiv v^2k_n^2, \tag{6.69}$$

with well-known (and similar) sinusoidal solutions

$$\begin{aligned} Z_n &= c_n \cos k_nz + s_n \sin k_nz, \\ T_n &= u_n \cos \omega_nt + v_n \sin \omega_nz \equiv \operatorname{Re}[a_n \exp\{-i\omega_nt\}], \end{aligned} \tag{6.70}$$

where c_n, v_n, u_n, and v_n (or, alternatively, $a_n = u_n + iv_n$) are constants. The general solution (6.67), with all ω_n different, may satisfy the boundary conditions at all times t only if for all $nc_n = 0$ and $\sin k_nl = 0$, giving the same wave number spectrum Eq. (6.64) and hence the own frequency spectrum (6.65), so that the general solution (6.67) of the so-called *boundary problem*, given by Eqs. (6.40) and Eq. (6.62), takes the form

$$q(z, t) = \operatorname{Re}\sum_n a_n \exp\{-i\omega_nt\} \sin k_nz, \tag{6.71}$$

where the complex amplitudes a_n are determined by the initial conditions.

Hence such sinusoidal standing waves (figure 6.10a) are not just an assumption, but a natural property of the 1D wave equation. It is also easy to verify that the result (6.71) is valid for the same system with different boundary conditions, although with a modified wave number spectrum. For example, if the rigid boundary conditions ($q = 0$) are implemented at $z = 0$, and the so-called *free* (or 'open') boundary conditions ($F = 0$, i.e. $\partial q/\partial z = 0$) at $z = l$, the spectrum becomes

$$k_n = \left(n - \frac{1}{2}\right)\frac{\pi}{l}, \qquad \text{with } n = 1, 2, 3, \ldots, \tag{6.72}$$

[21] This *variable separation method* is very general, and is discussed in all parts of this series, particularly in *Part EM* chapter 2.

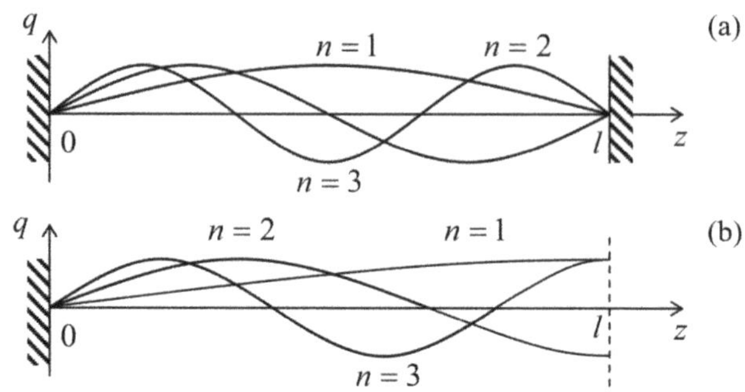

Figure 6.10. The lowest standing wave modes for the 1D systems with (a) two rigid boundaries and (b) one rigid and one open boundary.

so that the lowest standing waves look like those shown in figure 6.10b[22].

Note that the difference between the sequential values of k_n is still constant:

$$k_{n+1} - k_n = \frac{\pi}{l}, \tag{6.73}$$

i.e. the same as for the spectrum (6.64). This is natural, because in both cases the transfer from the nth mode to the $(n + 1)$th mode corresponds just to an addition of one mode half-wave—see figure 6.10. (This conclusion is valid for any combination of the rigid and free boundary conditions.) As was discussed above, for the discrete-particle chain we have started with (figure 6.4), the wave (6.40) and hence the above derivation of Eq. (6.71) are only valid in the acoustic wave limit, i.e. when the distance d between the particles is much less than the wavelengths $\lambda_n \equiv 2\pi/k_n$ of the mode under analysis. For a chain of length l, this means that the number of particles, $N \sim l/d$, has to be much larger than 1. However, a remarkable property of Eq. (6.71) is that it remains valid with the same wave number spectrum (6.64), not only in the acoustic limit, but also for arbitrary $N > 0$. Indeed, since $\sin k_n z \equiv (\exp\{+ik_n z\} - \exp\{-ik_n z\})/2$, each nth term of Eq. (6.71) may be represented as a sum of two traveling waves with equal but opposite wave vectors. As was discussed in section 6.3, such a wave is a solution of equation (6.24) describing the discrete-particle system for *any* k_n, with the only condition that its frequency obeys the general dispersion relation (6.30) rather than its acoustic limit Eq. (6.63).

Moreover, the expressions for k_n (with appropriate boundary conditions), such as Eq. (6.64) or Eq. (6.72), also survive the transition to arbitrary N, because their derivation above was based only on the sinusoidal form of the standing wave. The only new factor arising in the case of arbitrary N is that due to the equidistant property (6.73) of the wave number spectrum, as soon as n exceeds N, the waveforms Eq. (6.71) at particle locations $z_j = jd$ start to repeat. For example,

$$\begin{aligned} \sin k_{n+N} z_j &= \sin (k_n + N\Delta k) jd = \sin\left(k_n + N\frac{\pi}{d}\right) jd \\ &= \sin\left(k_n z_j + \pi j N\right) = \pm \sin k_n z_j. \end{aligned} \tag{6.74}$$

[22] The lowest standing wave of the system, with the smallest k_n and ω_n, is usually called its *fundamental mode*.

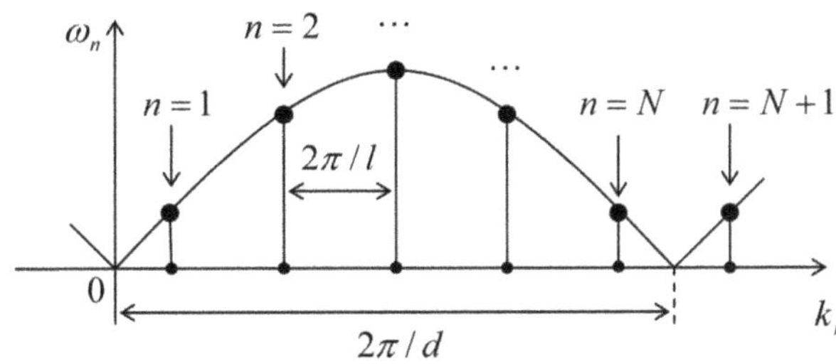

Figure 6.11. The wave numbers and own frequencies of a chain of a finite number N of particles in a chain with one rigid and one open boundary (schematically).

Hence the system has only N different (linearly independent) modes. But this result is in full compliance with the general conclusion made in section 6.2, that *any* system of N coupled 1D oscillators has exactly N own frequencies and corresponding oscillation modes. So, our analysis of a particular system, shown in figure 6.4, just exemplifies this general conclusion. Figure 6.11 illustrates this result for a particular finite value of N; the curve connecting the points shows exactly the same dispersion relation as was shown in figure 6.5, but now it is just a guide for the eye because for a system with a finite length l the wave number spectrum is discrete, and the intermediate values of k and ω do not have an immediate physical sense[23].

Note that the own frequencies of the system are not equidistant, while the wave numbers are. This insensitivity of the spacing (6.73) between the adjacent wave numbers to the particular physics of a macroscopically uniform system is a very general fact, common for waves of any nature, and is broadly used for the analysis of systems with very large number of particles (such as human-size crystals, with $N \sim 10^{23}$). For N so large, the effect of the boundary conditions, e.g. the difference between the spectra (6.64) and (6.72), is negligible, and they may be summarized as the following rule for the number of different standing waves within some interval $\Delta k \gg \pi/l$:

$$\Delta N \equiv \frac{\Delta k \big|_{\text{standing}}}{k_{n+1} - k_n} = \frac{l}{\pi} \Delta k \big|_{\text{standing}}. \tag{6.75a}$$

For such analyses, it is frequently more convenient to work with traveling waves rather than standing waves. In this case, we have to take into account that (as was just discussed above) each standing wave (6.66) may be decomposed to two

[23] Note that figure 6.11 shows the case of one rigid and one open boundary (see figure 6.10b), where $l = Nd$; for a conceptually simpler system with two rigid boundaries (figure 6.10a) we would need to take $l = (N + 1)d$, because neither of the end points can oscillate.

traveling waves with wave numbers $\pm k_n$, so that size of the interval Δk doubles and Eq. (6.75*a*) becomes[24]

$$\Delta N = \frac{l}{2\pi}\Delta k\Big|_{\text{traveling}} . \tag{6.75b}$$

Note that this state counting rule is valid for waves of just one type. As was discussed above, for the model system we have studied (figure 6.4), there are three types of such waves—one longitudinal and two transverse, so that if we need to count them all, ΔN should be multiplied by 3.

6.6 Wave decay and attenuation

Now let us discuss the effects of dissipation on the 1D waves, on the example of the same uniform system as shown in figure 6.4. The effects are simplest for a linear drag, which may be described (as was done for a single oscillator in section 5.1) by adding the term $\eta dq_j/dt$, to Eq. (6.24) for each particle:

$$m\ddot{q}_j + \eta\dot{q}_j - \kappa_{\text{ef}}(q_{j+1} - q_j) + \kappa_{\text{ef}}(q_j - q_{j-1}) = 0. \tag{6.76}$$

(In a uniform system, the drag coefficient η should be similar for all particles, although it may be different for the longitudinal and transverse oscillations.)

In order to analyze the dissipation effect on the *standing* waves, we may again use the variable separation method, i.e. look for the solution of Eq. (6.76) in a form similar to Eq. (6.67), naturally adjusting it for our current discrete system:

$$q(z_j, t) = \sum_n Z_n(z_j)T_n(t) . \tag{6.77}$$

After dividing all terms by $mZ_n(z_j)T_n(t)$, and separating the time-dependent and space-dependent terms, we obtain

$$\frac{\ddot{T}_n}{T_n} + \frac{\eta}{m}\frac{\dot{T}_n}{T_n} = \frac{\kappa_{\text{ef}}}{m}\left[\frac{Z_n(z_{j+1})}{Z_n(z_j)} + \frac{Z_n(z_{j+1})}{Z_n(z_j)} - 2\right] = \text{const.} \tag{6.78}$$

As we know from the previous section, the resulting equation for the function $Z_n(z_j)$ is satisfied if the variable separation constant is equal to $-\omega_n^2$, where ω_n obeys the dispersion relation (6.30) for the wave number k_n, properly calculated for the dissipation-free system, with the account of the given boundary conditions—see,

[24] Note that this simple, but very important relation is frequently derived using the so-called *Born–Carman boundary condition* $q_0(t) \equiv q_N(t)$, which implies bending the system of interest into a closed loop. For a 1D system with $N \gg 1$, such a mental exercise may be somehow justified, but for systems of higher dimension it is hardly physically plausible—and is unnecessary.

e.g. Eqs. (6.62) and (6.72). Hence for the function $T_n(t)$ we are obtaining the ordinary differential equation

$$\ddot{T}_n + 2\delta\dot{T}_n + \omega_n^2 T_n = 0, \qquad \text{with } \delta \equiv \frac{\eta}{2m}, \tag{6.79}$$

which is absolutely similar to Eq. (5.6*b*) for a single linear oscillator, which was studied in section 5.1. As we already know, it has the solution (5.9), describing the free-oscillation decay with the relaxation time given by Eq. (5.10), $\tau = 1/\delta$, and hence is similar for all modes[25].

Hence, the above analysis of the dissipation effect on the free standing waves has not brought any surprises, but it gives us a hint of how their forced oscillations, induced by some external forces $F_j(t)$ exerted on the particles, may be analyzed. Indeed, representing the force as a sum of spatial harmonics proportional to the system's modes,

$$F_j(t) = \frac{1}{m}\sum_n f_n(t)\, Z_n(z_j), \tag{6.80}$$

and using the variable separation (6.77), we arrive at the equation

$$\ddot{T}_n + 2\delta\dot{T}_n + \omega_n^2 T_n = f_n(t), \tag{6.81}$$

absolutely similar to Eq. (5.13*b*) for a single oscillator. This fact enables the use of all the methods discussed in section 5.1 for the forced oscillation analysis, except that the temporal Green's function, defined by either of the equivalent Eqs. (5.27) or (5.28), now acquires index n, i.e. becomes mode-dependent: $G(\tau) \to G_n(\tau)$. Performing the weighted summation similar to Eq. (6.80),

$$G_j(\tau) = \sum_n G_n(\tau)\, Z_n(z_j), \tag{6.82}$$

we obtain the *spatial-temporal Green's function* of the system—in this case, for a discrete, 1D set of spatial points $z_j = jd$. As in the single-oscillator case, it has a simple physical sense of the oscillations induced by a delta-functional force (i.e. a very short pulse), exerted on the jth particle. We will meet (and use) such spatial-temporal Green's functions in other parts of this series.

Now let us discuss the dissipation effects on the *traveling* waves, where they may take a completely different form of *attenuation*. Let us discuss this for a simple example, when one end (located at $z = 0$) of a very long chain ($l \to \infty$) is externally forced to perform sinusoidal oscillations of a certain frequency ω and a fixed

[25] Even an elementary experience with acoustic guitars shows that for their strings this particular conclusion is not valid: the higher modes ('overtones') decay substantially faster, leaving the fundamental mode oscillations for a slower decay. This is the result of an important energy loss (i.e. the wave decay) mechanism, not taken into account in Eq. (6.76), the radiation of the sound into the guitar's body through the string supports—mostly through the bridge. Such radiation may be described, after the proper modification of the boundary conditions (6.62), in terms of matching the wave impedance (6.47) of the string with the impedance of the support.

amplitude A_0. In this case, it is natural to look for the particular solution of Eq. (6.76) in a form very different from Eq. (6.77):

$$q_j(z, t) = \text{Re}\left[c_j e^{-i\omega t}\right], \tag{6.83}$$

with time-independent, but generally complex amplitudes c_j. As our discussion of a single oscillator in section 5.1 implies, this is not the general but rather a partial solution, which describes forced oscillations in the system to which it settles after some initial transient process. (At non-zero damping, we may be sure that this process fades after a finite time, and thus may be ignored.)

Plugging Eq. (6.83) into Eq. (6.76), we reduce it to an equation for the amplitudes c_j,

$$(-m\omega^2 - i\omega\eta + 2\kappa_{\text{ef}})\, c_j - \kappa_{\text{ef}} c_{j+1} - \kappa_{\text{ef}} c_{j-1} = 0, \tag{6.84}$$

which is a natural generalization of Eq. (6.25). As a result, partial solutions of the set of these equations (for $j = 0, 1, 2,\ldots$) may be looked for in the form (6.26) again, but now, because of the new, imaginary term in Eq. (6.84), we should be ready to obtain a complex phase shift α and hence a complex wave number $k \equiv \alpha/d$.[26] Indeed, the resulting characteristic equation for k,

$$\sin^2\frac{kd}{2} - \frac{\omega^2}{\omega_{\max}^2} + i\frac{2\omega\delta}{\omega_{\max}^2} \tag{6.85}$$

(where $\omega_{\max}$ is defined by Eq. (6.30), and the damping coefficient is defined just as in a single oscillator, $\delta \equiv \eta/2m$), does not have a real solution even at $\omega < \omega_{\max}$. Using the well-known expressions for the sine function of a complex argument[27], Eq. (6.85) may be readily solved in the most important low-damping limit $\delta \ll \omega$. In the linear approximation in δ, it does not affect the real part of k, but makes its imaginary part different from zero:

$$\begin{aligned} k &= \pm\frac{2}{d}\left(\arcsin\frac{\omega}{\omega_{\max}} + i\frac{\delta}{\omega_{\max}}\right) \\ &\equiv \pm\left(\frac{2}{d}\arcsin\frac{\omega}{\omega_{\max}} + i\frac{\delta}{v}\right), \qquad \text{for} \quad -\pi \leqslant \text{Re}\, k \leqslant \pi, \end{aligned} \tag{6.86}$$

with a periodic extension to other periods—see figure 6.5. Just as was done in Eq. (6.28), due to two values of the wave number, generally we have to take c_j in the form of not a single wave (6.26), but of a linear superposition of two partial solutions:

$$c_j = \sum_{\pm} c_{\pm} \exp\left\{\pm i\, \text{Re}\, k z_j \mp \frac{\delta}{v} z_j\right\}, \tag{6.87}$$

where the constants $c_{\pm}$ should be found from the boundary conditions. In our particular case, $|c_0| = A_0$ and $c_\infty = 0$, so that only one of these two waves, namely the one exponentially decaying at its penetration into the system, has a non-vanishing

[26] As a reminder, we have already met such a situation in the absence of damping, but at $\omega > \omega_{\max}$—see Eq. (6.38).

[27] See, e.g. Eq. (A.20).

magnitude: $|c_+| = A_0$, $c_- = 0$. Hence our solution describes a single wave, with the real amplitude and the oscillation energy decreasing as

$$A_j \equiv \left| c_j \right| = A_0 \exp\left\{ -\frac{\delta}{v} z_j \right\}, \quad E_j \propto A_j^2 \propto \exp\{-\alpha z_j\}, \quad \text{with } \alpha = \frac{2\delta}{v}, \tag{6.88}$$

i.e. with a frequency-independent *attenuation constant* $\alpha = 2\delta/v$, so that the spatial scale of wave penetration into a dissipative system is given by $l_d \equiv 1/\alpha$. Certainly, our simple solution (6.88) is only valid for a system of length $l \gg l_d$; otherwise, we would need the second term in the sum (6.87) to describe the wave reflected from its opposite end.

6.7 Nonlinear and parametric effects

Now let us discuss (because of the lack of time, very briefly, on a semi-quantitative level), the new nonlinear and parametric phenomena that appear in oscillatory systems with more than one degree of freedom—cf sections 5.4–5.8. One important new effect here is the *mutual phase-locking* of (two or more) weakly coupled self-excited oscillators with close frequencies: if the own frequencies of the oscillators are sufficiently close, their oscillation frequencies 'stick together' to become exactly equal. Although the dynamics of this process is very close to that of the phase-locking of a single oscillator by an external signal, which was discussed in section 5.4, it is rather counter-intuitive in the view of the results of section 6.1, and in particular the anti-crossing diagram shown in figure 6.2. The analysis of the effect using the van der Pol method (which is highly recommended to the reader) shows that the origin of the difference is the oscillator's nonlinearity, which makes the oscillation amplitude virtually independent of the phase evolution—see Eq. (5.68) and its discussion.

One more new effect is the so-called *non-degenerate parametric excitation*. It may be illustrated with the example of just two coupled oscillators—see section 6.1 above. Let us assume that the coupling constant κ, participating in Eq. (6.5), is not constant, but oscillates in time—say with frequency ω_p. In this case the forces acting on each oscillator from its counterpart, described by the right-hand sides of Eq. (6.5), will be proportional to $\kappa q_{2,1}(1 + \mu \cos\omega_p t)$. Assuming that the oscillations of q_1 and q_2 are close to sinusoidal ones, with certain frequencies $\omega_{1,2}$, we see that the force exerted on each oscillator will contain the so-called *combinational frequencies*

$$\omega_p \pm \omega_{2,1}. \tag{6.89}$$

If one of these frequencies is close to the own oscillation frequency of this oscillator, we can expect a substantial parametric interaction between the oscillators (on the top of the constant coupling effects discussed in section 6.1). According to Eq. (6.89), this may happen in two cases:

$$\omega_p = \omega_1 + \omega_2, \tag{6.90a}$$

$$\omega_p = \omega_1 - \omega_2. \tag{6.90b}$$

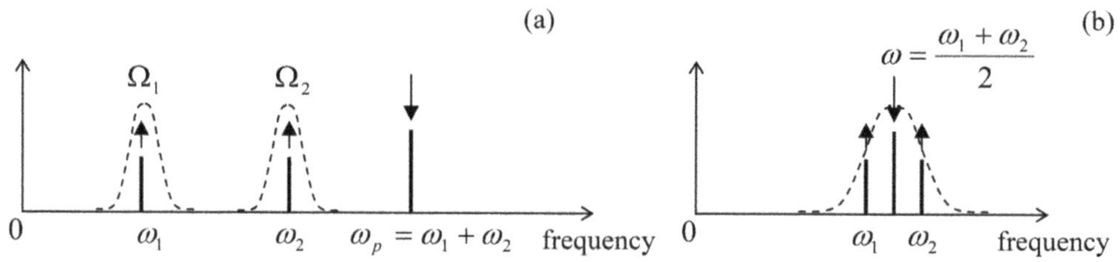

Figure 6.12. Spectra of oscillations at (a) the non-degenerate parametric excitation and (b) the four-wave mixing. The arrow directions symbolize the energy flowing into and out of the system.

The quantitative analysis (also highly recommended as a reader's exercise) shows that in the case (6.90*a*), the parameter modulation indeed leads to energy 'pumping' into oscillations. As a result, sufficiently large μ, at sufficiently small damping coefficients $\delta_{1,2}$ and detuning

$$\xi \equiv \omega_p - (\Omega_1 + \Omega_2), \tag{6.91}$$

may lead to a simultaneous self-excitation of two frequency components $\omega_{1,2}$. These frequencies, while being approximately equal to the corresponding own frequencies $\Omega_{1,2}$ of the system, are related to the *pumping frequency* ω_p by the exact relation (6.90*a*), but otherwise are arbitrary, e.g. may be incommensurate (figure 6.12a), thus justifying the term *non-degenerate* parametric excitation[28]. (The parametric excitation of a single oscillator, analyzed in section 5.5, is a particular, *degenerate* case of such excitation, with $\omega_1 = \omega_2 = \omega_p/2$.) On the other hand, for the case described by Eq. (6.90*b*), the parameter modulation always pumps energy from the oscillations, effectively increasing the system's damping.

Somewhat counter-intuitively, this difference between two cases (6.90) may be more simply interpreted using the notions of quantum mechanics. Namely, the equality $\omega_p = \omega_1 + \omega_2$ enables a decay of an external photon of energy $\hbar\omega_p$ into two photons of energies $\hbar\omega_1$ and $\hbar\omega_2$ of the oscillators. (In contrast, the complementary relation (6.90*b*), meaning that $\omega_1 = \omega_p + \omega_2$, results in a pumping-induced decay of oscillation photons.)

Note that even if the frequencies ω_1 and ω_2 of the parametrically excited oscillations are incommensurate, the oscillations are highly correlated. Indeed, the quantum-mechanical theory of this effect[29] shows that the generated photons are *entangled*. This fact makes the parametric excitation very popular for a broad class of experiments in several currently active fields including quantum computation and encryption, and Bell inequality/local reality studies[30].

Proceeding to nonlinear phenomena, let us note, first of all, that the simple reasoning that accompanied Eq. (5.109) in section 5.8 is also valid in the case when oscillations consist of two (or more) sinusoidal components with incommensurate frequencies. Replacing the notation 2ω with ω_p, we see that the non-degenerate

[28] Note that in some literature, the term *parametric down-conversion* (PDC) is used instead.
[29] Which is, surprisingly, not much more complex than the classical theory—see, e.g. *Part QM* section 5.5.
[30] See, e.g. *Part QM* chapter 10.

parametric excitation of the type Eq. (6.90*a*) is possible in a system of two coupled oscillators with a quadratic nonlinearity (of the type γq^2), 'pumped' by an intensive external signal at frequency $\omega_p \approx \Omega_1 + \Omega_2$. In optics, it is often more convenient to have all signals within the same, relatively narrow frequency range. A simple calculation, similar to the one made in Eqs. (5.107) and (5.108), shows that this may be done using the cubic nonlinearity[31] of the type αq^3, which allows a similar parametric energy exchange at the following frequency relation—see figure 6.12b:

$$2\omega = \omega_1 + \omega_2, \qquad \text{with } \omega \approx \omega_1 \approx \omega_2. \tag{6.92a}$$

This process is often called *four-wave mixing*, because it may be interpreted quantum-mechanically as the transformation of *two* externally delivered photons, each with energy $\hbar\omega_p$, into two other photons of energies $\hbar\omega_1$ and $\hbar\omega_2$. The word 'wave' in this term stems from the fact that at optical frequencies it is difficult to couple a sufficient volume of a nonlinear medium with lumped-type resonators. It is much easier to implement the parametric excitation of light (as well as other nonlinear phenomena such as higher harmonic generation) in *distributed systems* of a linear size much larger than the involved wavelengths. In such systems, the energy transfer from the incoming wave of frequency ω to generated waves of frequencies ω_1 and ω_2 is gradually accumulated at their joint propagation along the system. From the analogy between Eq. (6.85) (describing the evolution of wave's amplitude *in space*), and the usual equation of the linear oscillator (describing its evolution *in time*), it is clear that this energy transfer accumulation requires not only the frequencies ω, but also the wave numbers k be in similar relations. For example, the four-wave mixing requires that not only the frequency balance (6.92*a*), but also a similar relation

$$2k = k_1 + k_2, \tag{6.92b}$$

be *exactly* fulfilled. Since all three frequencies are close, this relation is easy to arrange. Unfortunately, due to the lack of time/space, for more discussion of this interesting subject, *nonlinear optics*, I have to refer the reader to special literature[32].

It may look like a dispersion-free media, with $\omega/k = v = \text{const}$, is the perfect solution for arranging the parametric interaction of waves, because in such media, for example, Eq. (6.92*b*) automatically follows from Eq. (6.92*a*). However, in such media not only the desirable three parametrically interacting waves, but also all their harmonics, have the same velocity. At these conditions, the energy transfer rates between all harmonics are of the same order. Perhaps the most important result of such a multi-harmonic interaction is that intensive incident traveling waves, interacting with a nonlinear medium, may develop sharply non-sinusoidal wave-forms, in particular those with an almost instant change of the field at a certain moment. Such *shock waves*, especially those of mechanical nature, are of great

[31] In optics, such nonlinearity is implemented using transparent crystals such as lithium niobate ($LiNbO_3$), with the cubic-nonlinear dependence of the electric polarization on the applied electric field: $\mathscr{P} \propto \mathscr{E} + \alpha\mathscr{E}^3$.

[32] See, e.g. [1] or the more modern treatment [2]. This field is currently very active. As just a single example, let me mention the recent experiments with parametric amplification of ultrashort (~20 fs) optical pulses to a peak power as high as $\sim 5 \times 10^{12}$ W—see [3].

interest for certain applications—some not quite innocent, e.g. the explosion of ordinary (chemical) and nuclear bombs[33].

To conclude this chapter, let me note that the above discussion of 1D acoustic waves will be extended in section 7.7 to elastic 3D media. There we will see that generally, the waves obey a more complex equation than the apparently natural generalization of Eq. (6.40):

$$\left(\frac{1}{v^2}\frac{\partial^2}{\partial t^2} - \nabla^2\right)q(\mathbf{r}, t) = 0, \tag{6.93}$$

where ∇^2 is the Laplace operator. This fact adds to the complexity of traveling-wave and standing-wave phenomena in higher dimensions. Moreover, in multi-dimensional systems, including such pseudo-1D systems as thin rods and pseudo-2D systems such as thin membranes, even static elastic deformations may be very non-trivial. A brief introduction to the general theory of small deformations, with a focus on elastic continua, will be the subject of the next chapter.

6.8 Problems

For the systems specified in problems 6.1–6.6:

(i) introduce convenient generalized coordinate(s) q_j of the system, and calculate own frequencies of its small harmonic oscillations near the equilibrium;

(ii) calculate the corresponding distribution coefficients, and sketch the oscillation modes.

Problem 6.1. Two elastically coupled pendula, confined to a vertical plane, with the parameters shown in the figure below (see also problems 1.8 and 2.9).

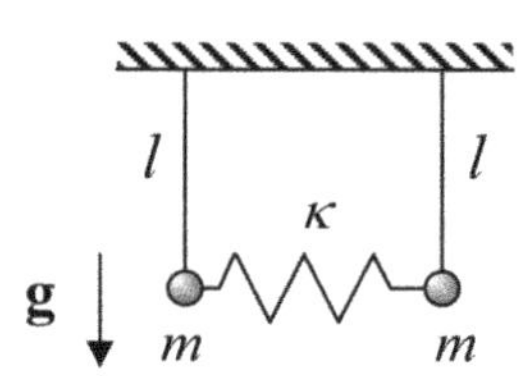

Problem 6.2. The double pendulum, confined to the vertical plane containing the support point (considered in problem 2.1), with $m' = m$ and $l = l'$—see the figure below.

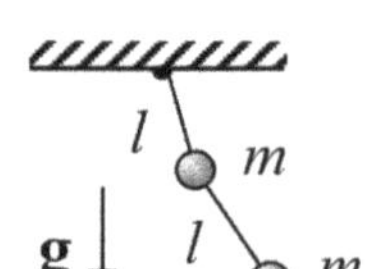

[33] The classical (and perhaps still the best) monograph on the subject is [4].

Problem 6.3. The chime bell (analyzed in problem 4.7, see the figure below), for the particular case $l = l'$.

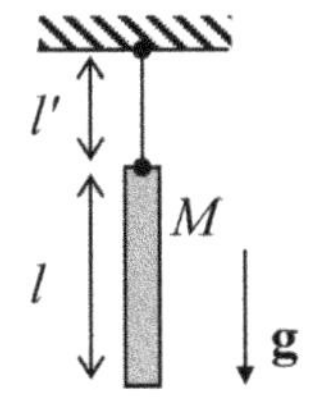

Problem 6.4. The triple pendulum shown in the figure below, with the motion confined to a vertical plane containing the support point.

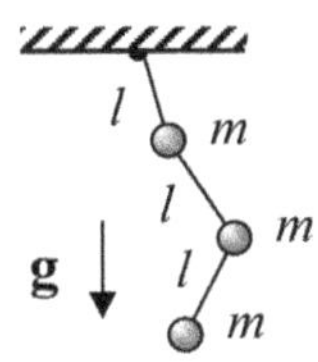

Hint: You may use any (e.g. numerical) method to calculate the characteristic equation roots.

Problem 6.5. A linear, symmetric system of three particles, shown in the figure below, where the connections between the particles not only act as ordinary usual elastic springs (giving potential energies $U = \kappa(\Delta l)^2/2$), but also resist the system's bending, giving an additional potential energy $U' = \kappa'\theta^2/2$, where θ is the (small) bending angle[34].

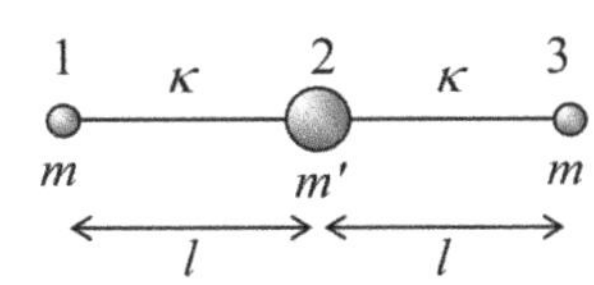

Problem 6.6. Three similar beads, which may slide along a circle of radius R without friction, connected with similar springs with elastic constants κ and equilibrium lengths l_0—see the figure below.

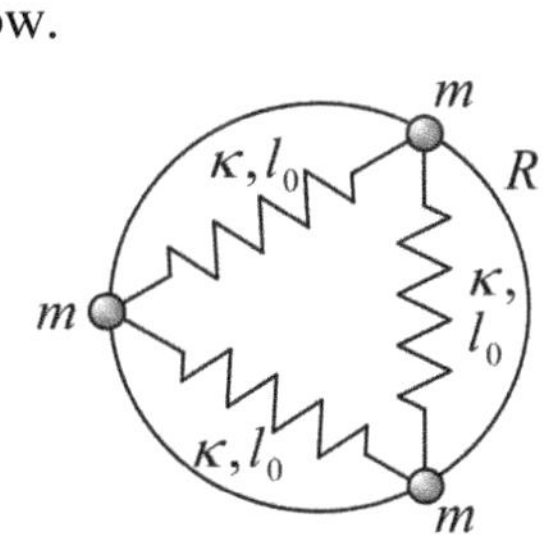

[34] This is a good model for small oscillations of linear molecules such as the now-infamous CO_2.

Problem 6.7. An external longitudinal force $F(t)$ is applied to the right particle of the system shown in figure 6.1, with $\kappa_L = \kappa_R = \kappa'$ and $m_1 = m_2 \equiv m$ (see the figure below), and the response $q_1(t)$ of the left particle to this force is being measured.

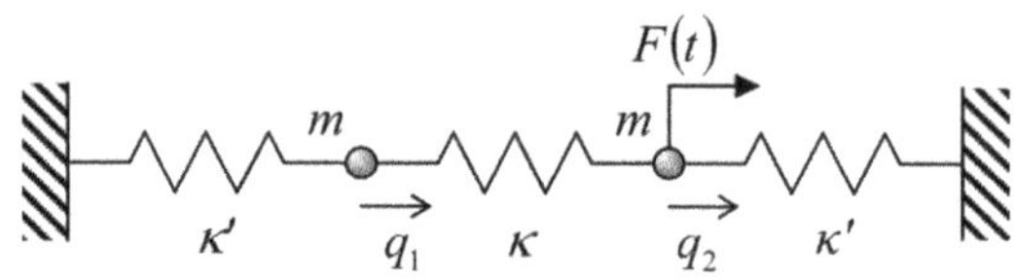

(i) Calculate the temporal Green's function for this response.
(ii) Use this function to calculate the response to force

$$F(t) = \begin{cases} 0, & \text{for } t < 0, \\ F_0 \sin \omega t, & \text{for } 0 \leqslant t, \end{cases}$$

with constant amplitude F_0 and frequency ω.

Problem 6.8. Use the Lagrangian formalism to re-derive Eq. (6.24) for both the longitudinal and the transverse oscillations in the system shown in figure 6.4a.

Problem 6.9. Calculate the energy (per unit length) of a sinusoidal traveling wave propagating in the 1D system shown in figure 6.4a. Use your result to calculate the average power flow created by the wave, and compare it to Eq. (6.49) valid in the acoustic wave limit.

Problem 6.10. Calculate the spatial distribution of the kinetic and potential energies in a standing, sinusoidal, 1D acoustic wave, and analyze its evolution in time.

Problem 6.11. A midpoint of a guitar string of length l has been slowly pulled up by height $h \ll l$, and then is let go. Neglecting dissipation, use two different approaches to calculate the midpoint displacement as a function of time.

Hint: You may like to use the following table series:

$$\sum_{m=1}^{\infty} \frac{\cos(2m-1)\xi}{(2m-1)^2} = \frac{\pi^2}{8}\left(1 - \frac{\xi}{\pi/2}\right), \quad \text{for } 0 \leqslant \xi \leqslant \pi.$$

Problem 6.12. Calculate the dispersion law $\omega(k)$ and the maximum and minimum frequencies of small longitudinal waves in a long chain of similar, spring-coupled pendula—see the figure below.

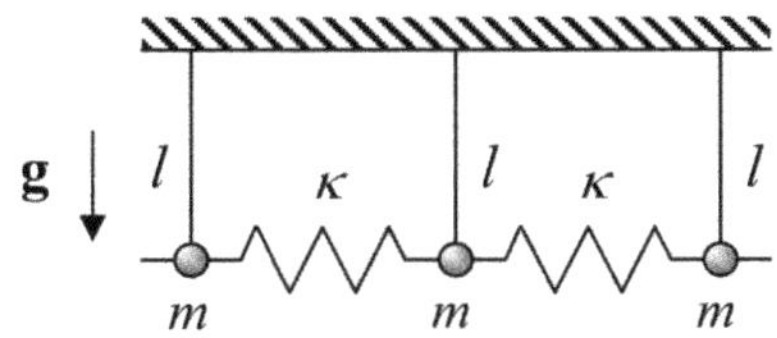

Problem 6.13. Calculate and analyze the dispersion relation $\omega(k)$ for small longitudinal waves in a long chain of elastically coupled particles with alternating masses—see the figure below. In particular, discuss the dispersion relation's period Δk, and its evolution at $m' \to m$.

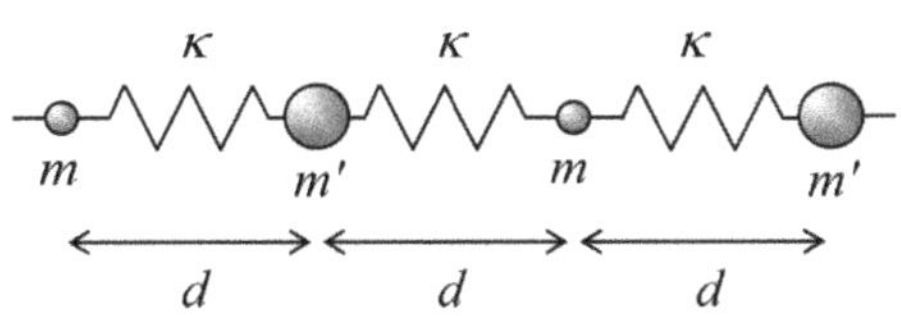

Problem 6.14. Analyze the traveling wave's reflection from a 'point inhomogeneity'-a single particle with a different mass $m_0 \neq m$, in an otherwise uniform 1D chain—see the figure below.

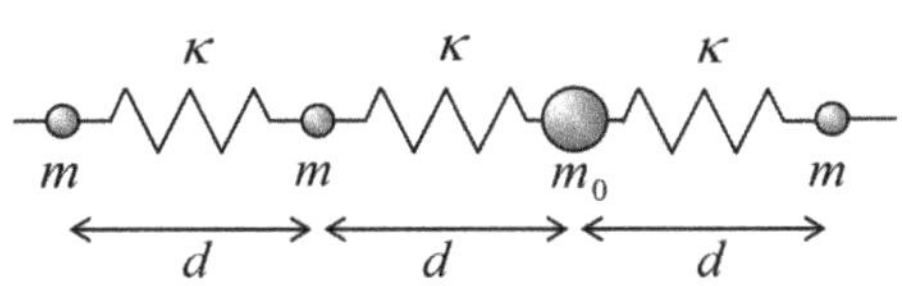

Problem 6.15.*

(i) Explore an approximate way to analyze waves in a continuous 1D system with parameters slowly varying along its length[35].

(ii) Apply this method to calculate the frequencies of transverse standing waves on a freely hanging heavy rope of length l, with constant mass μ per unit length—see the figure below.

(iii) For the three lowest standing wave modes, compare the results with those obtained in the solution of problem 6.4 for the triple pendulum.

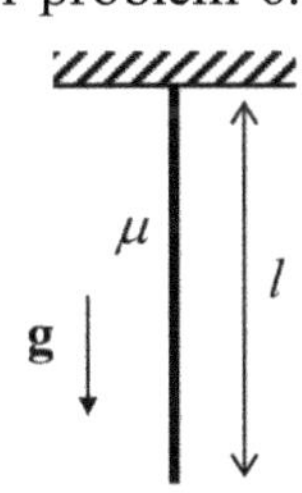

[35] The reader familiar with the WKB approximation in quantum mechanics (see, e.g. *Part QM* section 2.4) is welcome to adapt it for this classical application. Another possible starting point is the van der Pol approximation discussed in section 5.3, which should be translated from the time domain to the space domain.

Problem 6.16.* Use the van der Pol approximation to analyze the mutual phase-locking of two weakly coupled self-oscillators with the dissipative nonlinearity, for the cases of:

(i) direct coordinate coupling, described by Eq. (6.5), and

(ii) linear but otherwise arbitrary coupling of two similar oscillators.

Hint: In task (ii), describe the coupling by a linear operator, and express the result via its Fourier image.

Problem 6.17.* Extend the second task of the previous problem to the mutual phase-locking of N similar self-oscillators. In particular, explore the in-phase mode's stability for the case of the so-called *global coupling* via a single force F contributed equally by all oscillators.

Problem 6.18.* Find the condition of non-degenerate parametric excitation in a system of two coupled oscillators, described by Eq. (6.5), but with a time-dependent coupling: $\kappa \to \kappa(1 + \mu \cos \omega_p t)$, with $\omega_p \approx \Omega_1 + \Omega_2$ and $\Omega_2 - \Omega_1 \gg k/m$.

Hint: Assuming the modulation depth μ, static coupling κ, and detuning $\xi \equiv \omega_p - (\Omega_1 + \Omega_2)$ sufficiently small, use the van der Pol approximation for each of the coupled oscillators.

Problem 6.19. Show that the cubic nonlinearity of the type αq^3 indeed enables the parametric interaction ('four-wave mixing') of oscillations with incommensurate frequencies related by Eq. (6.92*a*).

Problem 6.20. Calculate the velocity of the transverse waves propagating on a thin, plane, elastic membrane, with mass $\mathscr{m}$ per unit area, pre-stretched with force τ per unit width.

References

[1] Bloembergen N 1996 *Nonlinear Optics* 4th edn (Singapore: World Scientific)

[2] Boyd R W 2008 *Nonlinear Optics* 3rd edn (New York: Academic)

[3] Zeng X *et al* 2017 *Opt. Lett.* **42** 2014

[4] Zeldovich Ya B 2002 *Physics of Shock Waves and High-Temperature Phenomena* (New York: Dover)

Chapter 7

Deformations and elasticity

The objective of this chapter is a brief discussion of small deformations of 3D continua, with a focus on the elastic properties of solids. The reader will see that the deformations of solids are non-trivial even in the absence of their time evolution, so that several key problems of statics will need to be discussed before proceeding to such dynamic phenomena as elastic waves in infinite media and thin rods.

7.1 Strain

As was discussed in the previous chapter, in *continua*, i.e. systems of particles so close to each other that the system discreteness may be neglected, the particle displacement $\mathbf{q}$ may be considered as a continuous function of space and time. The subject of this chapter is small deviations from the rigid-body approximation discussed in chapter 4, i.e. small *deformations*. The deformation smallness allows one to consider the displacement vector $\mathbf{q}$ as a function of the initial (pre-deformation) position of the particle, $\mathbf{r}$, and time t—just as was done in chapter 6 for 1D waves.

The first task of deformation theory is to exclude from consideration the types of motion considered in chapter 4, namely the translation and rotation unrelated to deformations. This means, first of all, that the variables describing deformations should not depend on the part of displacement distribution, which does not depend on position $\mathbf{r}$ (i.e. is common for the whole medium), because that part corresponds to a translational shift rather than to a deformation—see figure 7.1a. Moreover, even certain non-uniform displacements do not contribute to deformation. For example, Eq. (4.9) (with $d\mathbf{r}$ replaced with $d\mathbf{q}$ to comply with our current notation) shows that a small displacement of the type

$$d\mathbf{q}|_{\text{rotation}} = d\boldsymbol{\varphi} \times \mathbf{r}, \tag{7.1}$$

where $d\boldsymbol{\varphi} = \boldsymbol{\omega} dt$ is an infinitesimal vector common for the whole continuum, corresponds to its rotation about the direction of that vector, and has nothing to do with the body deformation—see figure 7.1b.

doi:10.1088/978-0-7503-1398-8ch7

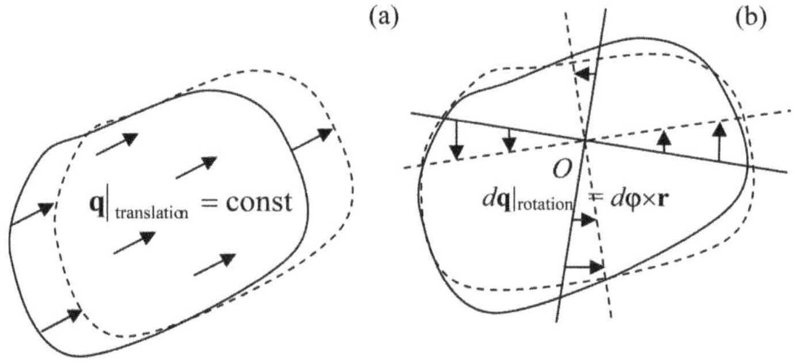

Figure 7.1. Two types of displacement vector distributions that are unrelated to deformation: (a) translation and (b) rotation.

This is why, in order to develop an adequate quantitative characterization of deformation, we should start by finding suitable functions of the spatial distribution of displacements, $\mathbf{q}(\mathbf{r})$, that exist only due to deformations. One such measure is the change of distance $dl = |d\mathbf{r}|$ between two close points:

$$(dl)^2|_{\text{after deformation}} - (dl)^2|_{\text{before deformation}} = \sum_{j=1}^{3}(dr_j + dq_j)^2 - \sum_{j=1}^{3}(dr_j)^2, \tag{7.2}$$

where dq_j is the jth Cartesian component of the difference $d\mathbf{q}$ between the displacements $\mathbf{q}$ of the points. If the deformation is small in the sense $|d\mathbf{q}| \ll |d\mathbf{r}| = dl$, we may keep in Eq. (7.2) only the terms proportional to the first power of the infinitesimal vector $d\mathbf{q}$:

$$\begin{aligned}(dl)^2|_{\text{after deformation}} - (dl)^2|_{\text{before deformation}} &= \sum_{j=1}^{3}[2dr_j dq_j + (dq_j)^2] \\ &\approx 2\sum_{j=1}^{3} dr_j dq_j.\end{aligned} \tag{7.3}$$

Since q_j is a function of three independent scalar arguments r_j, its full differential (at fixed time) may be represented as

$$dq_j = \sum_{j'=1}^{3} \frac{\partial q_j}{\partial r_{j'}} dr_{j'}. \tag{7.4}$$

Coefficients $\partial q_j/\partial r_{j'}$ may be considered as elements of a tensor[1] providing a linear relation between vectors $d\mathbf{r}$ and $d\mathbf{q}$. Plugging Eq. (7.4) into Eq. (7.2), we obtain

$$(dl)^2|_{\text{after deformation}} - (dl)^2|_{\text{before deformation}} = 2\sum_{j,j'=1}^{3} \frac{\partial q_j}{\partial r_{j'}} dr_j dr_{j'}. \tag{7.5}$$

[1] Since both $d\mathbf{q}$ and $d\mathbf{r}$ are legitimate physical vectors (whose Cartesian components are properly transformed as the transfer between reference frames), the 3×3 matrix with elements $\partial q_j/\partial r_{j'}$ is indeed a legitimate physical tensor—see the discussion in section 4.2.

A convenience of the tensor $\partial q_j/\partial r_{j'}$ for characterizing deformations is that it automatically excludes the translation displacement (figure 7.1a), which is independent of r_j. Its drawback is that its particular components are still affected by the rotation of the body (though the sum (7.5) is not). Indeed, according to the vector product definition, Eq. (7.1) may be represented in Cartesian coordinates as

$$dq_j|_{\text{rotation}} = (d\varphi_{j'}r_{j''} - d\varphi_{j''}r_{j'})\,\varepsilon_{jj'j''}, \tag{7.6}$$

where $\varepsilon_{jj'j''}$ is the *Levi–Civita symbol*[2]. Differentiating Eq. (7.6) over a particular Cartesian coordinate of vector $\mathbf{r}$, and taking into account that this partial differentiation (∂) is independent of (and hence may be swapped with) the differentiation (d) over the rotation angle φ, we obtain the amounts

$$d\left(\frac{\partial q_j}{\partial r_{j'}}\right)_{\text{rotation}} = -\varepsilon_{jj'j''}d\varphi_{j''}, \quad d\left(\frac{\partial q_{j'}}{\partial r_j}\right)_{\text{rotation}} = -\varepsilon_{j'jj''}d\varphi_{j''} = \varepsilon_{jj'j''}d\varphi_{j''}, \tag{7.7}$$

which may differ from 0. However, notice that the *sum* of these two differentials equals zero for any $d\boldsymbol{\varphi}$, which is possible only if[3]

$$\left(\frac{\partial q_{j'}}{\partial r_j} + \frac{\partial q_j}{\partial r_{j'}}\right)_{\text{rotation}} = 0, \quad \text{for } j \neq j'. \tag{7.8}$$

This is why it is convenient to rewrite Eq. (7.5) in a mathematically equivalent form,

$$(dl)^2|_{\text{after deformation}} - (dl)^2|_{\text{before deformation}} = 2\sum_{j,j'=1}^{3} s_{jj'}dr_j dr_{j'}, \tag{7.9a}$$

where $s_{jj'}$ are the elements of the so-called *symmetrized strain tensor* defined as

$$s_{jj'} \equiv \frac{1}{2}\left(\frac{\partial q_j}{\partial r_{j'}} + \frac{\partial q_{j'}}{\partial r_j}\right). \tag{7.9b}$$

(Note that this modification does not affect the diagonal elements: $s_{jj} = \partial q_j/\partial r_j$.) The advantage of the symmetrized tensor (7.9*b*) over the initial tensor $\partial q_j/\partial r_{j'}$ is that according to Eq. (7.8), at pure rotation all elements of the symmetrized strain tensor vanish.

Now let us discuss the physical meaning of this tensor. At was already mentioned in section 4.2, any symmetric tensor may be diagonalized by an appropriate selection of the reference-frame axes. In such principal axes, $s_{jj'} = s_{jj}\delta_{jj'}$, so that Eq. (7.4) takes the simple form

$$dq_j = \frac{\partial q_j}{\partial r_j}dr_j = s_{jj}dr_j. \tag{7.10}$$

---

[2] See, e.g. Eq. (A.83).

[3] As a result, the full sum (7.5), that includes three partial sums (7.8), is not affected by rotation—as we already know.

We may use this expression to calculate the change of each side of an elementary cuboid (parallelepiped) with sides dq_j parallel to the principal axes:

$$dr_j|_{\text{after deformation}} - dr_j|_{\text{before deformation}} \equiv dq_j = s_{jj}dr_j, \tag{7.11}$$

and of the cuboid's volume $dV = dr_1 dr_2 dr_3$:

$$\begin{aligned} dV|_{\text{after deformation}} - dV|_{\text{before deformation}} &= \prod_{j=1}^{3}(dr_j + s_{jj}dr_j) - \prod_{j=1}^{3} dr_j \\ &= dV\left[\prod_{j=1}^{3}(1 + s_{jj}) - 1\right], \end{aligned} \tag{7.12}$$

Since all our analysis is only valid in the linear approximation in small $s_{jj'}$, Eq. (7.12) is reduced to

$$dV|_{\text{after deformation}} - dV|_{\text{before deformation}} \approx dV \sum_{j=1}^{3} s_{jj} \equiv dV \, \text{Tr(s)}, \tag{7.13}$$

where Tr (*trace*)[4] of any matrix (in particular, any tensor) is the sum of its diagonal elements; in our case[5]

$$\text{Tr (s)} \equiv \sum_{j=1}^{3} s_{jj}. \tag{7.14}$$

So, the diagonal components of the tensor characterize the medium's compression/extension; then what is the meaning of its off-diagonal components? It may be illustrated on the simplest example of a purely *shear deformation*, shown in figure 7.2 (the geometry is assumed to be uniform along axis z, normal to the plane of drawing). In this case, all displacements (assumed small) have just one Cartesian component, in figure 7.2 along axis x, $\mathbf{q} = \mathbf{n}_x \alpha y$ (with $\alpha \ll 1$), so that the only non-vanishing component of the initial strain tensor $\partial q_j/\partial r_{j'}$ is $\partial q_x/\partial y = \alpha$, and the symmetrized tensor (7.9*b*) is

$$\text{s} = \begin{pmatrix} 0 & \alpha/2 & 0 \\ \alpha/2 & 0 & 0 \\ 0 & 0 & 0 \end{pmatrix}. \tag{7.15}$$

Evidently, the change of volume, given by Eq. (7.13), vanishes in this case. Thus, off-diagonal elements of the tensor characterize shear deformations.

To conclude this section, let me note that Eq. (7.9) is only valid in Cartesian coordinates. For the solution of some important problems with the axial or spherical symmetry, it is frequently convenient to express six different components of the symmetric strain tensor via three components of the displacement vector $\mathbf{q}$ in either cylindrical or spherical coordinates. A straightforward differentiation of the

[4] The traditional European notation for Tr is Sp (from the German *Spur* meaning 'trace' or 'track').

[5] Actually, the tensor theory shows that the trace does not depend on the particular choice of the coordinate axes.

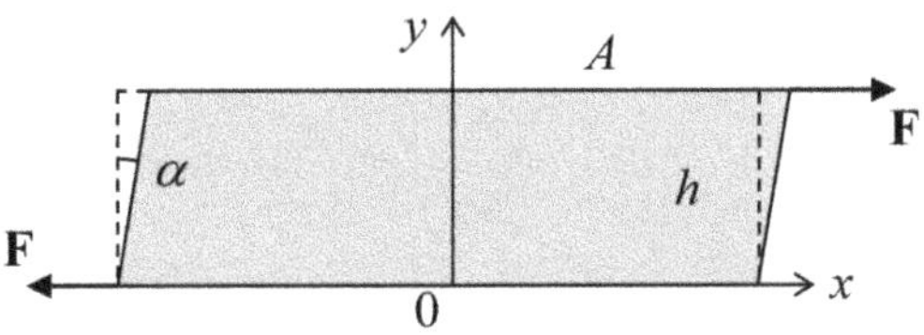

Figure 7.2. Example of a pure shear.

definitions of these curvilinear coordinates, similar to that used to derive the well-known expressions for spatial derivatives[6], yields, in particular, the following formulas for the diagonal elements of the tensor:

(i) in the cylindrical coordinates:

$$s_{\rho\rho} = \frac{\partial q_\rho}{\partial \rho}, \qquad s_{\varphi\varphi} = \frac{1}{\rho}\left(q_r + \frac{\partial q_\varphi}{\partial \varphi}\right), \qquad s_{zz} = \frac{\partial q_z}{\partial z}. \tag{7.16}$$

(ii) in the spherical coordinates:

$$s_{rr} = \frac{\partial q_r}{\partial r}, \quad s_{\theta\theta} = \frac{1}{r}\left(q_r + \frac{\partial q_\theta}{\partial \theta}\right), \quad s_{\varphi\varphi} = \frac{1}{r}\left(q_r + q_\theta \frac{\cos\theta}{\sin\theta} + \frac{1}{\sin\theta}\frac{\partial q_\varphi}{\partial \varphi}\right). \tag{7.17}$$

These expressions, which will be used below for the solution of some problems involving symmetrical geometries, may be a bit counter-intuitive. Indeed, Eq. (7.16) shows that even for a purely radial, axially symmetric deformation, $\mathbf{q} = \mathbf{n}_\rho q(\rho)$, the angular component of the strain tensor does not vanish: $s_{\varphi\varphi} = q/\rho$. (According to Eq. (7.17), in the spherical coordinates both angular components of the tensor exhibit the same property.) Note, however, that this relation describes a very simple geometric effect: the change of the lateral distance $\rho d\varphi \ll \rho$ between two close points with the same distance from the symmetry axis, at a small change of ρ that keeps the angle $d\varphi$ between the directions towards these two points constant.

7.2 Stress

Now let us discuss the forces that cause the strain—or, from an alternative point of view, are caused by the strain. Internal forces acting inside (i.e. between arbitrarily defined parts of) a continuum may be also characterized by a tensor. This *stress tensor*[7], with elements $\sigma_{jj'}$, relates the Cartesian components of the vector $d\mathbf{F}$ of the force acting on an elementary area dA of an (imaginary) interface between two parts of a continuum, with the components of the elementary vector $d\mathbf{A} = \mathbf{n}\, dA$ normal to the area—see figure 7.3:

$$dF_j = \sum_{j'=1}^{3} \sigma_{jj'}\, dA_{j'}. \tag{7.18}$$

[6] See, e.g. Eqs. (A.59)–(A.70).

[7] It is frequently called the *Cauchy stress tensor*, partly to honor A-L Cauchy (1789–1857) who introduced this notion, and partly to distinguish it from other possible definitions of the stress tensor, including the first and second *Piola–Kirchhoff tensors*. For the small deformations discussed in this course, all these notions coincide.

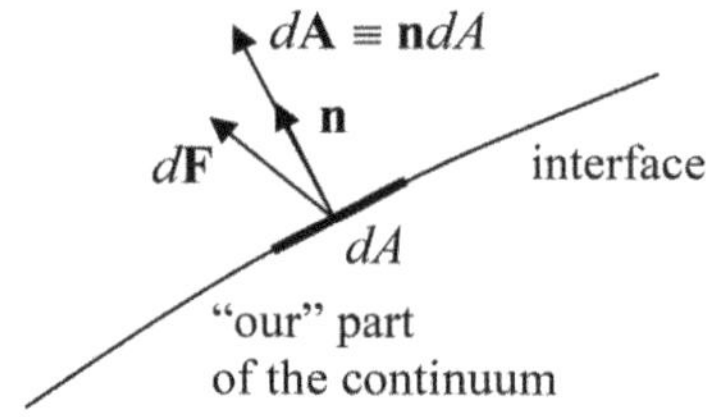

Figure 7.3. The definition of vectors $d\mathbf{A}$ and $d\mathbf{F}$.

The usual sign convention here is to take the *outer* normal $d\mathbf{n}$, i.e. to direct $d\mathbf{A}$ out of 'our' part of the continuum—the part on which the calculated force $d\mathbf{F}$ is exerted, by its complementary part.

In some cases the stress tensor's structure is very simple. For example, as will be discussed in detail in the next chapter, static and ideal *fluids* (i.e. liquids and gases) may only provide a force normal to any interface, and usually directed toward 'our' part of the body, so that

$$d\mathbf{F} = -\mathcal{P}d\mathbf{A}, \quad \text{i.e. } \sigma_{jj'} = -\mathcal{P}\delta_{jj'}, \tag{7.19}$$

where scalar $\mathcal{P}$ (in most cases positive) is called *pressure*, and generally may depend on both the spatial position and time. This type of stress, with $\mathcal{P} > 0$, is frequently called the *hydrostatic compression*—even if it takes place in solids.

However, in the general case the stress tensor also has off-diagonal terms, which characterize the shear stress. For example, if the shear strain shown in figure 7.2 is caused by a pair of forces $\pm\mathbf{F}$, they create internal forces $F_x\mathbf{n}_x$, with $F_x > 0$ if we speak about the force acting upon a part of the sample below the imaginary horizontal interface we are discussing. In order to avoid horizontal acceleration of each horizontal slice of the sample, the forces should not depend on y, i.e. $F_x =$ const $= F$. Superficially, it may look like the only non-vanishing component of the stress tensor is $dF_x/dA_y = F/A =$ const, so that tensor is asymmetric, in contrast to the strain tensor (7.15) of the same system. Note, however, that the pair of forces $\pm\mathbf{F}$ creates not only the shear stress, but also a non-vanishing rotating torque $\boldsymbol{\tau} = -Fh\mathbf{n}_z = -(dF_x/dA_y)Ah\mathbf{n}_z = -(dF_x/dA_y)V\mathbf{n}_z$, where $V = Ah$ is the sample's volume. So, if we want to perform a static stress experiment, i.e. avoid sample's rotation, we need to apply some other forces, e.g. a pair of vertical forces creating an equal and opposite torque $\boldsymbol{\tau}' = (dF_y/dA_x)V\mathbf{n}_z$, implying that $dF_y/dA_x = dF_x/dA_y = F/A$. As a result, the stress tensor becomes symmetric, and similar in structure to the symmetrized strain tensor (7.15):

$$\sigma = \begin{pmatrix} 0 & F_0/A & 0 \\ F_0/A & 0 & 0 \\ 0 & 0 & 0 \end{pmatrix}. \tag{7.20}$$

In many situations, the body may be stressed not only by forces applied to their surfaces, but also by some volume-distributed (*bulk*) forces $d\mathbf{F} = \mathbf{f}\,dV$, with a certain effective *bulk density* $\mathbf{f}$. (The most evident example of such forces is gravity. If its

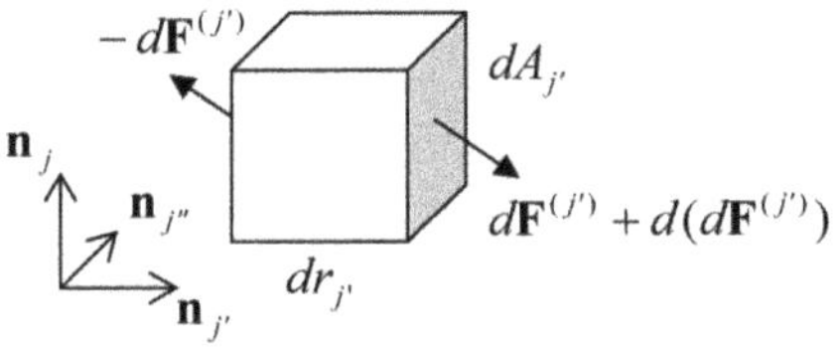

Figure 7.4. Deriving Eq. (7.23).

field is uniform as described by Eq. (1.16), then $\mathbf{f} = \rho\mathbf{g}$, where ρ is the mass density.) Let us derive the key formula describing the correct summation of the interface and bulk forces. For that, consider again an elementary cuboid with sides dr_j parallel to the corresponding coordinates axes (figure 7.4)—now not necessarily the principal axes of the stress tensor.

If elements $\sigma_{jj'}$ of the tensor do not depend on position, the force $d\mathbf{F}^{(j')}$ acting on the j'th face of the cuboid is exactly balanced by the equal and opposite force acting on its opposite face, because the vectors $d\mathbf{A}^{(j')}$ of these faces are equal and opposite. However, if $\sigma_{jj'}$ is a function of $\mathbf{r}$, then the net force $d(d\mathbf{F}^{(j')})$ does not vanish. (In this expression, the first differential sign refers to the elementary shift $dr_{j'}$, while the second one, to the elementary area $dA_{j'}$.) Using the expression $\sigma_{jj'}dA_{j'}$ for the j'th contribution to the sum (7.18), in the first order in $d\mathbf{r}$, the jth components of the vector $d(d\mathbf{F}^{(j')})$ are

$$d\left(dF_j^{(j')}\right) = d(\sigma_{jj'}dA_{j'}) = \frac{\partial \sigma_{jj'}}{\partial r_{j'}}dr_{j'}dA_{j'} \equiv \frac{\partial \sigma_{jj'}}{\partial r_{j'}}dV, \tag{7.21}$$

where the cuboid's volume $dV = dr_{j'}dA_{j'}$ evidently does not depend on the index j'. The addition of these force components for all three pairs of cuboid faces, i.e. the summation of Eq. (7.21) for all three values of the upper index j', yields the following relation for the jth Cartesian component of the net force exerted on the cuboid:

$$d(dF_j) = \sum_{j'=1}^{3} d\left(dF_j^{(j')}\right) = \sum_{j'=1}^{3} \frac{\partial \sigma_{jj'}}{\partial r_{j'}}dV. \tag{7.22}$$

Since any volume may be broken into such infinitesimal cuboids, Eq. (7.22) shows that the space-varying stress is equivalent to a volume-distributed force $d\mathbf{F}_{\text{ef}} = \mathbf{f}_{\text{ef}}\, dV$, whose *effective* (not real!) bulk density $\mathbf{f}_{\text{ef}}$ has the following Cartesian components

$$(f_{\text{ef}})_j = \sum_{j'=1}^{3} \frac{\partial \sigma_{jj'}}{\partial r_{j'}}, \tag{7.23}$$

so that in the presence of genuinely bulk forces $d\mathbf{F} = \mathbf{f}dV$, densities $\mathbf{f}_{\text{ef}}$ and $\mathbf{f}$ just add up.

Let us use this addition rule to spell out Newton's second law for a unit volume of a continuum:

$$\rho\frac{\partial^2 \mathbf{q}}{\partial t^2} = \mathbf{f}_{\text{ef}} + \mathbf{f}. \tag{7.24}$$

Using Eq. (7.23), the jth Cartesian component of Eq. (7.24) may be represented as

$$\rho \frac{\partial^2 q_j}{\partial t^2} = \sum_{j'=1}^{3} \frac{\partial \sigma_{jj'}}{\partial r_{j'}} + f_j. \tag{7.25}$$

This is the key equation of the continuum's dynamics (and statics), which will be repeatedly used below.

For the solution of some problems, it is also convenient to have a general expression for the work $\delta \mathscr{W}$ of the stress forces at a virtual deformation $\delta\mathbf{q}$—understood in the same variational sense as the virtual displacement $\delta\mathbf{r}$ in section 2.1. Using the equivalence between the stress forces and the effective bulk forces with density $\mathbf{f}_{\text{ef}}$, for any volume V of the medium we may write[8]

$$\delta \mathscr{W} = -\int_V \mathbf{f}_{\text{ef}} \cdot \delta\mathbf{q}\, d^3r = -\sum_{j=1}^{3} \int_V \left(f_{\text{ef}}\right)_j \delta q_j d^3r = -\sum_{j,j'=1}^{3} \int_V \frac{\partial \sigma_{jj'}}{\partial r_{j'}} \delta q_j d^3r. \tag{7.26}$$

Let us work out this integral by parts for a volume so large that the deformations δq_j on its surface are negligible. Then, swapping the operations of the variation and the spatial differentiation (just as was done with the time derivative in section 2.1), we obtain

$$\delta \mathscr{W} = \sum_{j,j'=1}^{3} \int_V \sigma_{jj'} \delta \frac{\partial q_j}{\partial r_{j'}} d^3r. \tag{7.27}$$

Assuming that the tensor $\sigma_{jj'}$ is symmetric, we may rewrite this expression as

$$\delta \mathscr{W} = \frac{1}{2} \sum_{j,j'=1}^{3} \int_V \left(\sigma_{jj'} \delta \frac{\partial q_j}{\partial r_{j'}} + \sigma_{j'j} \delta \frac{\partial q_j}{\partial r_{j'}} \right) d^3r. \tag{7.28}$$

Now, swapping indices j and j' in the second expression, we finally obtain

$$\delta \mathscr{W} = \frac{1}{2} \sum_{j,j'=1}^{3} \int_V \delta \left(\frac{\partial q_j}{\partial r_{j'}} \sigma_{jj'} + \frac{\partial q_{j'}}{\partial r_j} \sigma_{jj'} \right) d^3r = -\sum_{j,j'=1}^{3} \int_V \sigma_{jj'} \delta s_{jj'} d^3r, \tag{7.29}$$

where $s_{jj'}$ are the components of the strain tensor (7.9*b*). It is natural to rewrite this important formula as

$$\delta \mathscr{W} = \int_V \delta w(\mathbf{r}) d^3r, \quad \text{where} \quad \delta w(\mathbf{r}) \equiv \sum_{j,j'=1}^{3} \sigma_{jj'} \delta s_{j'j}, \tag{7.30}$$

and interpret the locally defined scalar function $\delta w(\mathbf{r})$ as the work of the stress forces per unit volume, at a small variation of the deformation.

---

[8] Here the sign corresponds to the work of the 'external' stress force $d\mathbf{F}$, exerted on 'our' part of the continuum by its counterpart—see figure 7.3. Note that some texts consider the alternative definition of $\delta \mathscr{W}$, leading to its opposite sign.

As a sanity check, for the pure pressure (7.19), Eq. (7.30) is reduced to the evident result $\delta\mathscr{W} = -\mathscr{P}\delta V$, where V is the volume of 'our' part of the continuum.

7.3 Hooke's law

In order to form a complete system of equations describing the continuum's dynamics, one needs to complement Eq. (7.25) with an appropriate *material equation* describing the relation between the forces, described by the stress tensor $\sigma_{jj'}$, and the deformations $\mathbf{q}$, described (in the small deformation limit) by the strain tensor $s_{jj'}$. This relation depends on the medium, and generally may be rather complex. Even leaving alone various anisotropic solids (e.g. crystals) and macroscopically inhomogeneous materials (such as ceramics or sand), strain typically depends not only on the current value of stress (possibly in a nonlinear way), but also on the previous history of stress application. Indeed, if strain exceeds a certain *plasticity threshold*, atoms (or nanocrystals) may slip to their new positions and never come back even if the strain is reduced. As a result, deformations become irreversible—see figure 7.5.

Only below the thresholds of nonlinearity and plasticity (which are typically close to each other), strain is nearly proportional to stress, i.e. obeys the famous *Hooke's law*[9]. However, even in this *elastic range* the law is not quite simple, and even for an isotropic medium is described not by one but by two constants, called the *elastic moduli*. The reason for that is that most elastic materials resist the strain accompanied by the volume change (say, the hydrostatic compression) differently from how they resist the shear deformation. In order to describe this difference, let us first represent the symmetrized strain tensor (7.9*b*) in the mathematically equivalent form

$$s_{jj'} = \left(s_{jj'} - \frac{1}{3}\delta_{jj'}\mathrm{Tr}(s)\right) + \left(\frac{1}{3}\delta_{jj'}\mathrm{Tr}(s)\right). \tag{7.31}$$

According to Eq. (7.13), the *traceless tensor* in the first parentheses of Eq. (7.31) does not give any contribution to the volume change, e.g. may be used to characterize a purely shear deformation, while the second term describes the hydrostatic compression alone. Hence we may expect that the stress tensor may be represented (again, in the elastic deformation range only!) as

$$\sigma_{jj'} = 2\mu\left(s_{jj'} - \frac{1}{3}\mathrm{Tr}(s)\delta_{jj'}\right) + 3K\left(\frac{1}{3}\mathrm{Tr}(s)\delta_{jj'}\right), \tag{7.32}$$

where K and μ are constants. (The inclusion of coefficients 2 and 3 into Eq. (7.32) is justified by the simplicity of some of its corollaries—see, e.g. Eqs. (7.36) and (7.41) below.) Indeed, experiments show that Hooke's law in this form is followed, at small strain, by all isotropic materials. In accordance with the above discussion, the constant μ (in some texts, denoted as G) is called the *shear modulus*, while the

[9] Named after R Hooke (1635–1703) who was the first to describe the law in its simplest, 1D version.

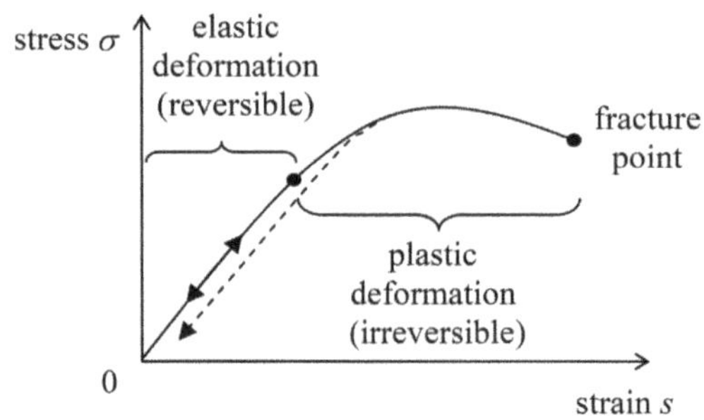

Figure 7.5. A typical relation between the stress and strain in solids (schematically).

constant K (sometimes called B), the *bulk modulus*. The two columns of table 7.1 show the approximate values of these moduli for typical representatives of several major classes of materials[10].

To better appreciate these values, let us first discuss the physical meanings of K and μ, using two simple examples of elastic deformation. In preparation for that, let us solve the set of nine (or rather six different) linear equations (7.32) for $s_{jj'}$. This is easy to do, due to the simple structure of these equations: they relate components $\sigma_{jj'}$ and $s_{jj'}$ with the same indices, besides the involvement of the tensor trace. This slight complication may be readily overcome by noticing that according to Eq. (7.32),

$$\text{Tr}(\sigma) \equiv \sum_{j=1}^{3} \sigma_{jj} = 3K \text{ Tr (s)}, \quad \text{i.e. Tr (s)} = \frac{1}{3K}\text{Tr}(\sigma). \tag{7.33}$$

Plugging this result into Eq. (7.32) and solving it for $s_{jj'}$, we readily get the reciprocal relation, which may be represented in a similar form:

$$s_{jj'} = \frac{1}{2\mu}\left(\sigma_{jj'} - \frac{1}{3}\text{Tr}(\sigma)\delta_{jj'}\right) + \frac{1}{3K}\left(\frac{1}{3}\text{Tr}(\sigma)\delta_{jj'}\right). \tag{7.34}$$

Now let us apply Hooke's law, in the form (7.32) or (7.34), to two simple situations in which the strain and stress tensors may be found without formulating the exact differential equations of the elasticity theory and boundary conditions for them. (That will be the subject of the next section.) The first experiment is the hydrostatic compression when the stress tensor is diagonal, and all its diagonal components are equal—see Eq. (7.19)[11]. For this case, Eq. (7.34) yields

$$s_{jj'} = -\frac{\mathcal{P}}{3K}\delta_{jj'}, \tag{7.35}$$

which means that regardless of the shear modulus, the strain tensor is also diagonal, with all diagonal components equal. According to Eqs. (7.11) and (7.13), this means

[10] Since the strain tensor elements, defined by Eq. (7.9), are dimensionless, while the strain, defined by Eq. (7.18), has the dimensionality of pressure (force per unit area), so do the elastic moduli K and μ.

[11] It may be proved that such a situation may be implemented not only in a fluid with pressure $\mathcal{P}$, but also by placing a solid sample of an *arbitrary* shape into a compressed fluid.

Table 7.1. Elastic moduli, density, and sound velocities of a few representative materials (approximate values).

Material	K (GPa)	μ (GPa)	E (GPa)	ν	ρ (kg m^{-3})	v_l (m s^{-1})	v_t (m s^{-1})
Diamond[a]	600	450	1100	0.20	3500	1830	1200
Hardened steel	170	75	200	0.30	7800	5870	3180
Water[b]	2.1	0	0	0.5	1000	1480	0
Air[b]	0.00010	0	0	0.5	1.2	332	0

[a]Averages over crystallographic directions (~10% anisotropy).
[b]At the so-called *ambient conditions* (T = 20 °C, $\mathcal{P}$ = 1 bar ≡ 10^5 Pa).

that all linear dimensions of the body are reduced by a similar fraction, so that its shape is preserved, while the volume is reduced by

$$\frac{\Delta V}{V} = \sum_{j=1}^{3} s_{jj} = -\frac{\mathcal{P}}{K}. \tag{7.36}$$

This equation clearly shows the physical sense of the bulk modulus K as the *reciprocal compressibility*.

As table 7.1 shows, the values of K may be dramatically different for various materials, and even for such 'soft stuff' as water this modulus in actually rather high. For example, even at the bottom of the deepest, 10 km ocean well ($\mathcal{P} \approx 10^3$ bar ≈ 0.1 GPa), the water's density increases by just about 5%. As a result, in most human-scale experiments, water may be treated as *incompressible*—a condition that will be widely used in the next chapter. Many solids are even much less compressible—see, for example, the first two rows of table 7.1.

The most compressible media are gases. For a gas, a certain background pressure $\mathcal{P}$ is necessary just for containing it within certain volume V, so that Eq. (7.36) is only valid for small increments of pressure, $\Delta\mathcal{P}$,

$$\frac{\Delta V}{V} = -\frac{\Delta \mathcal{P}}{K}. \tag{7.37}$$

Moreover, the compression of gases also depends on thermodynamic conditions. (In contrast, for most condensed media, the temperature effects are small.) For example, at ambient conditions most gases are reasonably well described by the equation of state for the model called the *ideal classical gas*:

$$\mathcal{P}V = Nk_BT, \quad \text{i.e. } \mathcal{P} = \frac{Nk_BT}{V}, \tag{7.38}$$

where N is the number of molecules in volume V, and $k_B \approx 1.38 \times 10^{-23}$ J K^{-1} is the Boltzmann constant[12]. For a small volume change ΔV at a constant temperature, this equation gives

$$\Delta\mathcal{P}|_{T=\text{const}} = -\frac{Nk_BT}{V^2}\Delta V = -\frac{\mathcal{P}}{V}\Delta V, \quad \text{i.e. } \frac{\Delta V}{V}|_{T=\text{const}} = -\frac{\Delta\mathcal{P}}{\mathcal{P}}. \tag{7.39}$$

[12] For the derivation and detailed discussion of Eq. (7.38) see, e.g. *Part SM* section 3.1.

Comparing this expression with Eq. (7.36), we obtain a remarkably simple result for the isothermal compression of gases,

$$K|_{T=\text{const}} = \mathcal{P}, \tag{7.40}$$

which means in particular that the bulk modulus listed in table 7.1 is actually valid, at the ambient conditions, for almost any gas. Note, however, that the change of thermodynamic conditions (say, from isothermal to adiabatic[13]) may affect the compressibility of the gas.

Now let us consider the second, rather different, fundamental experiment: a pure shear deformation shown in figure 7.2. Since the traces of matrices Eqs. (7.15) and (7.20), which describe this experiment, are equal to 0, for their off-diagonal elements, Eq. (7.32) gives merely $\sigma_{jj'} = 2\mu s_{jj'}$, so that the deformation angle α (see figure 7.2) is just

$$\alpha = \frac{1}{\mu}\frac{F}{A}. \tag{7.41}$$

Notice that the angle does not depend on the thickness h of the sample, though of course the maximal linear deformation $q_x = \alpha h$ is proportional to the thickness. Naturally, as table 7.1 shows, $\mu = 0$ for all fluids, because they do not resist static shear stress.

However, not all experiments, even the apparently simple ones, involve just either K or μ. Let us consider stretching a long and thin elastic rod of a uniform cross-section of area A—the so-called *tensile stress experiment* shown in figure 7.6.[14] Though the deformation of the rod near its clamped ends depends on the exact way forces $\mathbf{F}$ are applied (we will discuss this issue later on), we may expect that over most of its length the tension forces are directed virtually along the rod, $d\mathbf{F} = F_z\mathbf{n}_z$, and hence, with the coordinate choice shown in figure 7.6, $\sigma_{xj} = \sigma_{yj} = 0$ for all j, including the diagonal elements σ_{xx} and σ_{yy}. Moreover, due to the open lateral surfaces, on which, evidently, $dF_x = dF_y = 0$, there cannot be an internal stress force, of *any* direction, acting on any elementary internal boundary parallel to these surfaces. This means that $\sigma_{zx} = \sigma_{zy} = 0$. So, of all components of the stress tensor only one, σ_{zz}, is not equal to zero, and for a uniform sample, $\sigma_{zz} = \text{const} = F/A$. For this case, Eq. (7.34) shows that the strain tensor is also diagonal, but with different diagonal elements:

$$s_{zz} = \left(\frac{1}{9K} + \frac{1}{3\mu}\right)\sigma_{zz}, \tag{7.42}$$

$$s_{xx} = s_{yy} = \left(\frac{1}{9K} - \frac{1}{6\mu}\right)\sigma_{zz}. \tag{7.43}$$

[13] See, e.g. *Part SM* section 1.3.

[14] Although the analysis of compression in this situation gives similar results, in practical experiments a strong compression of a long sample may lead to the loss of horizontal stability—the so-called *buckling*—of the rod.

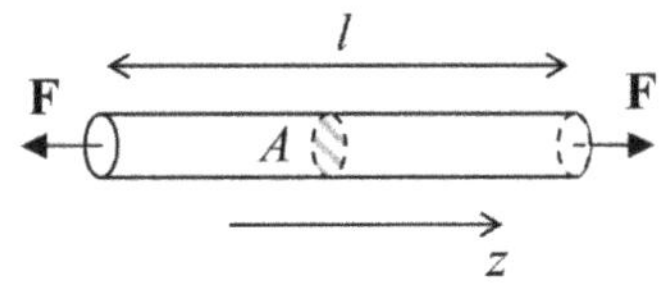

Figure 7.6. The tensile stress experiment.

Since the tensile stress is most common in the engineering (and physical experiment) practice, both combinations of the elastic moduli participating in these two relations have earned their own names. In particular, the constant in Eq. (7.42) is usually denoted as $1/E$ (but in many texts, as $1/Y$), where E is called the *Young's modulus*[15]:

$$\frac{1}{E} \equiv \frac{1}{9K} + \frac{1}{3\mu}, \quad \text{i.e. } E \equiv \frac{9K\mu}{3K + \mu}. \tag{7.44}$$

As figure 7.6 shows, in the tensile stress geometry $s_{zz} \equiv \partial q_z/\partial z = \Delta l/l$, so that the Young's modulus scales the linear relation between the relative extension of the rod and the force applied per unit area[16]:

$$\frac{\Delta l}{l} = \frac{1}{E}\frac{F}{A}. \tag{7.45}$$

The third column of table 7.1 shows the values of this modulus for two well-known solids: diamond (with the highest known value of E of all bulk materials[17]) and the steels (solid solutions of ~10% of carbon in iron) used in construction. Again, for all fluids the Young's modulus equals zero—as follows from Eq. (7.44) with $\mu = 0$.

I am confident that the reader of these notes is familiar with Eq. (7.42), in the form of Eq. (7.45), from his or her undergraduate studies. However, most probably this cannot be said about its counterpart Eq. (7.43), which shows that at the tensile stress, the rod's cross-section dimension also changes. This effect is usually characterized by the following dimensionless *Poisson's ratio*[18]:

$$-\frac{s_{xx}}{s_{zz}} = -\frac{s_{yy}}{s_{zz}} = -\left(\frac{1}{9K} - \frac{1}{6\mu}\right) \Big/ \left(\frac{1}{9K} + \frac{1}{3\mu}\right) = \frac{1}{2}\frac{3K - 2\mu}{3K + \mu} \equiv \nu, \tag{7.46}$$

According to this formula, for realistic materials with $K > 0$ and $\mu \geqslant 0$, values of ν may vary from (-1) to $(+1/2)$, but for the vast majority of materials[19] the values are

---

[15] Named after the polymath T Young (1773–1829)—somewhat unfairly, because his work on elasticity was predated by a theoretical analysis by L Euler in 1727 and detailed experiments by G Riccati in 1782.

[16] According to Eq. (7.45), E may be thought of as the force per unit area, which would double the initial sample's length, if only Hooke's law was valid for deformations that large—typically it is not.

[17] It is probably somewhat higher (up to 2000 GPa) in such nanostructures as carbon nanotubes and monatomic sheets (*graphene*), although there is still a substantial uncertainty in the experimental values of elastic moduli of these structures—see, e.g. [1, 2].

[18] In some older texts, the Poisson's ratio is denoted σ, but its notation as ν dominates the modern literature.

[19] The only known exceptions are certain exotic solids with very specific internal microstructures—see, e.g. [3] and references therein.

between 0 and 1/2—see table 7.1. The lower limit of this range is reached in porous materials such as cork, whose *lateral dimensions* almost do not change at the tensile stress. Some soft materials such as natural and synthetic rubbers represent the opposite case: $\nu \approx 1/2$.[20] Since according to Eqs. (7.13) and (7.42)–(7.46), the volume change is

$$\frac{\Delta V}{V} = s_{xx} + s_{yy} + s_{zz} = \frac{1}{E}\frac{F}{A}(1 - 2\nu) \equiv (1 - 2\nu)\frac{\Delta l}{l}, \tag{7.47}$$

such materials virtually do not change their *volume* at the tensile stress. The ultimate limit of this trend, $\Delta V/V = 0$, is provided by fluids and gases, because, as follows from Eq. (7.46) with $\mu = 0$, their Poisson ratio ν is exactly equal to 1/2. However, for most practicable construction materials such as steels (see table 7.1) the volume change (7.47) is as high as ~40% of that of the length.

Due to the clear physical sense of the coefficients E and ν, they are frequently used as a pair of independent elastic moduli, instead of K and μ. Solving Eqs. (7.44) and (7.46) for K and μ, we obtain

$$K = \frac{E}{3(1 - 2\nu)}, \qquad \mu = \frac{E}{2(1 + \nu)}. \tag{7.48}$$

Using these formulas, the two (equivalent) formulations of Hooke's law, expressed by Eqs. (7.32) and (7.34), may be rewritten as

$$\sigma_{jj'} = \frac{E}{1 + \nu}\left(s_{jj'} + \frac{\nu}{1 - 2\nu}\mathrm{Tr}(\mathrm{s})\,\delta_{jj'}\right), \tag{7.49a}$$

$$s_{jj'} = \frac{1 + \nu}{E}\left(\sigma_{jj'} - \frac{\nu}{1 + \nu}\,\mathrm{Tr}(\sigma)\,\delta_{jj'}\right). \tag{7.49b}$$

The linear relation between the strain and stress tensor in elastic continua enables the calculation of their potential energy U due to deformation. Indeed, to each infinitesimal part of this strain increase, we may apply Eq. (7.30), with the elementary work $\delta\mathscr{W}$ of the surface forces increasing the potential energy of 'our' part of the body by the equal amount δU. Let us slowly increase the deformation from a completely unstrained state (in which we may take $U = 0$) to a certain strained state, in the absence of bulk forces **f**, keeping the deformation type, i.e. the relation between the elements of the stress tensor, intact. In this case, all elements of the tensor $\sigma_{jj'}$ are proportional to the same single parameter characterizing the stress (say, the total applied force), and according to Hooke's law, all elements of the

[20] For example, *silicone rubbers* (synthetic polymers broadly used in engineering and physics experiment design) have, depending on their particular composition, synthesis, and thermal curing, $\nu = 0.47 \div 0.49$, and as a result combine respectable bulk moduli $K = (1.5 \div 2)$ GPa with a very low Young's modulus: $E = (0.0001 \div 0.05)$ GPa.

tensor $s_{j'j}$ are proportional to that parameter as well. In this case, integration over the variation yields the final value[21]

$$U = \int_V u(\mathbf{r}) d^3r, \quad u(\mathbf{r}) = \frac{1}{2} \sum_{j,j'=1}^{3} \sigma_{jj'} s_{j'j}. \tag{7.50}$$

Evidently, $u(\mathbf{r})$ may be interpreted as the volumic density of the potential energy of the elastic deformation.

7.4 Equilibrium

Now we are fully equipped to discuss the elastic deformation dynamics, but let us start with statics. The static (equilibrium) state may be described by requiring the right-hand side of Eq. (7.25) to equal zero. In order to find the elastic deformation, we need to plug $\sigma_{jj'}$ from the Hooke's law Eq. (7.49*a*), and then express elements $s_{jj'}$ via the displacement distribution—see Eq. (7.9). For a uniform material, the result is[22]

$$\frac{E}{2(1+\nu)} \sum_{j'=1}^{3} \frac{\partial^2 q_j}{\partial r_{j'}^2} + \frac{E}{2(1+\nu)(1-2\nu)} \sum_{j'=1}^{3} \frac{\partial^2 q_{j'}}{\partial r_j \partial r_{j'}} + f_j = 0. \tag{7.51}$$

Taking into account that the first sum in Eq. (7.51) is just the jth component of $\nabla^2 \mathbf{q}$, while the second sum is the jth component of $\nabla(\nabla \cdot \mathbf{q})$, we see that all three Eq. (7.51) for three Cartesian components ($j = 1, 2$, and 3) of the deformation vector $\mathbf{q}$, may be conveniently merged into one vector equation

$$\frac{E}{2(1+\nu)} \nabla^2 \mathbf{q} + \frac{E}{2(1+\nu)(1-2\nu)} \nabla(\nabla \cdot \mathbf{q}) + \mathbf{f} = 0. \tag{7.52}$$

For some applications, it is more convenient to recast this equation to a different form, using the well-known vector identity[23] $\nabla^2 \mathbf{q} = \nabla(\nabla \cdot \mathbf{q}) - \nabla \times (\nabla \times \mathbf{q})$. The result is

$$\frac{E(1-\nu)}{(1+\nu)(1-2\nu)} \nabla(\nabla \cdot \mathbf{q}) - \frac{E}{2(1+\nu)} \nabla \times (\nabla \times \mathbf{q}) + \mathbf{f} = 0. \tag{7.53}$$

It is interesting that in problems without volume-distributed forces ($\mathbf{f} = 0$), the Young's modulus E cancels out. Even more fascinating, in this case the equation may be re-written in a form not involving the Poisson ratio ν either. Indeed, acting by the del operator ∇ on the remaining terms of Eq. (7.53), we obtain a surprisingly simple equation

$$\nabla^2(\nabla \cdot \mathbf{q}) = 0. \tag{7.54}$$

---

[21] For clarity, let me reproduce this integration for the extension of a spring. In this case, $\delta U = \delta \mathcal{W} = F\delta x$, and if the spring's force is elastic, $F = \kappa x$, the integration yields $U = \kappa x^2/2 \equiv Fx/2$.

[22] As follows from Eq. (7.48), the coefficient before the first sum in Eq. (7.51) is just the shear modulus μ, while that before the second sum is equal to $(K + \mu/3)$.

[23] See, e.g. (A.73).

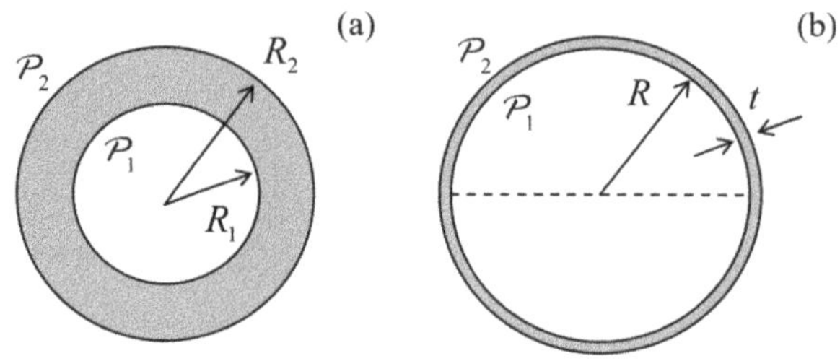

Figure 7.7. The spherical shell problem: (a) the general case and (b) the thin shell limit.

A natural question here is: how do the elastic moduli affect the deformation distribution if they do not participate in the differential equation describing it? The answer is two-fold. If what is fixed at the body's boundary are deformations, then the moduli are irrelevant, because the deformation distribution through the body does not depend on them. On the other hand, if the boundary conditions describe a fixed stress (or a combination of stress and strain), then the elastic constants creep into the solution via the recalculation of these conditions into the strain.

As a simple but representative example, let us calculate the deformation distribution in a (generally, thick) spherical shell under the different pressures inside and outside it—see figure 7.7a. Due to the spherical symmetry of the problem, the deformation is obviously spherically symmetric and radial, $\mathbf{q}(\mathbf{r}) = q(r)\mathbf{n}_r$, i.e. is completely described by one scalar function $q(r)$. Since the curl of such a radial vector field is zero[24], Eq. (7.53) is reduced to

$$\nabla(\nabla \cdot \mathbf{q}) = 0. \qquad (7.55)$$

This equation means that the divergence of the function $q(r)$ is constant within the shell. In spherical coordinates this means[25]

$$\frac{1}{r^2}\frac{d}{dr}(r^2 q) = \text{const.} \qquad (7.56)$$

Naming this constant $3a$ (with the numerical factor chosen for later notation convenience), and integrating Eq. (7.56) over r, we obtain its solution

$$q(r) = ar + \frac{b}{r^2}, \qquad (7.57)$$

[24] If this is not immediately evident, please have a look at Eq. (A.69) with $\mathbf{f} = f_r(r)\mathbf{n}_r$.

[25] See, e.g. Eq. (A.68) with $\mathbf{f} = q(r)\mathbf{n}_r$.

which also includes another integration constant, b. The constants a and b may be determined from the boundary conditions. According to Eq. (7.19),

$$\sigma_{rr} = \begin{cases} -\mathcal{P}_1, & \text{at } r = R_1, \\ -\mathcal{P}_2, & \text{at } r = R_2. \end{cases} \tag{7.58}$$

In order to relate this stress to strain, let us use Hooke's law, but for that we first need to calculate the strain tensor components for the deformation distribution Eq. (7.57). Using Eqs. (7.17), we obtain

$$s_{rr} = \frac{\partial q}{\partial r} = a - 2\frac{b}{r^3}, \qquad s_{\theta\theta} = s_{\varphi\varphi} = \frac{q}{r} = a + \frac{b}{r^3}, \tag{7.59}$$

so that $\text{Tr}(s) = 3a$. Plugging these relations into Eq. (7.49*a*) for σ_{rr}, we get

$$\sigma_{rr} = \frac{E}{1+\nu}\left[\left(a - 2\frac{b}{r^3}\right) + \frac{\nu}{1-2\nu}3a\right]. \tag{7.60}$$

Now plugging this relation into Eq. (7.58), we obtain a system of two linear equations for coefficients a and b. Solving this system, we obtain

$$a = \frac{1-2\nu}{E}\frac{\mathcal{P}_1 R_1^3 - \mathcal{P}_2 R_2^3}{R_2^3 - R_1^3}, \qquad b = \frac{1+\nu}{2E}\frac{(\mathcal{P}_1 - \mathcal{P}_2)R_1^3 R_2^3}{R_2^3 - R_1^3}. \tag{7.61}$$

Formulas Eqs. (7.57) and (7.61) give a complete solution of our problem. (Note that the elastic moduli are back, as was promised.) The result is rich in physical contents and deserves at least some analysis. First of all, note that according to Eq. (7.48), the coefficient $(1 - 2\nu)/E$ in the expression for a is just $1/3K$, so that the first term in Eq. (7.57) for the net deformation describes the hydrostatic compression. Now note that the second of Eqs. (7.61) yields $b = 0$ if $R_1 = 0$. Thus for a solid sphere we have only the hydrostatic compression, which was discussed in the previous section. Perhaps less intuitively, making two pressures equal also gives $b = 0$, i.e. the purely hydrostatic compression, for arbitrary $R_2 > R_1$.

However, in the general case $b \neq 0$, so that the second term in the deformation distribution Eq. (7.59), which describes the shear deformation[26], is also substantial. In particular, let us consider the important thin-shell limit, when $R_2 - R_1 \equiv t \ll R_{1,2} \equiv R$—see figure 7.7b. In this case, $q(R_1) \approx q(R_2)$ is just the change of the shell radius R, for which Eqs. (7.57) and (7.61) (with $R_2^3 - R_1^3 \approx 3R^2 t$) give

$$\begin{aligned} \Delta R \equiv q(R) \approx aR + \frac{b}{R^2} &\approx \frac{(\mathcal{P}_1 - \mathcal{P}_2)R^2}{3t}\left(\frac{1-2\nu}{E} + \frac{1+\nu}{2E}\right) \\ &= (\mathcal{P}_1 - \mathcal{P}_2)\frac{R^2}{t}\frac{1-\nu}{2E}. \end{aligned} \tag{7.62}$$

Naively, one could think that at least in this limit the problem could be analyzed by elementary means. For example, the total force exerted by the pressure

[26] Indeed, according to Eq. (7.48), the material-dependent factor in the second of Eq. (7.61) is just $1/4\mu$.

difference ($\mathcal{P}_1 - \mathcal{P}_2$) on the diametrical cross-section of the shell (see, e.g. the dashed line in figure 7.7b) is $F = \pi R^2(\mathcal{P}_1 - \mathcal{P}_2)$, giving the stress,

$$\sigma = \frac{F}{A} = \frac{\pi R^2(\mathcal{P}_1 - \mathcal{P}_2)}{2\pi Rt} = (\mathcal{P}_1 - \mathcal{P}_2)\frac{R}{2t}, \tag{7.63}$$

directed along the shell's walls. One can check that this simple formula may indeed be obtained, in this limit, from the strict expressions for $\sigma_{\theta\theta}$ and $\sigma_{\varphi\varphi}$ following from the general treatment carried out above. However, if we now tried to continue this approach by using the simple relation (7.45) to find the small change Rs_{zz} of the sphere's radius, we would arrive at a result with the structure of Eq. (7.62), but without the factor $(1 - \nu) < 1$ in the nominator. The reason for this error (which may be as significant as ~30% for typical construction materials—see table 7.1) is that Eq. (7.45), while being valid for *thin rods* of arbitrary cross-section, is invalid for *thin but broad sheets*, and in particular the thin shell in our problem. Indeed, while at the tensile stress both lateral dimensions of a thin rod may contract freely, in our problem all dimensions of the shell are under stress—actually, under much more tangential stress than radial stress[27].

7.5 Rod bending

The general approach to the static deformation analysis, outlined in the beginning of previous section, may be simplified not only for symmetric geometries, but also for uniform thin structures such as *thin plates* (also called 'membranes' or 'thin sheets') and *thin rods*. Due to the shortage of time, in this course I will demonstrate typical approaches to such systems only on the example of thin rods. (The theory of thin plates and shells is conceptually similar, but mathematically more complicated[28].)

In addition to the tensile stress analyzed in section 7.3, two other major types of rod deformation are *bending* and *torsion*. Let us start from a 'local' analysis of bending caused by a pair of equal and opposite external torques $\boldsymbol{\tau} = \pm\mathbf{n}_y\tau_y$ perpendicular to the rod axis z (figure 7.8), assuming that the rod is 'quasi-uniform', i.e. that on the longitudinal scale of this analysis (comparable with linear scale a of the cross-section) its material parameters and the cross-section A do not change substantially.

Just as in the tensile stress experiment (figure 7.6), the components of the stress forces $d\mathbf{F}$, normal to the rod length, have to equal zero on the surface of the rod. Repeating the arguments made for the tensile stress discussion, we arrive at the conclusion that only one diagonal component of the tensor (in figure 7.8, σ_{zz}) may differ from zero:

$$\sigma_{jj'} = \delta_{jz}\sigma_{zz}. \tag{7.64}$$

However, in contrast to the tensile stress, at pure static bending, the net force directed along the rod has to vanish:

$$F_z = \int_S \sigma_{zz} d^2r = 0, \tag{7.65}$$

[27] Strictly speaking, this is only true if the pressure difference is not too small, namely, if $|\mathcal{P}_1 - \mathcal{P}_2| \gg \mathcal{P}_{1,2}t/R$.

[28] For a review see, e.g. sections 11–15 in [4].

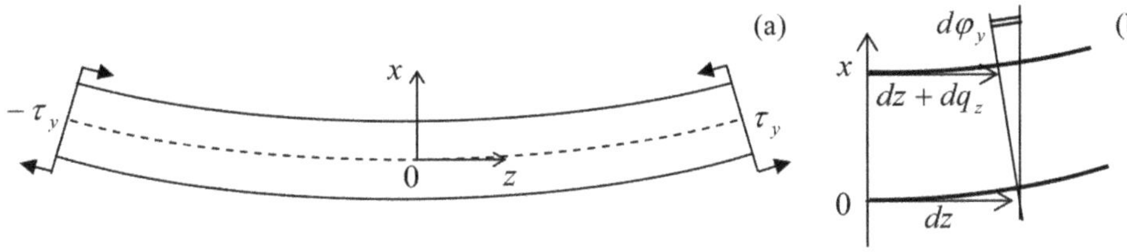

Figure 7.8. Rod bending, in a local reference frame (specific for each cross-section).

where S is the rod's cross-section, so that σ_{zz} has to change sign at some point of the axis x (in figure 7.8, selected to lie in the plane of the bent rod). Thus, the bending deformation may be viewed as a combination of stretching some layers of the rod (bottom layers in figure 7.8) with compression of other (top) layers.

Since it is difficult to make more immediate conclusions about the stress distribution, let us turn over to strain, assuming that the rod's cross-section is virtually constant on the length of our local analysis. From the above representation of bending as a combination of stretching and compression, it evident that the longitudinal deformation q_z has to vanish along some *neutral line* on the rod's cross-section—in figure 7.8, represented by the dashed line[29]. Selecting the origin of coordinate x on this line, and expanding the relative deformation in the Taylor series in x, due to the cross-section smallness, we may keep just the first, linear term:

$$s_{zz} \equiv \frac{dq_z}{dz} = -\frac{x}{R}. \tag{7.66}$$

The constant R has the sense of the *curvature radius* of the bent rod. Indeed, on a small segment dz the cross-section turns by a small angle $d\varphi_y = -\, dq_z/x$—see figure 7.8b. Using Eq. (7.66), we obtain $d\varphi_y = dz/R$, which is the usual definition of the curvature radius R in the differential geometry, for our special choice of the coordinate axes[30].

Expressions for other components of the strain tensor are more difficult to guess (as for the tensile stress, not all of them are equal to zero!), but what we already know about σ_{zz} and s_{zz} is already sufficient to start formal calculations. Indeed, plugging Eq. (7.64) into the Hooke's law in the form Eq. (7.49*b*), and comparing the result for s_{zz} with Eq. (7.66), we find

$$\sigma_{zz} = -E\frac{x}{R}. \tag{7.67}$$

From the same Eq. (7.49*b*), we could also find the transverse components of the strain tensor, conclude that they are related to s_{zz} exactly as at the tensile stress:

$$s_{xx} = s_{yy} = -\nu s_{zz}, \tag{7.68}$$

[29] Strictly speaking, that dashed line is the intersection of the *neutral surface* (the continuous set of such neutral lines for all cross-sections of the rod) with the plane of drawing.

[30] Indeed, for $(dx/dz)^2 \ll 1$, the general formula(A.24) for curvature (with the appropriate replacements $f \to x$ and $x \to z$) is reduced to $1/R = d^2x/dz^2 = d(dx/dz)/dz = d(\tan \varphi_y)/dz \approx d\varphi_y/dz$.

and then, integrating these relations along the cross-section of the rod, find the deformation of the cross-section's shape. More important for us, however, is the calculation of the relation between rod's curvature and the net torque acting on a given cross-section S (taking $dA_z > 0$):

$$\tau_y \equiv \int_S (\mathbf{r} \times d\mathbf{F})_y = -\int_S x\sigma_{zz} d^2r = \frac{E}{R}\int_S x^2 d^2r = \frac{EI_y}{R}, \tag{7.69}$$

where I_y is a geometric constant defined as

$$I_y \equiv \int_S x^2 dxdy. \tag{7.70}$$

Note that this factor, defining the bending rigidity of the rod, grows as fast as a^4 with the linear scale a of the cross-section[31].

In these expressions, x has to be counted from the neutral line. Let us see where exactly this line passes through the rod's cross-section. Plugging the result (7.67) into Eq. (7.65), we obtain the condition defining the neutral line:

$$\int_S xdxdy = 0. \tag{7.71}$$

This condition allows a simple interpretation. Imagine a thin sheet of some material, with a constant mass density σ per unit area, cut in the form of rod's cross-section. If we place a reference frame into its center-of-mass, then, by its definition,

$$\sigma\int_S \mathbf{r}\, dxdy = 0. \tag{7.72}$$

Comparing this condition with Eq. (7.71), we see that one of neutral lines has to pass through the center-of-mass of the sheet, which may be called the 'center-of-mass of the cross-section'.

Using the same analogy, we see that the integral I_y, given by Eq. (7.70), may be interpreted as the moment of inertia of the same imaginary sheet of material, with σ formally equal to 1, for its rotation about the neutral line—cf Eq. (4.24). This analogy is so convenient that the integral is usually called the *moment of inertia of the cross-section* and denoted similarly—just as has been done above. So, our basic result (7.69) may be re-written as

$$\frac{1}{R} = \frac{\tau_y}{EI_y}. \tag{7.73}$$

This relation is only valid if the deformation is small in the sense $R \gg a$. Still, since the deviations of the rod from its unstrained shape may accumulate along its length, Eq. (7.73) may be used for calculations of large 'global' deviations of the rod from equilibrium, on a length scale much larger than a. In order to describe such deformations, Eq. (7.73) has to be complemented by conditions of balance of the

[31] In particular, this is the reason why the usual electric wires are made not of a solid copper core, but rather a twisted set of thinner sub-wires, which may slip relative to each other, increasing wire flexibility.

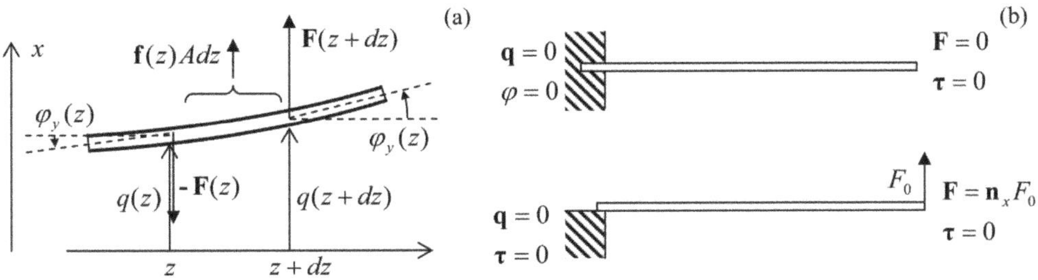

Figure 7.9. A global picture of rod bending: (a) the forces acting on a small fragment of a rod and (b) two bending problem examples, each with two typical but different boundary conditions.

bending forces and torques. Unfortunately, this requires a bit more differential geometry than I have time for, and I will only discuss this procedure for the simplest case of *relatively small* transverse deviations $q \equiv q_x$ of the rod from its initial straight shape, which will be used for axis z (figure 7.9a), for example by some bulk-distributed force $\mathbf{f} = \mathbf{n}_x f_x(z)$. (The simplest example is a uniform gravity field, for which $f_x = -\rho g = \text{const.}$) Note that in the forthcoming discussion the reference frame will be global, i.e. common for the whole rod, rather than local (pertaining to each cross-section) as in the previous analysis—cf figure 7.8.

First of all, we may write an evident differential equation for the average vertical force $\mathbf{F} = \mathbf{n}_x F_x(z)$ exerted on the part of the rod located to the left of its cross-section located at point z. This equation expresses the balance of vertical forces acting on a small fragment dz of the rod (figure 7.9a), necessary for the absence of its linear acceleration: $F_x(z + dz) - F_x(z) + f_x(z)Adz = 0$, giving

$$\frac{dF_x}{dz} = -f_x A, \qquad (7.74)$$

where A is the cross-section's area.

Note that this vertical component of the internal forces has been neglected at our derivation of Eq. (7.73), and hence our final results will be valid only if the ratio F_x/A is much smaller than the magnitude of σ_{zz} described by Eq. (7.67). However, these forces create the very torque $\boldsymbol{\tau} = \mathbf{n}_y \tau_y$ that causes the bending, and thus have to be taken into account at the analysis of the global picture. Such account may be made by writing the balance of torque components, acting on the same rod fragment of length dz, necessary for the absence of its angular acceleration:

$$\frac{d\tau_y}{dz} = -F_x. \qquad (7.75)$$

These two equations should be complemented by two geometric relations. The first of them is $d\varphi_y/dz = 1/R$, which has already been discussed above. We may immediately combine it with the basic result (7.73) of the local analysis, obtaining

$$\frac{d\varphi_y}{dz} = \frac{\tau_y}{EI_y}. \qquad (7.76)$$

The final equation is the geometric relation evident from figure 7.9a:

$$\frac{dq_x}{dz} = \varphi_y, \tag{7.77}$$

which is (as all expressions of our simple analysis) only valid for small bending angles, $|\varphi_y| \ll 1$.

The four differential equations Eqs. (7.74)–(7.77) are sufficient for the full solution of the weak bending problem, if complemented by appropriate boundary conditions. Figure 7.9b shows the most frequently met conditions. Let us solve, for example, the problem shown on the top panel of figure 7.9b: bending of a rod, 'clamped' at one end (say, immersed into a rigid wall), under its own weight. Considering, for the sake of simplicity, a uniform rod[32], we may integrate these equations one by one, each time using the appropriate boundary conditions. To start, Eq. (7.74), with $f_x = -\rho g$, yields

$$F_x = \rho g A z + \text{const} = \rho g A (z - l), \tag{7.78}$$

where the integration constant has been selected to satisfy the right-end boundary condition: $F_x = 0$ at $z = l$. As a sanity check, at the left wall ($z = 0$), $F_x = -\rho g A l = -mg$, meaning that the whole weight of the rod is exerted on the wall—fine.

Next, plugging Eq. (7.78) into Eq. (7.75) and integrating, we obtain

$$\tau_y = -\frac{\rho g A}{2}(z^2 - 2lz) + \text{const} = -\frac{\rho g A}{2}(z^2 - 2lz + l^2) = -\frac{\rho g A}{2}(z - l)^2, \tag{7.79}$$

where the integration constant's choice ensures the second right-boundary condition: $\tau_y = 0$ at $z = l$. Now proceeding in the same fashion to Eq. (7.76), we obtain

$$\varphi_y = -\frac{\rho g A}{2EI_y}\frac{(z - l)^3}{3} + \text{const} = -\frac{\rho g A}{6EI_y}[(z - l)^3 + l^3], \tag{7.80}$$

where the integration constant is selected to satisfy the *clamping condition* at the left end of the rod: $\varphi_y = 0$ at $z = 0$. (Note that this is different from the *support condition* illustrated on the lower panel of figure 7.9b, which allows the *angle* at $z = 0$ to be different from zero, but requires the *torque* to vanish.) Finally, integrating Eq. (7.77) with φ_y given by Eq. (7.80), we obtain the rod's global deformation law

$$q_x(z) = -\frac{\rho g A}{6EI_y}\left[\frac{(z - l)^4}{4} + l^3 z + \text{const}\right] = -\frac{\rho g A}{6EI_y}\left[\frac{(z - l)^4}{4} + l^3 z - \frac{l^4}{4}\right], \tag{7.81}$$

where the integration constant is selected to satisfy the second left-boundary condition: $q = 0$ at $z = 0$. So, the bending law is sort of complex even in this very

[32] As is clear from their derivation, Eqs. (7.74)–(7.77) are valid for any distribution of parameters A, E, I, and ρ over the rod's length, provided that the rod is *quasi-uniform*, i.e. its parameters' changes are so slow that the local relation (7.76) is still valid at any point.

simple problem. It is also remarkable how fast does the end's displacement grow with the increase of the rod's length:

$$q_x(l) = -\frac{\rho g A l^4}{8EI_y}. \tag{7.82}$$

To conclude the solution, let us discuss the validity of this result. First, the geometric relation (7.77) is only valid if $|\varphi_y(l)| \ll 1$, and hence if $|q_x(l)| \ll l$. Next, the local formula (7.76) is valid if $1/R = \tau(l)/EI_y \ll 1/a \sim A^{-1/2}$. Using the results Eqs. (7.79) and (7.82), we see that the latter condition is equivalent to $|q_x(l)| \ll l^2/a$, i.e. is weaker than the former one, because all our analysis has been based on the assumption that $l \gg a$. Another point of concern may be that the off-diagonal stress component $\sigma_{xz} \sim F_x/A$, that is created by the vertical gravity forces, has been ignored in our local analysis. For that approximation to be valid, this component must be much smaller than the diagonal component $\sigma_{zz} \sim aE/R = a\tau/I_y$ taken into account in that analysis. Using Eqs. (7.78) and (7.79), we obtain the following estimates: $\sigma_{xz} \sim \rho g l$, $\sigma_{zz} \sim a\rho g A l^2/I_y \sim a^3\rho g l^2/I_y$. According to its definition (7.70), I_y may be crudely estimated as a^4, so that we finally obtain the following simple condition: $a \ll l$, which has been assumed from the very beginning of our solution.

7.6 Rod torsion

One more class of analytically solvable elasticity problems is torsion of quasi-uniform, straight rods by a couple of axially oriented torques $\boldsymbol{\tau} = \mathbf{n}_z\tau_z$—see figure 7.10. This problem is simpler than the bending in the sense that due to its longitudinal uniformity, $d\varphi_z/dz = \text{const}$, it is sufficient to relate the torque τ_z to the *torsion parameter*

$$\kappa \equiv \frac{d\varphi_z}{dz}. \tag{7.83}$$

If the deformation is elastic and small (in the sense $\kappa a \ll 1$, where a is again the characteristic size of the rod's cross-section), κ is proportional to τ_z. Hence our task is to calculate their ratio,

$$C \equiv \frac{\tau_z}{\kappa} \equiv \frac{\tau_z}{d\varphi_z/dz}, \tag{7.84}$$

called the *torsional rigidity* of the rod.

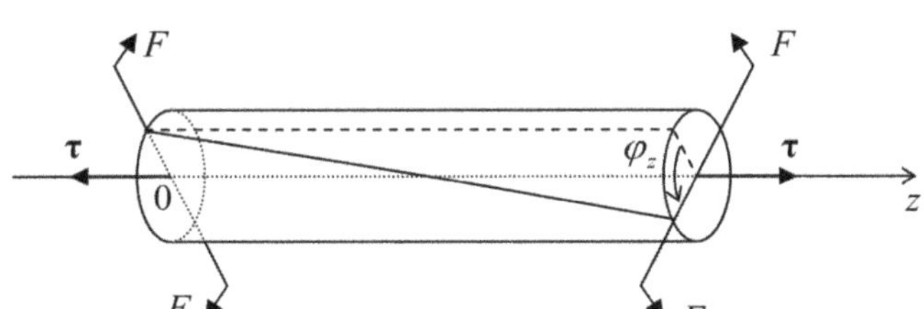

Figure 7.10. Rod torsion.

As a first guess (as we will see below, of limited validity), one may assume that the torsion does not change either the shape or size of the cross-section, but leads just to the mutual rotation of cross-sections about a certain central line. Using a reference frame with the origin on that line, this assumption immediately enables the calculation of components of the displacement vector $d\mathbf{q}$, by using Eq. (7.6) with $d\boldsymbol{\varphi} = \mathbf{n}_z d\varphi_z$:

$$dq_x = -yd\varphi_z = -\kappa ydz, \quad dq_y = xd\varphi_z = \kappa xdz, \quad dq_z = 0. \tag{7.85}$$

From here, we can calculate all Cartesian components (7.9) of the symmetrized strain tensor:

$$\begin{aligned} s_{xx} &= s_{yy} = s_{zz} = 0, \; s_{xy} = s_{yx} = 0, \\ s_{xz} &= s_{zx} = -\frac{\kappa}{2}y, \; s_{yz} = s_{zy} = \frac{\kappa}{2}x. \end{aligned} \tag{7.86}$$

The first of these equalities means that volume does not change, i.e. we are dealing with a pure shear deformation. As a result, all non-vanishing components of the stress tensor, calculated from Eq. (7.32), are proportional to the shear modulus alone:

$$\begin{aligned} \sigma_{xx} &= \sigma_{yy} = \sigma_{zz} = 0, \; \sigma_{xy} = \sigma_{yx} = 0, \\ \sigma_{xz} &= \sigma_{zx} = -\mu\kappa y, \; \sigma_{yz} = \sigma_{zy} = \mu\kappa x. \end{aligned} \tag{7.87}$$

(Note that for this problem, with a purely shear deformation, using alternative elastic moduli E and ν would be rather unnatural. If needed, we may always use the second of Eq. (7.48): $\mu = E/2(1 + \nu)$.)

Now it is straightforward to use this result to calculate the full torque as an integral over the cross-section's area A:

$$\tau_z \equiv \int_A (\mathbf{r} \times d\mathbf{F})_z = \int_A (xdF_y - ydF_x) = \int_A (x\sigma_{yz} - y\sigma_{xz})dxdy. \tag{7.88}$$

Using Eq. (7.87), we obtain $\tau_z = \mu\kappa I_z$, i.e.

$$C = \mu I_z, \quad \text{where} \quad I_z \equiv \int_A (x^2 + y^2)dxdy. \tag{7.89}$$

Again, just as in the case of thin rod bending, we have obtained an integral, I_z, similar to a moment of inertia, this time for the rotation about the axis z passing through a certain point of the cross-section. For any axially symmetric cross-section, this evidently should be the central point. Then, for example, for the practically important case of a round pipe with internal radius R_1 and external radius R_2, Eq. (7.89) yields

$$C = \mu \; 2\pi \int_{R_1}^{R_2} \rho^3 d\rho = \frac{\pi}{2}\mu(R_2^4 - R_1^4). \tag{7.90}$$

In particular, for the solid rod of radius R (which may be treated as a pipe with $R_1 = 0$ and $R_2 = R$), this gives torsional rigidity $C = (\pi/2)\mu R^4$, while for a hollow pipe of small thickness $t \ll R$, Eq. (7.90) is reduced to

$$C = 2\pi\mu R^3 t. \tag{7.91}$$

Note that per unit cross-section area A (and hence per unit mass) this rigidity is twice higher than that of a solid rod:

$$\frac{C}{A}|_{\text{thin round pipe}} = \mu R^2 > \frac{C}{A}|_{\text{solid round rod}} = \frac{1}{2}\mu R^2. \tag{7.92}$$

This fact is the basis of a broad use of thin pipes in construction and in physical experiment design.

However, for rods with axially asymmetric cross-sections, Eq. (7.89) gives *wrong* results. For example, for a narrow rectangle of area $A = w \times t$ with $t \ll w$, it yields $C = \mu t w^3/12$ (WRONG!), even functionally different from the correct result—cf Eq. (7.104) below. The reason for the failure of the above analysis is that it does not describe possible bending q_z of the rod's cross-section in the direction along the rod. (For axially symmetric rods, such bending is evidently forbidden by the symmetry, so that Eq. (7.89) is valid, and results (7.90)–(7.92) are absolutely correct.)

Let us describe[33] this, rather counter-intuitive effect by taking

$$q_z = \kappa\psi(x, y), \tag{7.93}$$

(where ψ is some function to be determined), but still keeping Eq. (7.85) for two other components of the displacement vector. The addition of ψ does not change the equality to zero of the diagonal components of the strain tensor, as well as of s_{xy} and s_{yx}, but contributes to other off-diagonal components:

$$s_{xz} = s_{zx} = \frac{\kappa}{2}\left(-y + \frac{\partial\psi}{\partial x}\right), \quad s_{yz} = s_{zy} = \frac{\kappa}{2}\left(x + \frac{\partial\psi}{\partial y}\right), \tag{7.94}$$

and hence to the corresponding elements of the stress tensor:

$$\sigma_{xz} = \sigma_{zx} = \mu\kappa\left(-y + \frac{\partial\psi}{\partial x}\right), \quad \sigma_{yz} = \sigma_{zy} = \mu\kappa\left(x + \frac{\partial\psi}{\partial y}\right). \tag{7.95}$$

Now let us find the requirement imposed on the function $\psi(x,y)$ by the fact that the stress force component parallel to the rod's axis,

$$dF_z = \sigma_{zx} dA_x + \sigma_{zy} dA_y = \mu\kappa dA\left[\left(-y + \frac{\partial\psi}{\partial x}\right)\frac{dA_x}{dA} + \left(x + \frac{\partial\psi}{\partial y}\right)\frac{dA_y}{dA}\right], \tag{7.96}$$

has to vanish at rod's surface(s), i.e. at each border of its cross-section. Coordinates $\{x, y\}$ of points at a border may be considered as unique functions, $x(l)$ and $y(l)$, of the arc l of that line—see figure 7.11.

[33] I would not be terribly shocked if reader skipped the balance of this section at the first reading. Although the following calculation is very elegant, instructive, and typical for the theory of elasticity, its results will not be used in other parts of this series.

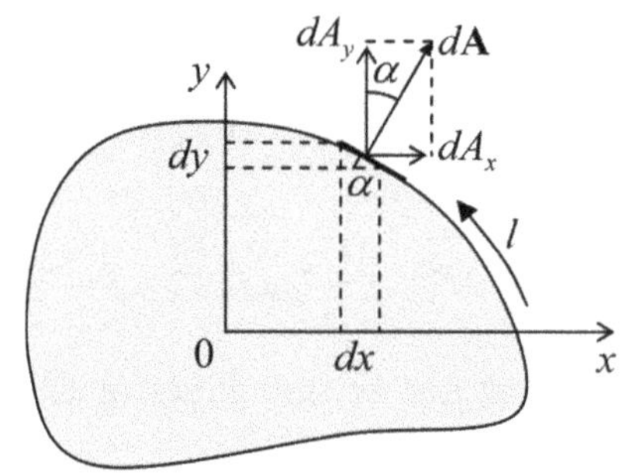

Figure 7.11. Deriving Eq. (7.99).

As this sketch shows, the elementary area ratios participating in Eq. (7.96) may be readily expressed via the derivatives of these functions: $dA_x/dA = \sin\alpha = dy/dl$, $dA_y/dA = \cos\alpha = -dx/dl$, so that we may write

$$\left[\left(-y + \frac{\partial\psi}{\partial x}\right)\left(\frac{dy}{dl}\right) + \left(x + \frac{\partial\psi}{\partial y}\right)\left(-\frac{dx}{dl}\right)\right]_{\text{border}} = 0. \tag{7.97}$$

Introducing, instead of ψ, a new function $\chi(x,y)$, defined by its derivatives as

$$\frac{\partial\chi}{\partial x} \equiv \frac{1}{2}\left(-x - \frac{\partial\psi}{\partial y}\right), \qquad \frac{\partial\chi}{\partial y} \equiv \frac{1}{2}\left(-y + \frac{\partial\psi}{\partial x}\right), \tag{7.98}$$

we may rewrite the condition (7.97) as

$$2\left(\frac{\partial\chi}{\partial y}\frac{dy}{dl} + \frac{\partial\chi}{\partial x}\frac{dx}{dl}\right)_{\text{border}} \equiv 2\frac{d\chi}{dl}\Big|_{\text{border}} = 0, \tag{7.99}$$

so that the function χ should be constant at each border of the cross-section.

In particular, for a singly connected cross-section, limited by just one continuous border line (as in figure 7.11), this constant is arbitrary, because according to Eq. (7.98), its choice does not affect the longitudinal deformation function $\psi(x,y)$ and hence the deformation as the whole. Now let us use the definition (7.98) of function χ to calculate the 2D Laplace operator of this function:

$$\nabla^2_{x,y}\chi \equiv \frac{\partial^2\chi}{\partial^2 x} + \frac{\partial^2\chi}{\partial^2 y} = \frac{1}{2}\frac{\partial}{\partial x}\left(-x - \frac{\partial\psi}{\partial y}\right) + \frac{1}{2}\frac{\partial}{\partial y}\left(-y + \frac{\partial\psi}{\partial y}\right) = -1. \tag{7.100}$$

This a 2D *Poisson equation* (frequently met, for example, in electrostatics), but with a very simple, constant right-hand side. Plugging Eq. (7.98) into Eq. (7.95), and those into Eq. (7.89), we may express the torque τ, and hence the torsional rigidity C, via the same function:

$$C \equiv \frac{\tau_z}{\kappa} = -2\mu\int_A\left(x\frac{\partial\chi}{\partial x} + y\frac{\partial\chi}{\partial y}\right)dxdy. \tag{7.101a}$$

Sometimes, it is easier to use this result in either of its two different forms. The first of them may be readily obtained from Eq. (7.101a) using integration by parts:

$$
\begin{aligned}
C &= -2\mu\left(\int dy \int x d\chi + \int dx \int y d\chi\right) \\
&= -2\mu\left[\int dy\left(x\chi_{\text{border}} - \int \chi dx\right) + \int dx\left(y\chi_{\text{border}} - \int \chi dy\right)\right] \\
&= 4\mu\left[\int_A \chi \, dxdy - \chi_{\text{border}} \int_A dxdy\right],
\end{aligned}
\qquad (7.101b)
$$

while the proof of one more form,

$$
C = 4\mu \int_A (\nabla_{x,y}\chi)^2 dxdy, \qquad (7.101c)
$$

is left for a reader's exercise.

Thus, if we need to know the rod's rigidity alone, it is sufficient to calculate the function $\chi(x,y)$ from Eq. (7.100) with the boundary condition $\chi|_{\text{border}} = \text{const}$, and then plug it into any of Eqs. (7.101). Only if we are also curious about the longitudinal deformation (7.93) of the cross-section, we may continue by using Eq. (7.98) to find the function $\psi(x,y)$ describing this deformation. Let us see how this general result works for the two examples discussed above. For the round cross-section of radius R, both the Poisson equation (7.100) and the boundary condition, $\chi = \text{const}$ at $x^2 + y^2 = R^2$, are evidently satisfied by the axially symmetric function

$$
\chi = -\frac{1}{4}(x^2 + y^2) + \text{const}. \qquad (7.102)
$$

For this case, either of Eq. (7.101) yields

$$
C = 4\mu \int_A \left[\left(-\frac{1}{2}x\right)^2 + \left(-\frac{1}{2}y\right)^2\right] dxdy = \mu \int_A (x^2 + y^2)\, d^2r, \qquad (7.103)
$$

i.e. the same result (7.89), which we had for $\psi = 0$. Indeed, plugging Eq. (7.102) into Eq. (7.98), we see that in this case $\partial\psi/\partial x = \partial\psi/\partial y = 0$, so that $\psi(x,y) = \text{const}$, i.e. the cross-section is not bent. (As was discussed in section 7.1, a uniform translation $q_z = \kappa\psi = \text{const}$ does not constitute a deformation.)

Now, turning to a rod with a narrow rectangular cross-section $A = w \times t$ with $t \ll w$, we may use this strong inequality to solve the Poisson equation (7.100) approximately, neglecting the second derivative of χ along the wider dimension (say, y). The remaining 1D differential equation $d^2\chi/d^2x = -1$, with boundary conditions $\chi|_{x=+t/2} = \chi|_{x=-t/2}$ has an evident solution $\chi = -x^2/2 + \text{const}$. Plugging this expression into any form of Eq. (7.101), we obtain the correct result for the torsional rigidity:

$$
C = \frac{1}{6}\mu w t^3. \qquad (7.104)
$$

Now let us have a look at the cross-section bending law (7.93) for this particular case. Using Eq. (7.98), we obtain

$$\frac{\partial \psi}{\partial y} = -x - 2\frac{\partial \chi}{\partial x} = x, \quad \frac{\partial \psi}{\partial x} = y + 2\frac{\partial \chi}{\partial y} = y. \tag{7.105}$$

Integrating these differential equations over the cross-section, and taking the integration constant (again, not contributing to the deformation) for zero, we obtain a beautifully simple result:

$$\psi = xy, \quad \text{i.e. } q_z = \kappa xy. \tag{7.106}$$

It means that the longitudinal deformation of the rod has a 'propeller bending' form: while the regions near the opposite corners (on the same diagonal) of the cross-section bend toward one direction of axis z, the corners on the other diagonal bend in the opposite direction. (This qualitative conclusion remains valid for rectangular cross-sections with any 'aspect ratio' t/w.)

For rods with several surfaces, i.e. with cross-sections limited by several boundaries (say, hollow pipes), the boundary conditions for function $\chi(x, y)$ requires a bit more care, and Eq. (7.101*b*) has to be modified, because the function may be equal to a different constant at each boundary. Let me leave the calculation of the torsional rigidity for this case for a reader's exercise.

7.7 3D acoustic waves

Now moving from the elastic statics to dynamics, we may start with Eq. (7.24), which may be transformed into the vector form exactly as this was done for the static case in the beginning of section 7.4. Comparing Eqs. (7.24) and (7.52), we immediately see that the result may be represented as

$$\rho\frac{\partial^2 \mathbf{q}}{\partial t^2} = \frac{E}{2(1+\nu)}\nabla^2\mathbf{q} + \frac{E}{2(1+\nu)(1-2\nu)}\nabla(\nabla\cdot\mathbf{q}) + \mathbf{f}(\mathbf{r}, t). \tag{7.107}$$

Let us use this general equation for the analysis of the perhaps most important type of time-dependent deformations: *acoustic waves*. First, let us consider the simplest case of a virtually infinite, uniform elastic medium, with no external forces **f**. In this case, due to the linearity and homogeneity of the resulting equation of motion, and taking clues from the analysis of the simple 1D model (see figure 6.4a) in sections 6.3–6.5[34], we may look for a particular time-dependent solution in the form of a sinusoidal, *linearly polarized, plane traveling wave*

$$\mathbf{q}(\mathbf{r}, t) = \text{Re}\,[\mathbf{a}e^{i(\mathbf{k}\cdot\mathbf{r}-\omega t)}], \tag{7.108}$$

where **a** is the constant complex amplitude of a wave (now a vector!), and **k** is the *wave vector*, whose magnitude is equal to the wave number k. The direction of these two vectors should be clearly distinguished: while **a** determines the wave's

[34] Note that Eq. (7.107) is more complex than the simple wave equation (6.40).

polarization, i.e. the direction of the particle displacements, the vector **k** is directed along the spatial gradient of the full phase of the wave

$$\Psi \equiv \mathbf{k} \cdot \mathbf{r} - \omega t + \arg a, \tag{7.109}$$

i.e. along the direction of the wave front propagation.

The importance of the angle between these two vectors may be readily seen from the following simple calculation. Let us point axis z of an (inertial) reference frame along the direction of vector **k,** and axis x in such direction that the vector **q**, and hence **a**, lie within the $\{x, z\}$ plane. In this case, all variables may change only along axis z, i.e. $\nabla = \mathbf{n}_z(\partial/\partial z)$, while the amplitude vector may be presented as the sum of just two Cartesian components:

$$\mathbf{a} = a_x\mathbf{n}_x + a_z\mathbf{n}_z. \tag{7.110}$$

Let us first consider a *longitudinal* wave, with the particle motion along the wave direction: $a_x = 0$, $a_z = a$. Then the vector **q** in Eq. (7.107), describing that wave, has only one (z-)component, so that $\nabla\cdot\mathbf{q} = dq_z/dz$ and $\nabla(\nabla\cdot\mathbf{q}) = \mathbf{n}_z(\partial^2\mathbf{q}/\partial z^2)$, and the Laplace operator gives the same expression: $\nabla^2\mathbf{q} = \mathbf{n}_z(\partial^2\mathbf{q}/\partial z^2)$. As a result, Eq. (7.107), with $\mathbf{f} = 0$, is reduced to a 1D acoustic wave equation

$$\rho\frac{\partial^2 q_z}{\partial t^2} = \left[\frac{E}{2(1+\nu)} + \frac{E}{2(1+\nu)(1-2\nu)}\right]\frac{\partial^2 q_z}{\partial z^2} = \frac{E(1-\nu)}{(1+\nu)(1-2\nu)}\frac{\partial^2 q_z}{\partial z^2}, \tag{7.111}$$

similar to Eq. (6.40). As we already know from section 6.4, this equation is indeed satisfied with the solution (7.108), provided that ω and k obey a linear dispersion relation, $\omega = v_l k$, with the following longitudinal wave velocity:

$$v_l^2 = \frac{E(1-\nu)}{(1+\nu)(1-2\nu)\rho} \equiv \frac{K + (4/3)\mu}{\rho}. \tag{7.112}$$

The last expression for v_l allows a simple interpretation. Let us consider a static experiment, similar to the tensile test experiment shown in figure 7.6, but with a sample much wider than l in both directions perpendicular to the force. Then the lateral contraction is impossible ($s_{xx} = s_{yy} = 0$), and we can calculate the only finite stress component, σ_{zz}, directly from Eq. (7.32) with $\mathrm{Tr(s)} = s_{zz}$:

$$\sigma_{zz} = 2\mu\left(s_{zz} - \frac{1}{3}s_{zz}\right) + 3K\left(\frac{1}{3}s_{zz}\right) = \left(K + \frac{4}{3}\mu\right)s_{zz}. \tag{7.113}$$

We see that the nominator in Eq. (7.112) is nothing more than the static elastic modulus for such a uniaxial deformation, and it is recalculated into the velocity exactly as the spring constant in the 1D waves considered in sections 6.3–6.4—cf Eq. (6.42).

Formula (7.112) becomes particularly simple in fluids, where $\mu = 0$, and the wave velocity is described by the well-known expression

$$v_l = \left(\frac{K}{\rho}\right)^{1/2}. \tag{7.114}$$

Note, however, that for gases, with their high compressibility and temperature sensitivity, the value of K participating in this formula may differ, at high frequencies, from that given by Eq. (7.40), because the fast compressions/extensions of gas are nearly adiabatic rather than isothermal. This difference is noticeable in table 7.1, one of whose columns lists the values of v_l for representative materials.

Now let us consider an opposite case of *transverse* waves with $a_x = a$, $a_z = 0$. In such a wave, the displacement vector is perpendicular to z, so that $\nabla\cdot\mathbf{q} = 0$, and the second term on the right-hand side of Eq. (7.107) vanishes. In contrast, the Laplace operator acting on such vector still gives the same non-zero contribution, $\nabla^2\mathbf{q} = n_z(\partial^2\mathbf{q}/\partial z^2)$, to Eq. (7.107), so that the equation yields

$$\rho\frac{\partial^2 q_x}{\partial t^2} = \frac{E}{2(1+\nu)}\frac{\partial^2 q_x}{\partial z^2}, \qquad (7.115)$$

and we again get the linear dispersion relation, $\omega = v_t k$, but with a different velocity:

$$v_t^2 = \frac{E}{2(1+\nu)\rho} = \frac{\mu}{\rho}. \qquad (7.116)$$

We see that the speed of transverse waves depends exclusively on the shear modulus μ of the medium[35]. This is also very natural: in such waves, the particle displacements $\mathbf{q} = \mathbf{n}_x q$ are perpendicular to the elastic forces $d\mathbf{F} = \mathbf{n}_z dF$, so that the only one component σ_{xz} of the stress tensor is involved. Also, the strain tensor $s_{jj'}$ has no diagonal components, Tr(s) = 0, so that μ is the only elastic modulus actively participating in Hooke's law (7.32). In particular, fluids cannot carry transverse waves at all (formally, their velocity (7.116) vanishes), because they do not resist shear deformations. For all other materials, the longitudinal waves are faster than the transverse ones[36]. Indeed, for all known natural materials the Poisson ratio is positive, so that the velocity ratio that follows from Eqs. (7.112) and (7.116),

$$\frac{v_l}{v_t} = \left(\frac{2-2\nu}{1-2\nu}\right)^{1/2}, \qquad (7.117)$$

is above $\sqrt{2} \approx 1.4$. For the most popular construction materials, with $\nu \approx 0.3$, the ratio is about 2—see table 7.1.

Let me emphasize again that for both the longitudinal and the transverse waves, the dispersion relation between the wave number and frequency is linear: $\omega = vk$. As was already discussed in chapter 6, in this case of *acoustic waves* (or just 'sound'), the phase and group velocities are equal, and waves of more complex form, consisting of several (or many) Fourier components of the type (7.108), preserve their form during

[35] Because of that, one can frequently meet the term *shear waves*. Note also that in contrast to the traversal waves in the simple 1D model analyzed in chapter 6 (see figure 6.4a), those in a 3D continuum do not need a pre-stretch tension $\mathscr{T}$. We will return to the effect of tension in the next section.

[36] Because of this difference between v_l and v_t, in geophysics the longitudinal waves are known as *P-waves* (with the letter P for "primary"), because they arrive at the detection site (say from an earthquake) first, before the transverse waves, called the *S-waves*, with S standing for "secondary".

propagation. This means that both Eqs. (7.111) and (7.115) are satisfied by solutions of the type (6.41):

$$q_{\pm}(z, t) = f_{\pm}\left(t \mp \frac{z}{v}\right), \tag{7.118}$$

where the functions $f_{\pm}$ describe the propagating waveforms. (However, if the initial wave is a mixture, of the type (7.107), of longitudinal and transverse components, these components, propagating with different velocities, will 'run from each other'.) As one may infer from the analysis of a periodic system model in chapter 6, the wave dispersion would be back at very high (*hypersound*) frequencies where the wave number k becomes close to the reciprocal distance d between the particles of the medium (e.g. atoms or molecules), and hence the approximation of the medium as a continuum, used through this chapter, becomes invalid.

As we already know from chapter 6, in addition to the velocity, the waves of each type are characterized by one more important parameter, the wave impedance Z of the continuum, for an acoustic waves frequently called its *acoustic impedance*. Generalizing Eq. (6.46) to the 3D case, we may define the impedance as the magnitude of the ratio of the force *per unit area* (i.e. the corresponding component of the stress tensor) exerted by the wave, and the particles' velocity. For example, for the longitudinal waves,

$$Z_l \equiv \left| \frac{\sigma_{zz}}{\partial q_z/\partial t} \right| = \left| \frac{\sigma_{zz}}{s_{zz}} \frac{s_{zz}}{\partial q_z/\partial t} \right| = \left| \frac{\sigma_{zz}}{s_{zz}} \frac{\partial q_z/\partial z}{\partial q_z/\partial t} \right|. \tag{7.119}$$

Plugging in Eqs. (7.108), (7.112), and (7.113), we obtain

$$Z_l = \left[\left(K + \frac{4}{3}\mu\right)\rho\right]^{1/2}, \tag{7.120}$$

in a clear analogy with the first of Eqs. (6.48). Similarly, for the transverse waves, the appropriately modified definition, $Z_t \equiv |\sigma_{xz}/(\partial q_x/\partial z)|$, yields

$$Z_t = (\mu\rho)^{1/2}. \tag{7.121}$$

Just like in the 1D model studied in chapter 6, one role of the wave impedance is to scale the power carried by the wave. For plane 3D waves in infinite media, with their infinite wave front area, it is more appropriate to speak about *power density*, i.e. power $\wp = d\mathscr{P}/dA$ per unit area of the front, and characterize it by not only its magnitude,

$$\wp = \frac{d\mathbf{F}}{dA} \cdot \frac{\partial \mathbf{q}}{\partial t}, \tag{7.122}$$

but also the direction of the energy propagation, that (for a plane wave in an isotropic medium) coincides with the direction of the wave vector $\mathbf{k}$: $\boldsymbol{\wp} \equiv \wp \mathbf{n}_\mathbf{k}$. Using the definition (7.18) of the stress tensor, we may represent the Cartesian components of this *Umov vector*[37] as

$$\mathscr{p}_j = \sum_{j'} \sigma_{jj'} \frac{\partial q_{j'}}{\partial t}. \tag{7.123}$$

Returning to plane waves propagating along axis z, and acting exactly like in section 7.6, for both the longitudinal and transverse waves we again arrive at Eq. (6.49), but for $\mathscr{p}$ rather than $\mathscr{P}$ (due to a different definition of the wave impedance—per unit area rather than per each particle chain). For the sinusoidal waves of the type (7.108), it yields

$$\mathscr{P}_z = \frac{\omega^2 Z}{2} a a^*, \tag{7.124}$$

with Z being the corresponding impedance—either Z_l or Z_t.

Just as in the 1D case, one more important effect in which the notion of impedance is crucial, is *wave reflection* from at an interface between two media. The two boundary conditions, necessary for the analysis of these processes, may be obtained from the continuity of vectors $\mathbf{q}$ and $d\mathbf{F}$. (The former condition is evident, while the latter one may be obtained by applying Newton's second law to the infinitesimal volume $dV = dAdz$, where segment dz straddles the interface.) Let us start from the simplest case of the *normal incidence* on a plane interface between two uniform media with different elastic moduli and mass densities. Due to the symmetry, it is evident that the longitudinal/transverse incident wave may only excite the longitudinal/transverse reflected and transferred waves, but not the counterpart wave type. Thus we can literally repeat the calculations of section 6.4, again arriving at the fundamental relations Eqs. (6.55) and (6.56), with the only replacement of Z and Z' with the corresponding values of either Z_l (7.120) or Z_t (7.121). Thus, at the normal incidence the wave reflection is determined solely by the acoustic *impedances* of the media, while the sound *velocities* are not involved.

The situation, however, becomes more involved at a non-vanishing, incidence angle $\theta^{(i)}$ (figure 7.12), where the transmitted wave is generally also *refracted*, i.e. propagates under a different angle, $\theta^{(r)} \neq \theta^{(i)}$, beyond the interface. Moreover, at $\theta^{(i)} \neq 0$ the directions of particle motion (vector $\mathbf{q}$) and of the stress forces (vector $d\mathbf{F}$) in the incident wave are neither exactly parallel nor exactly perpendicular to the interface. Thus this wave serves as an actuator for the reflected and refracted waves of both types—see figure 7.12, drawn for the particular case when the incident wave is transverse. The corresponding four angles, $\theta_t^{(r)}$, $\theta_t^{(r)}$, θ_l', θ_t', may be readily related to $\theta^{(i)}$ by the 'kinematic' condition that the incident wave, as well as the reflected and refracted waves of both types, should have the same spatial distribution along the

[37] Named after N Umov who introduced this concept in 1874. Ten years later, a similar concept for electromagnetic waves (see, e.g. *Part EM* section 6.4) was suggested by J Poynting, so that some textbooks use the term 'Umov–Poynting vector'. In a dissipation-free, elastic medium, the Umov vector obeys the following continuity equation: $\partial(\rho v^2/2 + u)/\partial t + \nabla \cdot \mathscr{p} = 0$, with u given by Eq. (7.50), which expresses the conservation of the total (kinetic plus potential) energy of the elastic deformation.

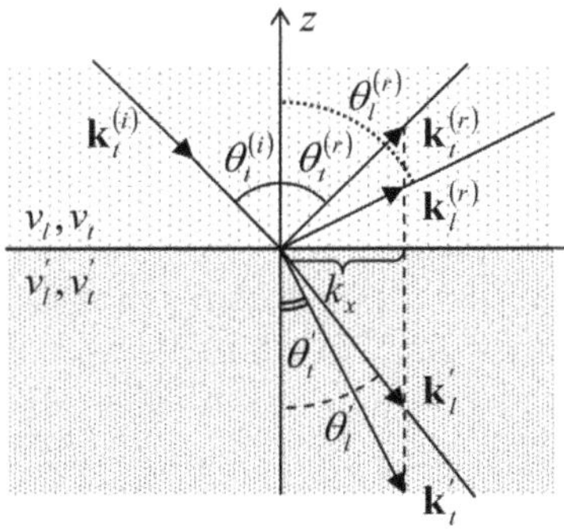

Figure 7.12. Deriving the 'kinematic' conditions (7.126) of the acoustic wave reflection and refraction (for the transverse incident wave case).

interface plane, i.e. for the interface particles participating in all five waves. According to Eq. (7.108), the necessary boundary condition is the equality of the tangential components (in figure 7.12, k_x), of all five wave vectors:

$$k_t \sin \theta_t^{(r)} = k_l \sin \theta_l^{(r)} = k_l' \sin \theta_l' = k_t' \sin \theta_t' = k_x \equiv k_t \sin \theta_t^{(i)}. \qquad (7.125)$$

Since the acoustic wave vector magnitudes k, at fixed frequency ω, are inversely proportional to the corresponding wave velocities, we immediately obtain the following relations:

$$\theta_t^{(r)} = \theta_t^{(i)}, \qquad \frac{\sin \theta_l^{(r)}}{v_l} = \frac{\sin \theta_l'}{v_l'} = \frac{\sin \theta_t'}{v_t'} = \frac{\sin \theta_t^{(i)}}{v_t}, \qquad (7.126)$$

so that generally all four angles are different. (This is of course the analog of the well-known *Snell law* in optics—where, however, only transverse waves are possible.) These relations show that, just like in optics, the direction of a wave propagating into a medium with lower velocity is closer to the normal (in figure 7.12, axis z). In particular, this means that if $v' > v$, the acoustic waves, at larger angles of incidence, may exhibit the effect of total internal reflection, so well known from optics[38], when the *refracted* wave vanishes. In addition, Eq. (7.126) show that in acoustics, a *reflected* longitudinal wave, with velocity $v_l > v_t$, may vanish at sufficiently large angles of the transverse wave incidence.

All these facts automatically follow from general expressions for amplitudes of the reflected and refracted waves via the amplitude of the incident wave. These relations are straightforward to derive (again, from the continuity of the vectors $\mathbf{q}$ and $d\mathbf{F}$), but since they are much bulkier then those in the electromagnetic wave theory (where they are called the *Fresnel formulas*[39]), I do not have time/space for spelling them out. Let me only note that, in contrast to the case of normal incidence, these relations involve eight medium parameters: the values of impedances Z, Z', and velocities v, v' on both sides of the interface, and for both the longitudinal and transverse waves.

[38] See, e.g. *Part EM* section 7.5.

[39] Discussion of these may be also found in *Part EM* section 7.5.

There is another factor that makes boundary acoustic effects more complex. Within certain frequency ranges, interfaces (and in particular surfaces) of elastic solids may sustain so-called *surface acoustic waves* (SAW), in particular, the *Rayleigh waves* and the *Love waves*[40]. The main feature that distinguishes such waves from their *bulk* (longitudinal and transverse) counterparts, discussed above, is that the displacement amplitudes are largest at the interface and decay exponentially into the bulk of both adjacent media. The characteristic depth of this penetration is of the order of the wavelength, although not exactly equal to it.

In the Rayleigh waves, the particle displacement vector $\mathbf{q}$ has two components: one longitudinal (and hence parallel to the interface) and another transverse (perpendicular to the interface). In contrast to the bulk waves, the components are coupled (via their interaction with the interface) and hence propagate with a single velocity v_R. As a result, the trajectory of each particle in the Rayleigh wave is an ellipse in the plane perpendicular to the interface. A straightforward analysis[41] of the Rayleigh waves on the *surface* of an elastic solid (i.e. its interface with vacuum) yields the following equation for v_R:

$$\left(2 - \frac{v_\mathrm{R}^2}{v_t^2}\right)^4 = 16\left(1 - \frac{v_\mathrm{R}^2}{v_t^2}\right)^2\left(1 - \frac{v_\mathrm{R}^2}{v_l^2}\right)^2. \tag{7.127}$$

According to this formula, and Eqs. (7.112) and (7.116), for realistic materials with the Poisson index between 0 and 1/2, the Rayleigh waves are slightly (by 4%–13%) slower than the bulk transverse waves—and hence substantially slower than the bulk longitudinal waves.

In contrast, the Love waves are purely transverse, with vector $\mathbf{q}$ oriented parallel to the interface. However, the interaction of these waves with the interface reduces their velocity v_L in comparison with that (v_t) of the bulk transverse waves, keeping it in the narrow interval between v_t and v_R:

$$v_\mathrm{R} < v_\mathrm{L} < v_t < v_l. \tag{7.128}$$

The practical importance of surface acoustic waves is that their amplitude decays very slowly with distance r from their point-like source: $a \propto 1/r^{1/2}$, while any bulk waves decay much faster, as $a \propto 1/r$. (Indeed, in the latter case the power $\mathscr{P} \propto a^2$, emitted by such source, is distributed over a spherical surface area proportional to r^2, while in the former case all the power goes into a thin surface circle whose length scales as r.) At least two areas of applications of the surface acoustic waves have to be mentioned: in geophysics (for earthquake detection and Earth crust seismology) and electronics (for signal processing, with a focus on frequency filtering). Unfortunately, I cannot dwell on these interesting topics and I have to refer the reader to special literature[42].

---

[40] Named, respectively, after Lord Rayleigh (born J Strutt, 1842–1919) who theoretically predicted the very existence of surface acoustic waves, and A Love (1863–1940).

[41] Unfortunately, I do not have time/space to reproduce this calculation either; see, e.g. section 24 in [4].

[42] See, e.g. [5, 6].

7.8 Elastic waves in restricted geometries

From what we have discussed at the end of the last section, it should be pretty clear that generally the propagation of acoustic waves in elastic bodies of finite size may be very complicated. There is, however, one important limit in which several important results may be readily obtained. This is the limit of (relatively) low frequencies, where the wavelength is much larger than at least one dimension of a system. Let us consider, for example, various waves that may propagate along thin rods, in this case 'thin' meaning that the characteristic size a of rod's cross-section is much smaller than not only the length of the rod, but also the wavelength $\lambda = 2\pi/k$. In this case there is a considerable range of distances z along the rod,

$$a \ll \Delta z \ll \lambda, \tag{7.129}$$

in which we can neglect the dynamic effects due to medium inertia, and apply results of our earlier static analyses.

For example, for a longitudinal wave of stress, which is essentially a wave of periodic tensile extensions and compressions of the rod, within the range (7.129) we may use the static relation (7.42):

$$\sigma_{zz} = Es_{zz}. \tag{7.130}$$

For what follows, it is easier to use the general equation of elastic dynamics not in its vector form (7.107), but rather in the precursor, Cartesian-component form (7.25), with $f_j = 0$. For plane waves propagating along axis z, only one component (with $j' \to z$) of the sum on the right-hand side of this equation is non-vanishing, and it is reduced to

$$\rho\frac{\partial^2 q_j}{\partial^2 t^2} = \frac{\partial \sigma_{jz}}{\partial z}. \tag{7.131}$$

In our current case of longitudinal waves, all components of the stress tensor apart from σ_{zz} are equal to zero. With σ_{zz} from Eq. (7.130), and using the definition $s_{zz} = \partial q_z/\partial z$, Eq. (7.131) is reduced to a simple wave equation,

$$\rho\frac{\partial^2 q_z}{\partial^2 t^2} = E\frac{\partial^2 q_z}{\partial z^2}, \tag{7.132}$$

which shows that the velocity of such *tensile waves* is

$$v = \left(\frac{E}{\rho}\right)^{1/2}. \tag{7.133}$$

Comparing this result with Eq. (7.112), we see that the tensile wave velocity, for any realistic continuum with a positive Poisson ratio, is lower than the velocity v_l of longitudinal waves in the bulk of the same material. The reason for this difference is simple: in thin rods, the cross-section is free to oscillate (e.g. shrink in the longitudinal extension phase of the passing wave)[43], so that the effective force

[43] For this reason, the tensile waves can be called longitudinal only in a limited sense: while the stress wave is purely longitudinal: $\sigma_{xx} = \sigma_{yy} = 0$, the strain wave is not: $s_{xx} = s_{yy} = -\sigma s_{zz} \neq 0$, i.e. $\mathbf{q}(\mathbf{r}, t) \neq \mathbf{n}_z q_z$.

resisting the longitudinal deformation is smaller than in a border-free space. Since (as is clearly visible from the wave equation), the scale of the force gives the scale of v^2, this difference translates into slower waves in rods. Of course as wave frequency is increased, at $ka \sim 1$ there is a (rather complex and cross-section-dependent) crossover from Eqs. (7.133) to (7.112).

Proceeding to transverse waves in rods, let us first have a look at long *bending waves*, for which the condition (7.129) is satisfied, so that with vector $\mathbf{q} = \mathbf{n}_x q_x$ (with axis x along the bending direction—see figure 7.8) is virtually constant in the whole cross-section. In this case, the only component of the stress tensor contributing to the net transverse force F_x is σ_{xz}, so that the integral of Eq. (7.131) over the cross-section is

$$\rho A \frac{\partial^2 q_x}{\partial t^2} = \frac{\partial F_x}{\partial z}, \quad \text{with} \quad F_x = \int_A \sigma_{xz} d^2 r. \tag{7.134}$$

Now, if Eq. (7.129) is satisfied, we again may use static local relations Eqs. (7.75)—(7.77), with all derivatives d/dz duly replaced with their partial form $\partial/\partial z$, to express the force F_x via the bending deformation q_x. Plugging these relations into each other one by one, we arrive at a rather unusual differential equation

$$\rho A \frac{\partial^2 q_x}{\partial t^2} = -EI_y \frac{\partial^4 q_x}{\partial z^4}. \tag{7.135}$$

Looking for its solution in the form of a sinusoidal wave (7.108), we obtain a nonlinear dispersion relation[44]:

$$\omega = \left(\frac{EI_y}{\rho A} \right)^{1/2} k^2. \tag{7.136}$$

Such relation means that the bending waves are not acoustic at *any* frequency, and cannot be characterized by a single velocity that would be valid for all wave numbers k, i.e. for all spatial Fourier components of a waveform. According to our discussion in section 6.3, such *strongly dispersive* systems cannot pass non-sinusoidal waveforms too far without changing their waveform considerably.

This situation changes, however, if the rod is pre-stretched with a tension force $\mathscr{T}$—just as in the discrete 1D model that was analyzed in section 6.3. The calculation of the effect of this force is essentially similar; let us repeat it for the continuous case, for a minute neglecting the bending stress—see figure 7.13.

Still sticking to the limit of small angles φ, the additional vertical component $d\mathscr{T}_x$ of the net force acting on a small rod fragment of length dz is $\mathscr{T}_x(z - dz) - \mathscr{T}_x(z) = \mathscr{T}\,\varphi_y(z + dz) - \mathscr{T}\,\varphi_y(z) \approx \mathscr{T}\,(\partial\varphi_y/\partial z)dz$, so that $\partial F_x/\partial z = \mathscr{T}(\partial\varphi_y/\partial z)$. With the geometric relation (7.77) in its partial-derivative form $\partial q_x/\partial z = \varphi_y$, this additional

[44] Note that since the 'moment of inertia' I_y, defined by Eq. (7.70), may depend on the bending direction (unless the cross-section is sufficiently symmetric), the dispersion relation (7.136) may give different results for different directions of the bending wave polarization.

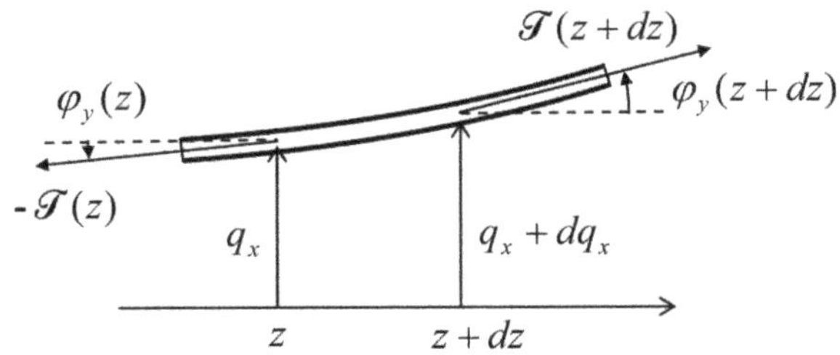

Figure 7.13. Additional forces in a thin rod ('string'), due to the background tension $\mathscr{T}$

term becomes $T(\partial^2 q_x/\partial z^2)$. Now adding it to the right-hand side of Eq. (7.135), we obtain the following dispersion relation

$$\omega^2 = \frac{1}{\rho A}\left(EI_y k^4 + \mathscr{T} k^2\right). \tag{7.137}$$

Since the product ρA in the denominator of this expression is just the rod's mass per unit length (which was denoted by μ in chapter 6), at low k (and hence low frequencies), this expression is reduced to the linear dispersion law, with the velocity given by Eq. (6.43):

$$v = \left(\frac{\mathscr{T}}{\rho A}\right)^{1/2}. \tag{7.138}$$

So Eq. (7.137) describes a smooth crossover from the 'guitar-string' acoustic waves to the highly dispersive bending waves (7.136).

Now let us consider another type of transverse waves in thin rods—the so-called *torsional waves*, which are essentially the dynamic propagation of the torsional deformation discussed in section 7.6. The easiest way to describe these waves, again within the limits (7.129), is to write the equation of rotation of a small segment dz of the rod about axis z, passing through the 'center-of-mass' of its cross-section, under the difference of torques $\boldsymbol{\tau} = \mathbf{n}_z\tau_z$ applied on its ends—see figure 7.10:

$$\rho I_z dz \frac{\partial^2 \varphi_z}{\partial t^2} = d\tau_z, \tag{7.139}$$

where I_z is the 'moment of inertia' defined by Eq. (7.89), which now, after its multiplication by ρdz, i.e. by the mass per unit area, has turned into the real moment of inertia of a dz-thick slice of the rod. Dividing both sides of Eq. (7.139) by dz, and using the static local relation (7.84), $\tau_z = C\kappa = C(\partial\varphi_z/\partial z)$, we obtain the following differential equation

$$\rho\, I_z \frac{\partial^2 \varphi_z}{\partial t^2} = C\frac{\partial^2 \varphi_z}{\partial z^2}. \tag{7.140}$$

Just as Eqs. (7.111), (7.115), and (7.132), this equation describes an acoustic (dispersion-free) wave, which propagates with the following frequency-independent velocity

$$v = \left(\frac{C}{\rho\, I_z}\right)^{1/2}. \tag{7.141}$$

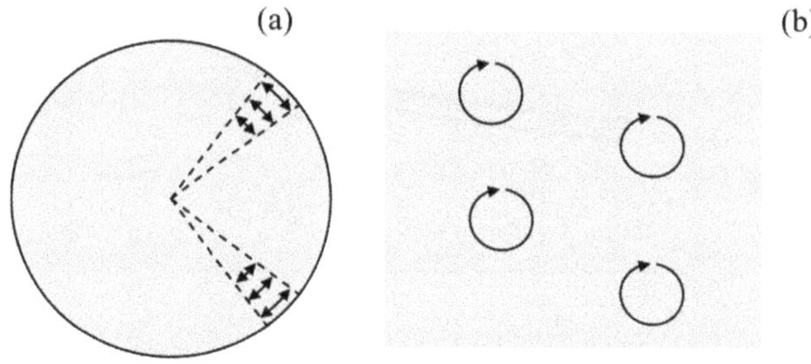

Figure 7.14. Particle trajectories in two different transverse waves with the same velocity: (a) torsional waves in a thin round rod and (b) circularly polarized waves in an infinite (or very broad) sample.

As we have seen in section 7.6, for rods with axially symmetric cross-sections, the torsional rigidity C is described by the simple relation (7.89), $C = \mu I_z$, so that relation (7.141) is reduced to Eq. (7.116) for the transverse waves in infinite media. The reason for this similarity is simple: in a torsional wave, particles oscillate along small arcs (figure 7.14a), so that if the rod's cross-section is round, its surface is stress-free, and does not perturb or modify the motion in any way, and hence does not affect the transverse velocity.

This fact raises the interesting issue of the relation between the torsional and *circularly polarized* waves. Indeed, in section 7.7 I have not emphasized enough that Eq. (7.116) is valid for a transverse wave polarized in any direction perpendicular to vector **k** (in our notation, directed along axis z). In particular, this means that such waves are doubly degenerate: any isotropic elastic continuum can carry simultaneously two non-interacting transverse waves propagating in the same direction with the same velocity (7.116), with mutually perpendicular linear polarizations (directions of vector **a**), for example, directed along axes x and y[45]. If both waves are sinusoidal (7.108), with the same frequency, each point of the medium participates in two simultaneous sinusoidal motions within the $[x,y]$ plane:

$$q_x = \text{Re}\left[a_x e^{i(kz-\omega t)}\right] = A_x \cos\Psi, \quad q_y = \text{Re}\left[a_y e^{i(kz-\omega t)}\right] = A_y \cos(\Psi + \varphi), \quad (7.142)$$

where $\Psi \equiv kz - \omega t + \varphi_x$, and $\varphi \equiv \varphi_y - \varphi_x$. The elementary geometry tells us that the trajectory of such a motion on the $[x, y]$ plane is an ellipse (figure 7.15), so that such waves are called *elliptically polarized*. The most important particular cases of such polarization are:

(i) $\varphi = 0$ or π: a *linearly polarized* wave, with vector **a** directed at angle $\theta = \tan^{-1}(A_y/A_x)$ relative to axis x; and

(ii) $\varphi = \pm\pi/2$ and $A_x = A_y$: two possible *circularly polarized* waves, with *right* or *left* polarization, respectively[46].

---

[45] As was shown in section 6.3, this is true even in the simple 1D model shown in figure 6.4a.

[46] The circularly polarized waves play an important role in quantum mechanics, where they may be most naturally quantized, with their elementary excitations (in the case of mechanical waves we are discussing, called *phonons*) having either positive or negative angular momentum $L_z = \pm\hbar$.

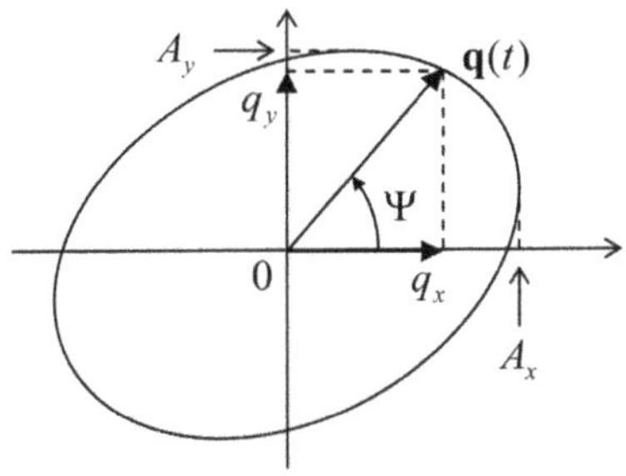

Figure 7.15. Trajectory of a particle in an elliptically polarized transverse wave, within the plane perpendicular to the direction of wave propagation.

Now comparing the trajectories of particles in the torsional wave in a thin round rod (or pipe) and the circularly polarized wave in a broad sample (figure 7.14), we see that, despite the same wave propagation velocity, these transverse waves are rather different. In the former case (figure 7.14a) each particle moves back and forth along an arc, with the arc length different for different particles (and vanishing at rod's center), so that the waves are *not* planar. On the other hand, in a circularly polarized wave all particles move along similar, circular trajectories, so that such wave *is* planar.

To conclude this chapter, let me briefly mention the opposite limit, when the size of the body, from whose boundary the waves are completely reflected[47], is much larger than the wavelength. In this case, the waves propagate almost as in an infinite 3D continuum (which was analyzed in section 7.7), and the most important new effect is the finite numbers of wave modes in the body. Repeating the 1D analysis at the end of section 6.5 for each dimension of a 3D cuboid of volume $V = l_1 l_2 l_3$, and taking into account that the numbers k_n in each of the three dimensions are independent, we obtain the following generalization of Eq. (6.75) for the number ΔN of different traveling waves with wave vectors within a relatively small volume d^3k of the wave vector space:

$$dN = g\frac{V}{(2\pi)^3}d^3k \gg 1, \quad \text{for} \quad \frac{1}{V} \ll d^3k \ll k^3, \qquad (7.143)$$

where $k >> 1/l_{1,2,3}$ is the center of this volume, and g is the number of different possible wave modes with the same wave vector $\mathbf{k}$. For the mechanical waves analyzed above, with one longitudinal mode, and two transverse modes with different polarizations, $g = 3$.

[47] For acoustic waves, such a condition is easy to implement. Indeed, from section 7.7 we already know that the strong inequality of wave impedances Z is sufficient for such reflection. The numbers in table 7.1 show that, for example, the impedance of a longitudinal wave in a typical metal (say, steel) is almost two orders of magnitude higher than that in air, ensuring their virtually full reflection from the surface.

Note that since the derivation of Eqs. (6.75) and (7.143) does not use other properties of the waves (in particular, their dispersion relations), the mode counting rule is ubiquitous in physics, being valid, in particular, for electromagnetic waves (where $g = 2$) and quantum 'de Broglie waves' (i.e. wavefunctions), whose degeneracy factor g is usually determined by the particle's spin[48].

7.9 Problems

Problem 7.1. A uniform thin sheet of an isotropic, elastic material, of area $A \gg t^2$, is compressed by two planar, parallel, broad, rigid surfaces—see the figure below. Assuming no slippage between the sheet and the compressing surfaces, calculate the relative compression $(-\Delta t/t)$ as a function of the compressing force. Compare the result with that for the tensile stress, calculated in section 7.3.

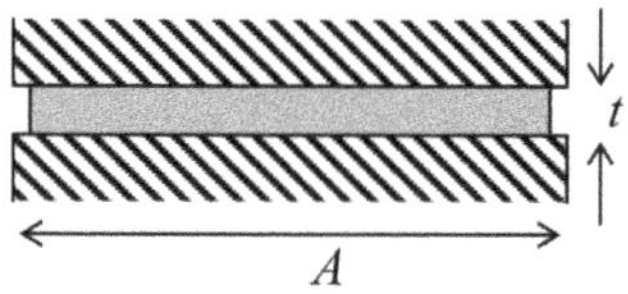

Problem 7.2. A thin, very wide sheet of an isotropic, elastic material is clamped in two rigid, planar, parallel walls that are pulled apart with force F, along the sheet's length l. Find the relative extension $\Delta l/l$ of the sheet in the direction of the force, and its relative compression $\Delta t/t$ in the perpendicular direction, and compare the results with Eqs. (7.45)–(7.46) for the tensile stress, and the solution of the previous problem.

Problem 7.3. Calculate the radial extension ΔR of a thin, long, round cylindrical pipe, due to its rotation with a constant angular velocity ω about its symmetry axis (see the figure below), in terms of the elastic moduli E and ν. The external pressure is negligible both inside and outside the pipe.

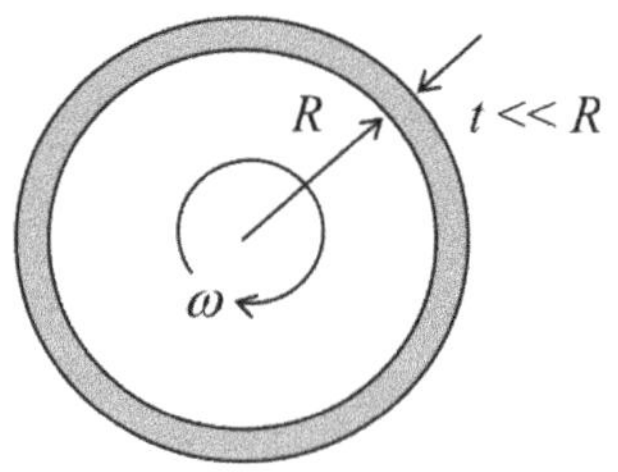

Problem 7.4. A long, uniform rail with the cross-section shown in the figure below, is being bent with the same (small) torque twice: first within the plane xz and then within the plane yz. Assuming that $t \ll l$, find the ratio of the rail bending deformations in these two cases.

48 See, e.g. *Part EM* section 7.8 and *Part QM* section 1.7.

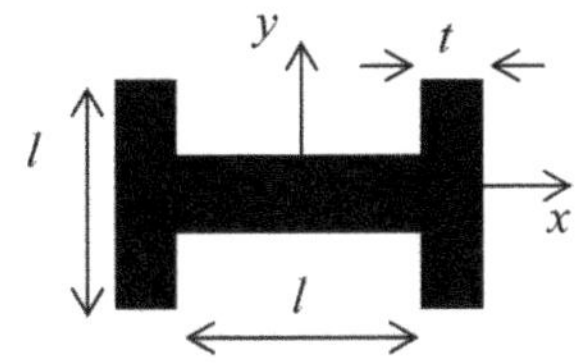

Problem 7.5. Two thin rods of the same length and mass are made of the same elastic, isotropic material. The cross-section of one of them is a circle, while another one is an equilateral triangle—see the figure below. Which of the rods is more stiff against bending along its length? Quantify the relation. Does the result depend on the bending plane orientation?

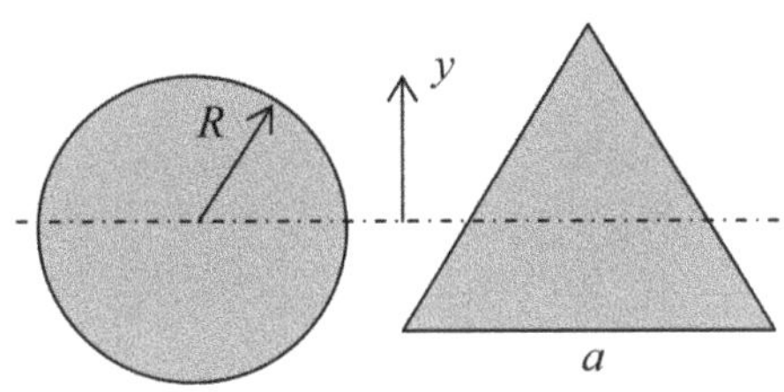

Problem 7.6. A thin, elastic, uniform, initially straight beam is placed on two point supports at the same height—see the figure below. What support point placement minimizes the largest deviation of the beam from the horizontal line, under its own weight?

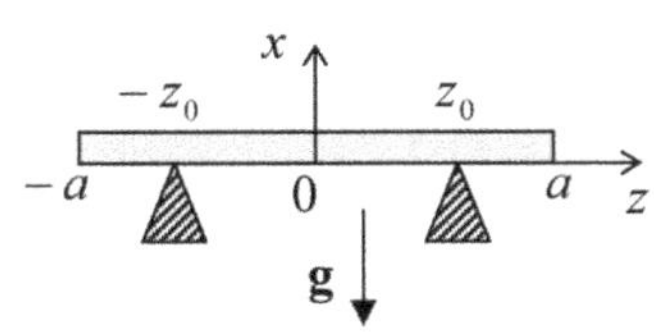

Hint: An approximate answer (with an accuracy better than 1%) is acceptable.

Problem 7.7. Calculate the largest longitudinal compression force $\mathscr{T}$ that may be withstood by a thin, straight, elastic rod without bucking (see the figures below) for the two cases shown:

(i) the rod's ends are clamped, and

(ii) the rod is free to turn about the support points.

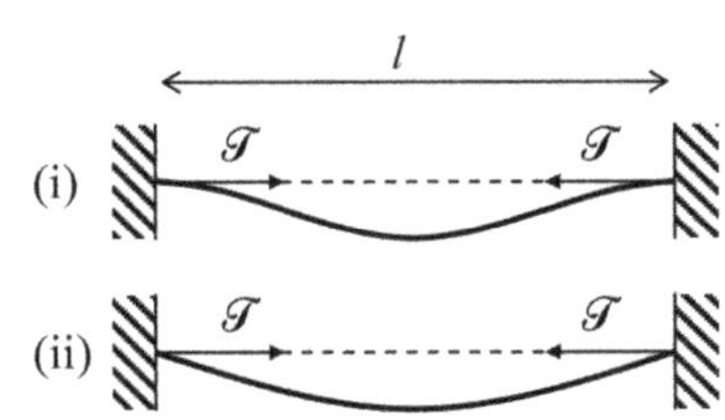

Problem 7.8. An elastic, light, thin pole with square cross-section of area $A = a \times a$, has been firmly dug into the ground in a vertical position, sticking out by height $h \gg a$. What largest compact mass M may be placed straight on the top of the pole without loss of stability?

Problem 7.9. Calculate the potential energy of a small and slowly changing, but otherwise arbitrary bending deformation of a uniform, elastic, initially straight rod. Can the result be used to derive the dispersion relation (7.136)?

Problem 7.10. Calculate the torsional rigidity of a thin, uniform rod whose cross-section is an ellipse with semi-axes a and b.

Problem 7.11. Calculate the potential energy of a small but otherwise arbitrary torsional deformation $\varphi_z(z)$ of a uniform, straight, elastic rod.

Problem 7.12. Calculate the spring constant $\kappa \equiv dF/dl$ of a coil spring made of a uniform, elastic wire, with a circular cross-section of diameter d, wound as a dense round spiral of $N \gg 1$ turns of radius $R \gg d$—see the figure below.

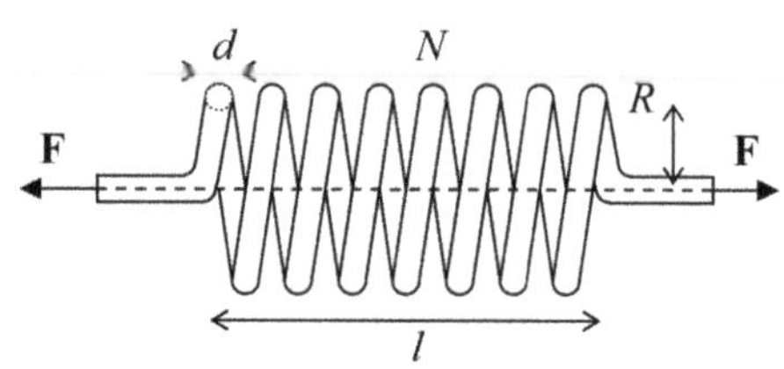

Problem 7.13. The coil spring, described in the previous problem, is now used as what is sometimes called the *torsion spring*—see the figure below. Find the corresponding spring constant $d\tau/d\varphi$, where τ is the torque of the external forces **F** relative to the center of the coil (point 0).

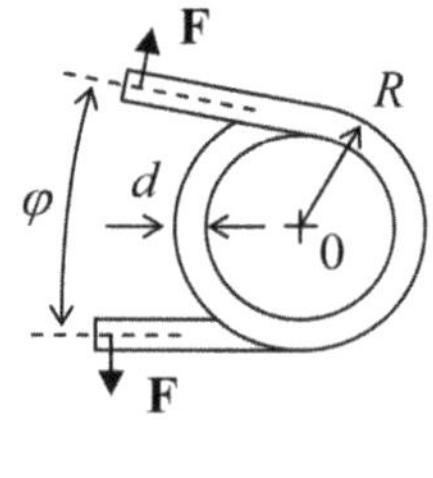

Problem 7.14. Use Eqs. (7.99) and (7.100) to recast Eq. (7.101b) for the torsional rigidity C of a thin rod into the form given by Eq. (7.101c).

Problem 7.15.* Generalize Eq. (7.101b) to the case of rods with more than one cross-section's boundary. Use the result to calculate the torsional rigidity of a thin round pipe, and compare it to Eq. (7.91).

Problem 7.16. A long steel wire has a circular cross-section with a 3 mm diameter, and is pre-stretched with a constant force of 10 N. Which of the longitudinal and transverse waves with frequency 1 kHz has the largest group velocity in the wire? Accept the following parameters for the steel (see table 7.1): $E = 170$ GPa, $\nu = 0.30$, and $\rho = 7.8$ g cm^{-3}.

Problem 7.17. Define and calculate the wave impedances for (i) tensile and (ii) torsional waves in a thin rod, appropriate in the long wave limit. Use the results to calculate the fraction of each wave's power reflected from the connection of a long rod with a round cross-section to a similar rod, but with a twice smaller diameter—see the figure below.

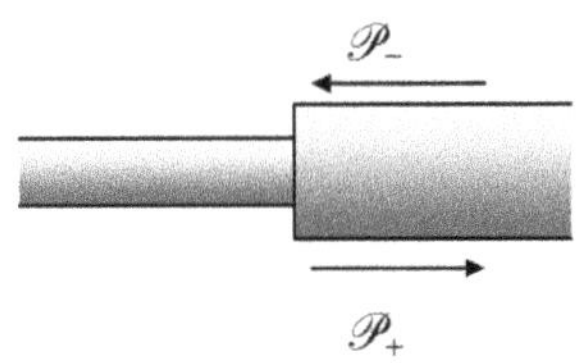

References

[1] Lee C *et al* 2008 *Science* **321** 5887
[2] Lee J-U *et al* 2012 *Nano Lett.* **12** 4444
[3] Lakes R 1987 *Science* **235** 1038
[4] Landau L and Lifshitz E 1986 *Theory of Elasticity* 3rd edn (Oxford: Butterworth-Heinemann)
[5] Aki K and Richards P G 2003 *Quantitative Seismology* 2nd edn (University Science Books)
[6] Morgan D 2007 *Surface Acoustic Waves* 2nd edn (New York: Academic)

IOP Publishing

Classical Mechanics
Lecture notes
Konstantin K Likharev

Chapter 8

Fluid mechanics

This chapter describes the basic notions of the mechanics of fluids, discusses a few core problems of the statics and dynamics of ideal and viscous fluids, and gives a very brief, rather superficial, review of such a complicated phenomenon as turbulence. In addition, the viscous fluid flow discussion is used as a platform for an elementary introduction to numerical methods of the partial differential equation solution—whose importance extends well beyond this particular field.

8.1 Hydrostatics

The mechanics of *fluids* (defined as the materials that cannot keep their geometric form on their own, and include both liquids and gases) is both simpler and more complex than that of elastic solids, with the simplicity falling squarely in the domain of *statics*—often called *hydrostatics*, because water has always been the main fluid for the human race and hence for science and engineering. Indeed, fluids, by definition, cannot resist static shear deformations. There are two ways to express this fact. First, we can formally take the shear modulus μ, describing this resistance, to be equal to zero. Then Hooke's law (7.32) shows that the stress tensor is diagonal:

$$\sigma_{jj'} = \delta_{jj'}\sigma_{jj}. \tag{8.1}$$

Alternatively, the same conclusion may be reached just by looking at the stress tensor definition (7.19) and/or figure 7.3, and saying that in the absence of shear stress, the elementary interface $d\mathbf{F}$ has to be perpendicular to the area element dA, i.e. parallel to the vector $d\mathbf{A}$.

Moreover, in fluids at equilibrium, all three diagonal components σ_{jj} of the stress tensor have to be equal. To prove that, it is sufficient to single out (mentally rather than physically) from a static fluid a small volume in the shape of a right prism, with mutually perpendicular faces normal to the two directions we are interested in—in figure 8.1, along axes x and y.

doi:10.1088/978-0-7503-1398-8ch8

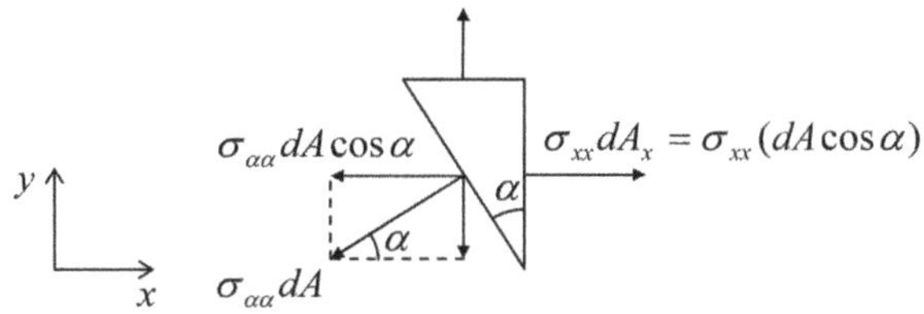

Figure 8.1. Proving the pressure isotropy.

The prism is in equilibrium if each Cartesian component of the vector of the total force exerted on all its faces equals zero. For the x-component this balance may be expressed as $\sigma_{xx} dA_x - (\sigma_{\alpha\alpha} dA) \cos\alpha = 0$. However, from the geometry (figure 8.1), $dA_x = dA \cos\alpha$, and so that the above expression yields $\sigma_{\alpha\alpha} = \sigma_{xx}$. A similar argument for the y-component gives $\sigma_{\alpha\alpha} = \sigma_{yy}$, so that $\sigma_{xx} = \sigma_{yy}$. Changing the orientation of the prism, we can obtain such equalities for any pair of diagonal components of the stress tensor, σ_{jj}, so that all three of them have to be equal.

This common diagonal element of the stress matrix is usually represented as $(-\mathcal{P})$, because in the vast majority of cases, the parameter $\mathcal{P}$, called *pressure*, is positive. Thus we arrive at the key relation (which has already been mentioned in section 7.2):

$$\sigma_{jj'} = -\mathcal{P}\delta_{jj'}. \tag{8.2}$$

In the absence of bulk forces, pressure should be constant through the volume of fluid, due to the translational symmetry. Let us see how this result is affected by bulk forces. With the simple stress tensor (8.2), the general condition of equilibrium of a continuous medium, expressed by Eq. (7.25) with a zero left-hand side, becomes just

$$-\frac{\partial\mathcal{P}}{\partial r_j} + f_j = 0, \tag{8.3}$$

and may be re-written in the following convenient vector form:

$$-\nabla\mathcal{P} + \mathbf{f} = 0. \tag{8.4}$$

In the simplest case of a heavy fluid with mass density ρ, in a uniform gravity field, $\mathbf{f} = \rho\mathbf{g}$, the equation of equilibrium becomes

$$-\nabla\mathcal{P} + \rho\mathbf{g} = 0, \tag{8.5}$$

with only one non-vanishing component (vertical, near the Earth's surface). If, in addition, the fluid may be considered *incompressible*, with its density ρ constant[1],

[1] As was discussed in section 7.3 in the context of table 7.1, this is an excellent approximation, for example, for human-scale experiments with water.

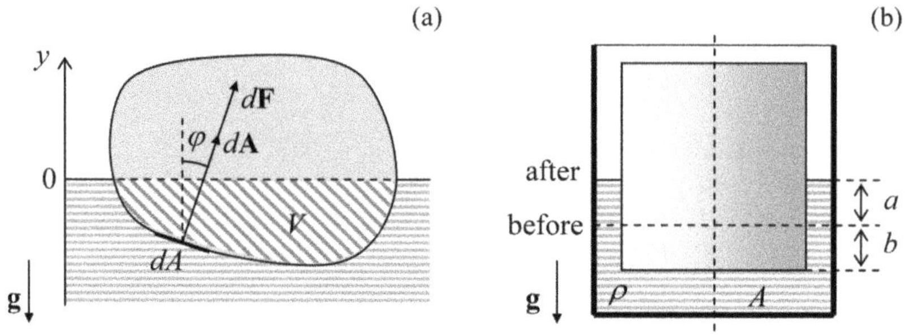

Figure 8.2. Calculating the buoyant force.

this equation may be readily integrated over the vertical coordinate (say, y) to give the so-called *Pascal equation*[2]:

$$\mathcal{P} + \rho g y = \text{const}, \tag{8.6}$$

where the direction of axis y is taken opposite to that of vector **g**.

Two manifestations of this key equation are most common, and well known. The first one is the fact that in interconnected vessels filled with a fluid, its pressure is equal at all points at equal height (y), regardless of the vessel shape, provided that the fluid is in equilibrium.

The second result is the *buoyant force* **F** exerted by a liquid on a (possibly partly) submerged body, i.e. the vector sum of the elementary pressure forces $d\mathbf{F} = \mathcal{P}d\mathbf{A}$ exerted on all elementary areas dA of the submerged part of the body's surface—see figure 8.2. According to Eq. (8.6), with the constant equal to zero (corresponding to zero pressure at the liquid's surface taken for $y = 0$, see figure 8.2a), the vertical component of this elementary force is

$$dF_y = dF\cos\varphi = \mathcal{P}dA\cos\varphi = -\rho g y\cos\varphi dA = -\rho g y dA_h. \tag{8.7}$$

where $dA_h = \cos\varphi\, dA$ is the horizontal footprint of the elementary area dA (say, $dxdz$). Now integrating this relation over all the surface, we get the total vertical buoyant force[3]:

$$F_b = \rho g\int_S(-y)dA_h \equiv \rho g V, \tag{8.8}$$

where V is the volume of the *submerged* part of the body's volume, while ρ is the liquid's density, so that by magnitude, F_b equals the weight of the liquid which *would* fill the submerged volume.

This well-known *Archimedes principle* may be proved even more simply using the following argument: the fluid's pressure forces, and hence the resulting buoyant force, cannot depend on what is inside the body's volume. Hence F_b would be the same if we

[2] The equation, and the SI unit of pressure 1 Pa = 1 N m^{-2}, are named after B Pascal (1623–62) who not only pioneered hydrostatics, but also invented the first mechanical calculator and made several other important contributions to mathematics—and Christian philosophy!

[3] The force is strictly vertical, because the horizontal components of the elementary forces $d\mathbf{F}$ exerted on opposite elementary areas dA, at the same height y, cancel.

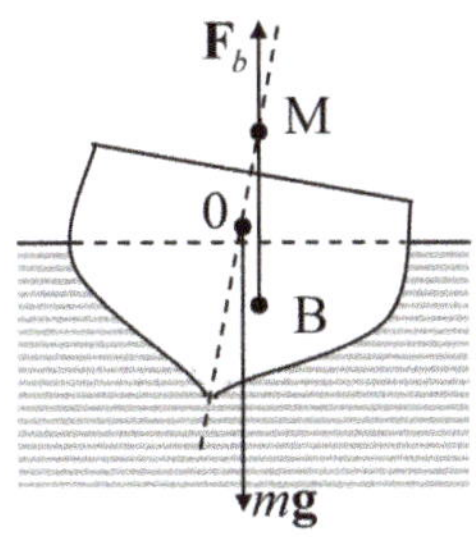

Figure 8.3. Angular stability of a floating body.

filled the volume V in question with a fluid similar to the surrounding one. But in this case the surface does not play any role, and the fluid should be still in equilibrium, so that both forces acting on it, the buoyant force $\mathbf{F}_b$ and the internal liquid's weight $m\mathbf{g} = \rho V\mathbf{g}$, have to be equal and opposite, thus proving Eq. (8.8) again.

Despite the simplicity of the Archimedes principle, its different, erroneous formulations, such as 'The buoyant force's magnitude is equal to the weight of the displaced liquid' (WRONG!) creeps from one undergraduate textbook to another, leading to application errors. A typical example is shown in figure 8.2b, where a solid vertical cylinder with the base area A is pressed into a liquid inside a container of a comparable size, pushing the liquid's level up. The correct answer for the buoyant force, following from Eq. (8.8), is

$$F_b = \rho g V = \rho g A(a + b), \tag{8.9a}$$

because the volume V of the *submerged* part of the inner cylinder is evidently $A(a + b)$. But the wrong formulation cited above, using the term *displaced liquid*, would give a different answer:

$$F_b = \rho g V_{\text{displaced}} = \rho g A b. \quad \text{(WRONG!)} \tag{8.9b}$$

Evidently, the latter result is correct only asymptotically, in the limit of a very large container.

Another frequent error in hydrostatics concerns the angular stability of a freely floating body—a problem of evident vital importance for boat/ship design. It is sometimes claimed that the body is stable only if the effective point of buoyant force application (in figure 8.3, point B) is *above* the center-of-mass (0) of the whole floating body[4]. However, as figure 8.3 shows, this is unnecessary; indeed in the shown case, point B (which is just the center-of-mass the liquid would have in the submerged part) is *below* point 0, even at a vanishingly small tilt. Still, the in this case the torque created by the pair of forces $\mathbf{F}_b$ and $m\mathbf{g}$, evidently tries to return the body to the equilibrium position, which is therefore stable. As figure 8.3 shows, the actual condition of the angular stability may be expressed as the requirement for the point M (in shipbuilding, called the *metacenter* of the ship's hull) to be above the center-of-mass 0[5].

[4] Please note the crucial difference between the whole body and that of its submerged part.

[5] A small tilt of the body leads to a small lateral displacement of point B, but does not affect the position of the metacenter M.

To conclude this section, let me note that the integration of Eq. (8.4) may be more complex in the case that the bulk forces **f** depend on position[6], and/or if the fluid is substantially compressible. In the latter case, Eq. (8.4) should be solved together with the media-specific *equation of state* $\rho = \rho(\mathcal{P})$ describing its compressibility law—whose example is given by Eq. (7.38) for ideal gases: $\rho \equiv mN/V = m\mathcal{P}/k_B T$, where m is the mass of one gas molecule.

8.2 Surface tension effects

Besides the bulk (volume-distributed) forces, one more possible source of pressure is *surface tension*. This effect results from the difference between the potential energy of atomic interactions on the interface between two different fluids and that in their bulks, and thus may be described by an additional potential energy

$$U_i = \gamma A, \tag{8.10}$$

where A is the interface area, and γ is called the *surface tension constant* (or just the 'surface tension'). For a stable interface of any two fluids, γ is always positive[7]. For surfaces of typical liquids (or their interfaces with air), at the room temperature, the surface tension equals a few 10^{-2} J m^{-2}, corresponding to the potential energy U_i of a few 10^{-2} eV per surface molecule, i.e. just a fraction of the full *binding* (or 'cohesive') *energy* of the same liquid, which is typically of the order of 10^{-1} eV per molecule.

In the absence of other forces, the surface tension makes a liquid drop spherical to minimize its surface area A at fixed volume. For the analysis of the surface tension effects for more complex geometries, and in the presence of other forces, it is convenient to reduce it to a certain additional effective pressure drop $\Delta\mathcal{P}_{ef}$ at the interface. In order to calculate $\Delta\mathcal{P}_{ef}$, let us consider the condition of equilibrium of a small part dA of a smooth interface between two fluids (figure 8.4), in the absence of bulk forces.

If the pressures $\mathcal{P}_{1,2}$ on the two sides of the interface are different, the work of stress forces on fluid 1 at a small virtual displacement $\delta\mathbf{r} = \mathbf{n}\delta r$ of the interface (where $\mathbf{n} = d\mathbf{A}/dA$ is the unit vector normal to the interface) equals[8]

$$\delta\mathcal{W} = dA\delta r(\mathcal{P}_1 - \mathcal{P}_2). \tag{8.11}$$

For equilibrium, this work has to be compensated by an equal change of the interface energy, $\delta U_i = \gamma\delta(dA)$. Differential geometry tells us that in the linear

[6] A simple example of such a problem is given by the fluid equilibrium in a container rotating with a constant angular velocity $\boldsymbol{\omega}$. If we solve such a problem in a reference frame rotating together with the container, the real bulk forces should be complemented by the centrifugal force Eq. (4.93), depending on $\mathbf{r}$.

[7] If γ of the interface of two certain fluids is negative, it self-reconfigures to decrease U_s by increasing the interface area, i.e. fragments the system into a solution.

[8] This equality readily follows from the general equation (7.30), with the stress tensor elements expressed by Eq. (8.2), but in this simple case of the net stress force $d\mathbf{F} = (\mathcal{P}_1 - \mathcal{P}_2)d\mathbf{A}$ parallel to the interface element vector $d\mathbf{A}$, it may be even more simply obtained just from the definition of work $\delta\mathcal{W} = d\mathbf{F}\cdot\delta\mathbf{r}$ at the virtual displacement $\delta\mathbf{r} = \mathbf{n}\delta r$.

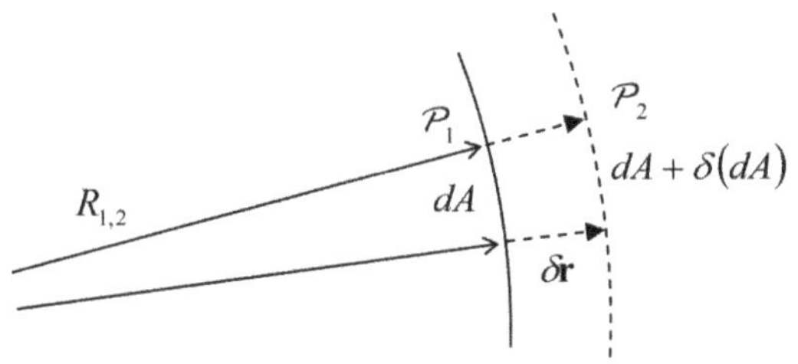

Figure 8.4. Deriving the Young–Laplace formula (8.13).

approximation in δr, the relative change of the elementary surface area, corresponding to a fixed solid angle $d\Omega$, may be expressed as

$$\frac{\delta(dA)}{dA} = \frac{\delta r}{R_1} + \frac{\delta r}{R_2}, \tag{8.12}$$

where $R_{1,2}$ are the so-called *principal radii* of the interface curvature[9]. Combining Eqs. (8.10)–(8.12), we obtain the *Young–Laplace formula*[10]

$$\mathcal{P}_1 - \mathcal{P}_2 = \Delta\mathcal{P}_{\text{ef}} \equiv \gamma\left(\frac{1}{R_1} + \frac{1}{R_2}\right). \tag{8.13}$$

In particular, this formula shows that the additional pressure created by surface tension inside a spherical drop of a liquid, of radius R, equals $2\gamma/R$, i.e. decreases with R. In contrast, according to Eq. (8.5)–(8.6), the effects of bulk forces, for example gravity, grow as $\rho g R$. The comparison of these two pressure components shows that if the drop radius (or more generally, the characteristic linear size of a fluid sample) is much larger than the so-called *capillary length*

$$a_c \equiv \left(\frac{2\gamma}{\rho g}\right)^{1/2}, \tag{8.14}$$

the surface tension may be safely ignored—as will be done in the following sections of this chapter, apart from a brief discussion at the end of section 8.4. For the water surface, or more exactly its interface with air at ambient conditions, $\gamma \approx 0.073$ J m^{-2} (i.e. N m^{-1}), while $\rho \approx 1000$ kg m^{-3}, so that $a_c \approx 4$ mm.

On the other hand, in very narrow tubes, such as blood capillary vessels with radius $a \sim 1$ μm, i.e. $a \ll a_c$, the surface tension effects are very important. The key notion for the analysis of these effects is the *contact angle* θ_c (also called the 'wetting angle') at an equilibrium edge of a liquid wetting a solid—see figure 8.5.

[9] This general formula may be verified by elementary means for a sphere of radius r (for which $R_1 = R_2 = r$ and $dA = r^2 d\Omega$, so that $\delta(dA)/dA = \delta(r^2)/r^2 = 2\delta r/r$), and a round cylindrical interface of radius R (for which $R_1 = r$, $R_2 = \infty$, and $dA = rd\varphi dz$, so that $\delta(dA)/dA = \delta r/r$). For more on curvature, see, for example, [1].

[10] This result (not to be confused with Eq. (8.15), called the *Young's equation*) was derived in 1806 by P-S Laplace (of the Laplace operator/equation fame) on the basis of the first analysis of the surface tension effects by T Young a year earlier.

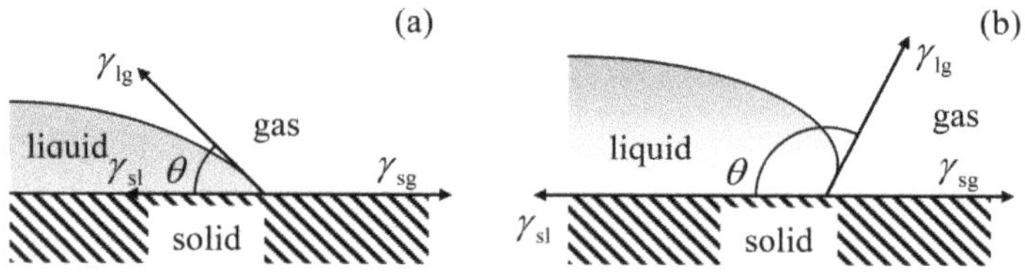

Figure 8.5. Contact angles for (a) hydrophilic and (b) hydrophobic surfaces.

According to its definition (8.7), the constant γ may be interpreted as a force (per unit length of the interface boundary) directed normally to the boundary, and 'trying' to reduce the interface area. As a result, the balance of horizontal components of the three such forces, shown in figure 8.5, immediately yields

$$\gamma_{sl} + \gamma_{lg} \cos \theta_c = \gamma_{sg}, \tag{8.15}$$

where the indices at constants γ correspond to three possible interfaces between the liquid, solid and gas. For the so-called *hydrophilic* surfaces that 'like to be wet' by this particular liquid (not necessarily water), meaning that $\gamma_{sl} < \gamma_{sg}$, this relation yields $\cos\theta_c > 0$, i.e. $\theta_c < \pi/2$—the situation shown in figure 8.5a. On the other hand, for *hydrophobic* surfaces with $\gamma_{sl} > \gamma_{sg}$, Young's equation (8.15) yields larger contact angles, $\theta_c > \pi/2$—see figure 8.5b.

Let us use this notion to solve the simplest but perhaps the most important problem of this field—finding the height h of the fluid column lifted by the surface tension forces in a narrow vertical tube made of a hydrophilic material, assuming its internal surface to be a round cylinder of radius a—see figure 8.6. Inside an incompressible fluid, pressure drops with height according to the Pascal equation (8.6), so that just below the surface, $\mathcal{P} \approx \mathcal{P}_0 - \rho g h$, where $\mathcal{P}_0$ is the background (e.g. atmospheric) pressure. This means that at $a \ll h$ the pressure variation along the concave surface (called the *meniscus*) of the liquid is negligible, so that according to the Young–Poisson equation (8.13), the sum $(1/R_1 + 1/R_2)$ has to be virtually constant along the surface. Due to the axial symmetry of the problem, this means that the surface has to be a part of a sphere[11]. From the contact angle definition, radius R of the sphere is equal to $a/\cos \theta_c$—see figure 8.6.

Plugging this relation into Eq. (8.13) with $\mathcal{P}_1 - \mathcal{P}_2 = \rho g h$, we get the following equation for h:

$$\rho g h = \frac{2\gamma \cos \theta_c}{a}. \tag{8.16a}$$

In hindsight, this result might be obtained more directly—by requiring the total weight $\rho g V = \rho g(\pi a^2 h)$ of the lifted liquid's column to be equal to the vertical component $F \cos \theta_c$ of the full surface tension force $F = \gamma p$, acting on the perimeter

[11] Note that this is not true for tubes with different shapes of their cross-section.

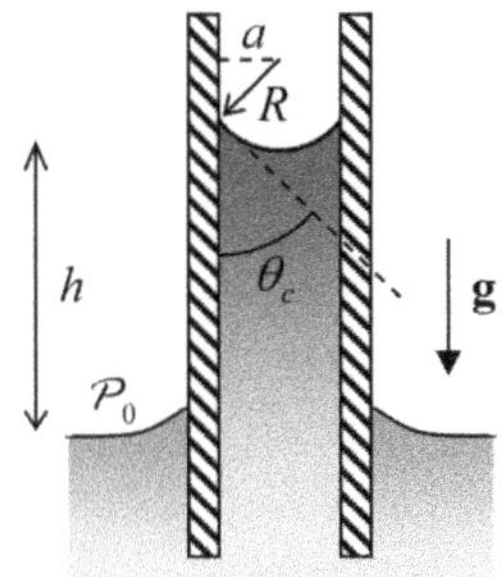

Figure 8.6. Liquid's rise in a vertical capillary tube.

$p = 2\pi a$ of the meniscus. Using the definition Eq. (8.14) of the capillary length a_c, Eq. (8.16*a*) may be represented as the so-called *Jurin rule:*

$$h = \frac{a_c^2}{a}\cos\theta_c \leqslant \frac{a_c^2}{a}; \tag{8.16b}$$

according to our initial assumption $h \gg a$, Eq. (8.16) is only valid for narrow tubes, with radius $a \ll a_c$.

This capillary rise is the basic mechanism of lifting water with nutrients from roots to the branches and leaves of plants, so that the tallest tree height corresponds to the Jurin rule (8.16), with $\cos\theta_c \approx 1$, and the pore radius a limited from below by a few microns, because of the viscosity effects restricting the fluid discharge—see section 8.5 below.

8.3 Kinematics

In contrast to the stress tensor (which is useful and simple, see Eq. (8.2)), the strain tensor is *not* a very useful notion in fluid mechanics. Indeed, apart from very few situations[12], typical problems of this field involve fluid *flow*, i.e. a state when the velocity of fluid particles has some non-zero time average. This means that the trajectory of each individual particle is a long line, and the notion of its displacement $\mathbf{q}$ becomes impracticable. However, particle's velocity $\mathbf{v} \equiv d\mathbf{q}/dt$ is a much more useful notion, in particular if it is considered as a function the *observation point* $\mathbf{r}$ and (generally) time t. In an important class of fluid dynamics problems, the so-called *stationary* (or 'steady', or 'static') *flow*, the velocity defined in this way does not depend on time, $\mathbf{v} = \mathbf{v}(\mathbf{r})$.

There is, however, a price to pay for the convenience of this notion: namely, due to the difference between the vectors $\mathbf{q}$ and $\mathbf{r}$, a particle's acceleration $\mathbf{a} = d^2\mathbf{q}/dt^2$ (which participates, in particular, in Newton's second law) cannot be calculated just as the time derivative of the velocity $\mathbf{v}(\mathbf{r}, t)$. This fact is evident, for example, for the

[12] One of them is the sound propagation, where the particle displacements $\mathbf{q}$ are typically small, so that results of section 7.7 are applicable. As a reminder, they show that in fluids, with $\mu = 0$, the transverse sound cannot propagate (formally, has zero velocity and impedance), while the longitudinal sound's velocity is finite—see Eq. (7.114).

static flow case, in which the acceleration of individual fluid particles may be very significant even if $\mathbf{v}(\mathbf{r})$ does not depend on time—just think about the acceleration of a drop of water flowing over the Niagara Falls' rim, first accelerating fast and then virtually stopping below, while the water velocity $\mathbf{v}$ at every particular point, as measured from a bank-based reference frame, is nearly constant. Thus the main task of fluid kinematics is to express $\mathbf{a}$ via $\mathbf{v}(\mathbf{r}, t)$; let us do this.

Since each Cartesian component v_j of the velocity has to be considered as a function of four *independent* scalar variables: three Cartesian components r_j of vector $\mathbf{r}$ and time t, its full time derivative may be represented as

$$\frac{dv_j}{dt} = \frac{\partial v_j}{\partial t} + \sum_{j'=1}^{3} \frac{\partial v_j}{\partial r_{j'}} \frac{dr_{j'}}{dt}. \tag{8.17}$$

Let us apply this *general* relation to a *specific* set of infinitesimal changes $\{dr_1, dr_2, dr_3\}$ that follows a small displacement $d\mathbf{q}$ of a certain particular particle of the fluid, $d\mathbf{r} = d\mathbf{q} = \mathbf{v}dt$, i.e.

$$dr_j = v_j dt. \tag{8.18}$$

In this case dv_j/dt is the jth component a_j of the particle's acceleration $\mathbf{a}$, so that Eq. (8.17) yields the following key relation of fluid kinematics:

$$a_j = \frac{\partial v_j}{\partial t} + \sum_{j'=1}^{3} v_{j'} \frac{\partial v_j}{\partial r_{j'}}. \tag{8.19a}$$

Using the del operator ∇, this result may be rewritten in the following compact vector form[13]:

$$\mathbf{a} = \frac{\partial \mathbf{v}}{\partial t} + (\mathbf{v} \cdot \nabla)\mathbf{v}. \tag{8.19b}$$

This relation already signals the main technical problem of the fluid dynamics: many equations involving particle's acceleration are nonlinear in velocity, excluding such a powerful tool as the linear superposition principle (which was used so frequently in the previous chapters of this course) from the applicable mathematical arsenal.

One more basic relation of the fluid kinematics is the so-called *continuity equation*, which is essentially just the differential version of the mass conservation law. Let us mark, inside a fluid flow, an arbitrary volume V limited by a stationary (time-independent) surface S. The total mass of the fluid inside the volume may change only due to its flow though the boundary:

$$\frac{dM}{dt} \equiv \frac{d}{dt}\int_V \rho d^3r = -\int_S \rho v_n d^2r \equiv -\int_S \rho \mathbf{v} \cdot d\mathbf{A}, \tag{8.20a}$$

where the elementary area vector $d\mathbf{A}$ is defined just as in section 7.2—see figure 8.7.

[13] Note that the operator relation $d/dt = \partial/\partial t + (\mathbf{v}\cdot\nabla)$ is applicable to an arbitrary (scalar or vector) function; it is frequently called the *convective derivative*. (Alternative adjectives, such as 'Lagrangian', 'substantial', or 'Stokes', are sometimes used for this derivative as well.) The relation has numerous applications well beyond the fluid dynamics—see, e.g. *Part EM* chapter 9 and *Part QM* chapter 1.

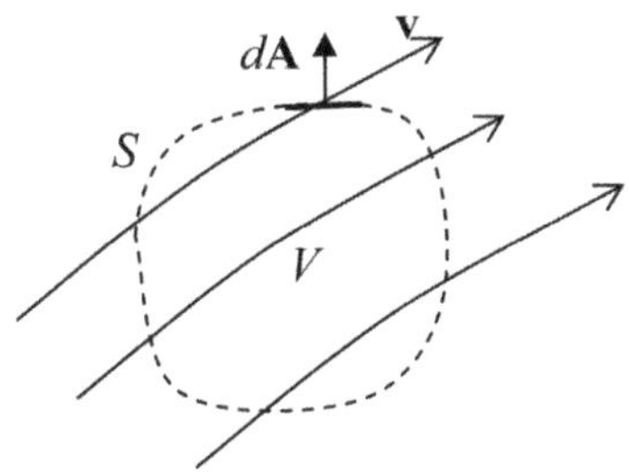

Figure 8.7. Deriving the continuity equation.

Using the same the same divergence theorem that has been used several times in this course[14], the surface integral in Eq. (8.20*a*) may be transformed into the integral of $\nabla(\rho\mathbf{v})$ over volume V, so that this equation may be rewritten as

$$\int_V \left(\frac{\partial \rho}{\partial t} + \nabla \cdot \mathbf{j}\right) d^3r = 0, \qquad (8.20b)$$

where vector $\mathbf{j} \equiv \rho\mathbf{v}$ defined is called either the *mass flux density* or the *mass current*. Since Eq. (8.20*b*) is valid for an arbitrary stationary volume V, the function under the integral has to vanish at any point:

$$\frac{\partial \rho}{\partial t} + \nabla \cdot \mathbf{j} = 0. \qquad (8.21)$$

Note that such continuity equation is valid not only for mass, but for other conserved physics quantities (e.g. the electric charge, probability, etc), with the proper re-definition of ρ and $\mathbf{j}$[15].

8.4 Dynamics: ideal fluids

Let us start our discussion of fluid dynamics from the simplest case when the stress tensor obeys the simple expression (8.2) even at the fluid motion. Physically, this means that the fluid viscosity effects, leading to the mechanical energy loss, are negligible. (We will discuss the conditions of this assumption in the next section.) Then the equation of motion of such an *ideal fluid* (essentially Newton's second law for its unit volume) may be obtained from Eq. (7.25) using the simplifications of its right-hand side, discussed in section 8.1:

$$\rho\mathbf{a} = -\nabla\mathcal{P} + \mathbf{f}. \qquad (8.22)$$

Now using the basic kinematic relation (8.19), we arrive at the following *Euler equation*[16]:

$$\rho\frac{\partial \mathbf{v}}{\partial t} + \rho(\mathbf{v}\cdot\nabla)\mathbf{v} = -\nabla\mathcal{P} + \mathbf{f}. \qquad (8.23)$$

[14] If the reader still needs a reminder, see Eq. (A.78).

[15] See, e.g. *Part EM* section 4.1, *Part QM* section 1.4, and *Part SM* section 5.6.

[16] It was derived in 1755 by the same L Euler whose name has already been (reverently) mentioned several times in this volume.

Generally this equation has to be solved together with the continuity equation (8.21) and the equation of state of the particular fluid, $\rho = \rho(\mathcal{P})$. However, as we have already discussed, in many situations the compressibility of water and other important liquids is very low and may be ignored, so that ρ may be treated as a given constant. Moreover, in many cases the bulk forces **f** are conservative and may be represented as a gradient of a certain potential function $u(\mathbf{r})$—the potential energy per unit volume:

$$\mathbf{f} = -\nabla u; \tag{8.24}$$

for example, for a uniform gravity field, $u(\mathbf{r}) = \rho g y$, where y is referred to some (arbitrary) horizontal level. In this case, the right-hand side of Eq. (8.23) becomes $-\nabla(\mathcal{P} + u)$. For these cases, it is beneficial to recast the left-hand side of that equation as well, using the following well-know identity of vector algebra[17]

$$(\mathbf{v} \cdot \nabla)v = \nabla\left(\frac{v^2}{2}\right) - \mathbf{v} \times (\nabla \times \mathbf{v}). \tag{8.25}$$

As a result, the Euler equation takes the form

$$\rho\frac{\partial \mathbf{v}}{\partial t} - \rho\mathbf{v} \times (\nabla \times \mathbf{v}) + \nabla\left(\mathcal{P} + u + \rho\frac{v^2}{2}\right) = 0. \tag{8.26}$$

In a stationary flow, the first term of this equation vanishes. If the second term, describing fluid's *vorticity*, is zero as well, then equation (8.26) has the first integral of motion,

$$\mathcal{P} + u + \frac{\rho}{2}v^2 = \text{const}, \tag{8.27}$$

called the *Bernoulli equation*. Numerous examples of application of equation (8.27) to simple problems of stationary flow in pipes, in the Earth gravity field, should be well known to the reader, so I hope I can skip their discussion without much harm.

In the general case an ideal fluid may have vorticity, so that Eq. (8.27) is not always valid. Moreover, due to the absence of viscosity in an ideal fluid, the vorticity, once created, does not decrease along the *streamline*—the fluid particle's trajectory, to which the velocity is tangential in every point[18]. Mathematically, this fact[19] is expressed by the following *Kelvin theorem*: $(\nabla \times \mathbf{v})\cdot d\mathbf{A}$ = const along any small contiguous group of streamlines crossing an elementary area dA[20].

However, in many important cases the vorticity of fluid is negligible. For example, if the vorticity exits in some part of the fluid volume (say, induced by local turbulence, see section 8.6 below), but decays due to the fluid's viscosity (to be discussed in section

[17] It readily follows, for example, from Eq. (A.76) with $\mathbf{g} = \mathbf{f} = \mathbf{v}$.

[18] Perhaps the most spectacular manifestation of the vorticity conservation are the famous *toroidal vortex rings* (see, e.g. a nice photo and a movie at https://en.wikipedia.org/wiki/Vortex_ring), predicted in 1858 by H von Helmholtz, and then demonstrated by P Tait in a series of spectacular experiments with smoke in air. The persistence of such a ring, once created, is only limited by the fluid's viscosity—see the next section.

[19] This theorem was first formulated (verbally) by H von Helmholtz.

[20] Its proof may be found, e.g. in section 8 of [2].

8.5), well before it reaches the region of our interest. If this viscosity is sufficiently small, its effects on the fluid's flow in the region of interest are negligible, i.e. the ideal fluid approximation is still acceptable. Another important case is when a solid body of arbitrary shape is embedded into an ideal fluid whose flow is uniform (meaning, by definition, that $\mathbf{v}(\mathbf{r},t) = \mathbf{v}_0 = \text{const}$) at large distances[21], its vorticity is zero everywhere. Indeed, since $\nabla \times \mathbf{v}$ at the uniform flow, the vorticity is zero at distant points of any streamline, and according to the Kelvin theorem, should equal zero everywhere.

In such cases, the velocity distribution, as any curl-free vector field, may be represented as a gradient of some effective potential function,

$$\mathbf{v} = -\nabla\phi. \tag{8.28}$$

Such *potential flow* may be described by a simple differential equation. Indeed, the continuity equation (8.21) for a steady flow of an incompressible fluid is reduced to $\nabla\cdot\mathbf{v} = 0$. Plugging Eq. (8.28) into this relation, we obtain the scalar Laplace equation,

$$\nabla^2\phi = 0, \tag{8.29}$$

which should be solved with appropriate boundary conditions. For example, the flow may be limited by solid bodies inside which that the fluid cannot penetrate. Then the fluid velocity $\mathbf{v}$ at the solid-body boundaries should not have a normal component; according to Eq. (8.28), this means

$$\left.\frac{\partial\phi}{\partial n}\right|_{\text{surfaces}} = 0. \tag{8.30}$$

On the other hand, if at large distances the fluid flow is known, e.g. uniform, then:

$$\nabla\phi = -\mathbf{v}_0 = \text{const}, \qquad \text{at } r \to \infty. \tag{8.31}$$

As the reader may already know (for example, from a course of electrodynamics[22]), the Laplace equation (8.29) is readily solvable analytically in several simple (symmetric) but important situations. Let us consider, for example, the case of a round cylinder, with radius R, immersed into a flow with the initial velocity $\mathbf{v}_0$ perpendicular to the cylinder's axis (figure 8.8). For this problem, it is natural to use the cylindrical coordinates, with axis z parallel to the cylinder's axis. In this case the velocity distribution is evidently independent of z, so that we may simplify the general expression of the Laplace operator in cylindrical coordinates[23] by taking $\partial/\partial z = 0$. As a result, Eq. (8.29) is reduced to[24]

$$\frac{1}{\rho}\frac{\partial}{\partial\rho}\left(\rho\frac{\partial\phi}{\partial r}\right) + \frac{1}{\rho^2}\frac{\partial^2\phi}{\partial\theta^2} = 0, \qquad \text{at} \quad \rho \geqslant R. \tag{8.32}$$

---

[21] This case is very important, because the motion of a solid body, with a constant velocity $\mathbf{u}$, in the otherwise stationary fluid, gives exactly the same problem (with $\mathbf{v}_0 = -\mathbf{u}$), in a reference frame bound to the body.

[22] See, e.g. *Part EM* sections 2.3–2.8.

[23] See, e.g. Eq. (A.61).

[24] Let me hope that letter ρ, used here to denote the magnitude of the 2D radius-vector $\boldsymbol{\rho} = \{x, y\}$, will not be confused with the fluid's density—which does not participate in this boundary problem.

The general solution of this equation may be obtained using the variable separation method, similar to that used in section 6.5—see Eq. (6.67). The result is[25]

$$\phi = a_0 + b_0 \ln \rho + \sum_{n=1}^{\infty}(c_n \cos n\varphi + s_n \sin n\varphi)(a_n\rho^n + b_n\rho^{-n}), \qquad (8.33)$$

where coefficients a_n and b_n have to be found from the boundary conditions (8.30) and (8.31). Choosing axis $x = r \cos \varphi$ to be parallel to vector $\mathbf{v}_0$ (figure 8.8a) we may spell out these conditions in the form

$$\frac{\partial \phi}{\partial \rho} = 0, \qquad \text{at} \quad \rho = R, \qquad (8.34)$$

$$\phi \to -v_0\rho \cos \varphi + \phi_0, \qquad \text{at} \quad \rho \gg R, \qquad (8.35)$$

where ϕ_0 is an arbitrary constant, which does not affect the velocity distribution, and may be taken for zero. The condition (8.35) is incompatible with all terms of the sum (8.33), except the term with $n = 1$ (with $s_1 = 0$ and $c_1a_1 = -v_0$), so that Eq. (8.33) is reduced to

$$\phi = \left(-v_0\rho + \frac{c_1b_1}{\rho}\right)\cos \varphi. \qquad (8.36)$$

Now, plugging this solution into equation (8.34), we obtain $c_1b_1 = -v_0R^2$, so that, finally,

$$\phi = -v_0\left(\rho + \frac{R^2}{\rho}\right)\cos \varphi. \qquad (8.37a)$$

Figure 8.8a shows the surfaces of constant velocity potential ϕ. In order to find the fluid velocity, it is easier to rewrite Eq. (8.37) in the Cartesian coordinates $x = \rho \cos \varphi$, $y = \rho \sin \varphi$:

$$\phi = -v_0x\left(1 + \frac{R^2}{\rho^2}\right) = -v_0x\left(1 + \frac{R^2}{x^2 + y^2}\right). \qquad (8.37b)$$

From this equation, we may readily calculate the Cartesian components $v_x = -\partial\phi/\partial x$ and $v_y = -\partial\phi/\partial y$ of the fluid velocity. Figure 8.8b shows the flow streamlines[26]. One can see that the largest potential gradient, and hence the maximum fluid's speed, is achieved at the vertical diameter's ends ($\rho = R$, $\phi = \pm\, \pi/2$), where

[25] See, e.g. *Part EM* Eq. (2.112). Note that the most general solution of Eq. (8.32) also includes a term proportional to ϕ, but this term should be zero for such a single-valued function as the velocity potential.

[26] They may be found by integration of the evident equation $dy/dx = v_y(x,y)/v_x(x,y)$. For our simple problem this integration may be done analytically, giving the relation $y[1 - R^2/(x^2 + y^2)] = \text{const}$, where the constant is specific for each streamline.

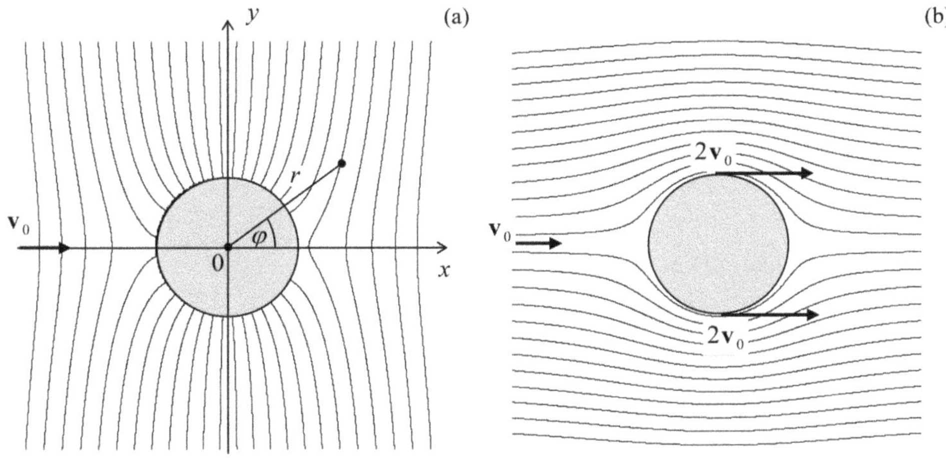

Figure 8.8. The flow of an ideal, incompressible fluid around a cylinder: (a) equipotential surfaces and (b) streamlines.

$$v = v_x = -\left.\frac{\partial\phi}{\partial x}\right|_{\substack{r=R\\x=0}} = 2v_0. \tag{8.38}$$

Now the pressure distribution may be calculated by plugging Eq. (8.37) into the Bernoulli equation (8.27) with $u(\mathbf{r}) = 0$. The result shows that the pressure reaches maximum at the ends of the longitudinal diameter $y = 0$, while at the ends of the transverse diameter $x = 0$, where the velocity is largest, it is lower by $2\rho v_0^2$ (where ρ is the fluid density again—apologies for the notation jitters!). Note that the distributions of both the velocity and the pressure are symmetric with respect to the transverse axis $x = 0$, so that the fluid flow does not create any net drag force in its direction. It may be shown that this result, which stems from the conservation of the mechanical energy of an ideal fluid, remains valid for a solid body of arbitrary shape moving inside an infinite volume of an ideal fluid—the so-called *D'Alambert paradox*. However, if a body moves near an ideal fluid's surface, its energy may be transformed into that of surface waves, and drag becomes possible.

Speaking about the *surface waves* in a gravity field[27], their description is one more classical problem of the ideal fluid dynamics. Let us consider an open surface of an ideal fluid of density ρ in a uniform gravity field $\mathbf{f} = \rho\mathbf{g} = -\rho g\mathbf{n}_y$—see figure 8.9. If the wave amplitude A is sufficiently small, we can neglect the nonlinear term $(\mathbf{v}\cdot\nabla)\mathbf{v} \propto A^2$ in the Euler equation (8.23) in comparison with the first term, $\partial\mathbf{v}/\partial t$, which is linear in A. For a wave with frequency ω and wavenumber k, particle's velocity $\mathbf{v} = d\mathbf{q}/dt$ is of the order of ωA, so that this approximation is legitimate if $\omega^2 A \gg k(\omega A)^2$, i.e. when

$$kA \ll 1, \tag{8.39}$$

[27] The alternative, historic term 'gravity waves' for this phenomenon may nowadays lead to confusion with the relativistic effect of gravity waves (which may propagate in free space), whose recent direct observation by the LIGO collaboration [3] has become a major physics sensation.

i.e. when the wave's amplitude A is much smaller than its wavelength $\lambda = 2\pi/k$. Due to this assumption, we may neglect the fluid vorticity effects, and (for an incompressible fluid) again use the Laplace equation (8.29) for the wave's analysis.

Looking for its solution in the natural form of a sinusoidal wave, uniform in one of the horizontal directions (x),

$$\phi = \text{Re}\,[\Phi(y)e^{i(kz-\omega t)}], \tag{8.40}$$

we obtain a very simple equation

$$\frac{d^2\Phi}{dy^2} - k^2\Phi = 0, \tag{8.41}$$

with an exponential solution (decaying, as it has to, at $y \to -\infty$), $\Phi = \Phi_A \exp\{ky\}$, so that Eq. (8.40) becomes

$$\phi = \text{Re}\,[\Phi_A e^{ky} e^{i(kz-\omega t)}] = \Phi_A e^{ky} \cos(kz - \omega t), \tag{8.42}$$

where the last form is valid if Φ_A is real—which may always be arranged by a proper selection of the origins of z and/or t. Note that the rate k of the wave's decay in the vertical direction is exactly equal to the wavenumber of its propagation in the horizontal direction—along the fluid's surface. Because of that, the trajectories of fluid particles are exactly circular. Indeed, using Eqs. (8.28) and (8.42) to calculate velocity components,

$$\begin{aligned} v_x = 0, \quad v_y &= -\frac{\partial \phi}{\partial y} = -k\Phi_A e^{ky}\cos(kz - \omega t), \\ v_z = -\frac{\partial \phi}{\partial z} &= -k\Phi_A e^{ky}\sin(kz - \omega t), \end{aligned} \tag{8.43}$$

we see that v_y and v_z, at the same height y, have equal real amplitudes, and are phase-shifted by $\pi/2$. This result becomes even more clear if we use the velocity definition $\mathbf{v} = d\mathbf{q}/dt$ to integrate Eq. (8.43) over time to recover the particle displacement law $\mathbf{q}(t)$. Due to the strong inequality equation (8.39), the integration may be done at fixed y and z:

$$q_y = q_A e^{ky}\sin(kz - \omega t), \qquad q_z = q_A e^{ky}\cos(kz - \omega t), \qquad \text{with } q_A \equiv \frac{k}{\omega}\Phi_A. \tag{8.44}$$

Note that the phase of oscillations of v_z coincides with that of q_y. It means, in particular, that at a wave's top ('crest'), fluid particles are moving in the direction of the wave's propagation—see the arrows in figure 8.9.

It is remarkable that all this picture follows from the Laplace equation alone! The 'only' remaining feature to calculate is the dispersion law $\omega(k)$, and for that we need to combine Eq. (8.42) with what remains, in our linear approximation, of the Euler equation (8.23). In this approximation, and with the bulk force potential $u = \rho g y$, this equation is reduced to

$$\nabla\left(-\rho\frac{\partial\phi}{\partial t} + \mathcal{P} + \rho g y\right) = 0. \tag{8.45}$$

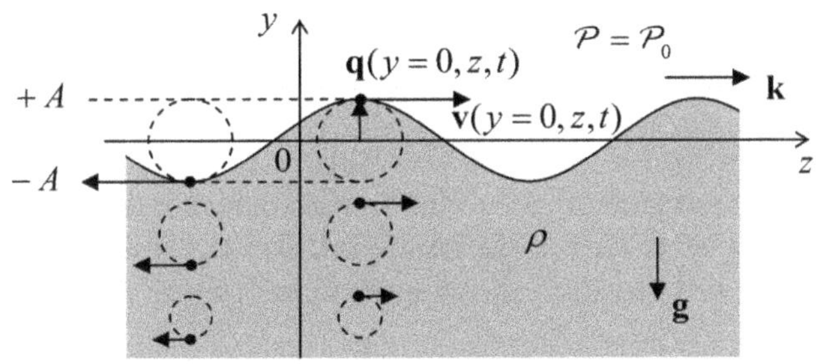

Figure 8.9. Small surface wave on a deep heavy fluid. Dashed lines show fluid particle trajectories. (For clarity, the displacement amplitude A is strongly exaggerated.)

This equation means that the function in the parentheses is constant in space; at the surface, and at negligible surface tension, it should be equal to pressure $\mathcal{P}_0$ above the surface (say, the atmospheric pressure), which we assume to be constant. This means that on the surface, the contributions to $\mathcal{P}$ that come from the first and the third term in Eq. (8.45) should compensate each other. Let us take the average surface position for $y = 0$; then the surface with waves is described by the relation $y\,(z,\,t) = q_y(y,\,z,\,t)$—see figure 8.9. Due to the strong relation (8.39), we can use Eqs. (8.42) and (8.44) with $y = 0$, so that the above compensation condition yields

$$-\rho\omega\Phi_A \sin(kz - \omega t) + \rho g\frac{k}{\omega}\Phi_A \sin(kz - \omega t) = 0. \tag{8.46}$$

This condition is identically satisfied on the whole surface (and for any Φ_A) as soon as

$$\omega^2 = gk, \tag{8.47}$$

this equality giving the dispersion relation we were looking for.

Looking at this surprisingly simple result (which includes just one constant, g) note, first of all, that it does not involve the fluid's density. This is not too surprising, because due to the weak equivalence principle, particle masses always drop out of the results of problems involving gravitational forces alone. Second, the dispersion law (8.47) is strongly nonlinear, and in particular does not have the acoustic wave limit at all. This means that the surface wave propagation is strongly dispersive, with both the phase velocity $v_{ph} \equiv \omega/k = g/\omega$ and the group velocity $v_{gr} \equiv d\omega/dk = g/2\omega = v_{ph}/2$ diverging at $\omega \to 0$.

This divergence is an artifact of our assumption of the infinite fluid thickness. A rather straightforward generalization of the above calculations to a fluid's layer of a finite thickness h, using the additional boundary condition $v_y|_{y=-h} = 0$, yields the following modified dispersion relation,

$$\omega^2 = gk\tanh\ kh. \tag{8.48}$$

It shows that relatively long waves, with $\lambda \gg h$, i.e. with $kh \ll 1$, propagate without dispersion (i.e. have $\omega/k = \text{const} \equiv v$), with the following velocity:

$$v = (gh)^{1/2}. \tag{8.49}$$

For the Earth oceans, this velocity is rather high, approaching 300 m s^{-1} (!) for h = 10 km. This result explains, in particular, the very fast propagation of tsunami waves.

In the opposite limit of very short waves (large k), Eq. (8.47) also does not give a good description of experimental data, due to the surface tension effects—see section 8.2. Using Eq. (8.13), it is easy (and hence left for the reader) to show that their account leads (at $kh \gg 1$) to the following modification of Eq. (8.47):

$$\omega^2 = gk + \frac{\gamma k^3}{\rho}. \tag{8.50}$$

According to this formula, the surface tension is important at wavelengths smaller than the capillary constant a_c given by Eq. (8.14). Much shorter waves, for whom Eq. (8.50) yields $\omega \propto k^{3/2}$, are called the *capillary waves*—or just 'ripples'.

8.5 Dynamics: viscous fluids

Fluid viscosity of many fluids, at not too high velocities, may be described surprisingly well by adding, to the static stress tensor (8.2), additional components proportional to the velocity $\mathbf{v} \equiv d\mathbf{q}/dt$:

$$\sigma_{jj'} = -\mathcal{P}\delta_{jj'} + \tilde{\sigma}_{jj'}(\mathbf{v}). \tag{8.51}$$

In the view of our experience with Hooke's law (7.32), expressing a stress tensor proportional to particle displacements $\mathbf{q}$, we may expect a similar expression with the replacement $\mathbf{q} \to \mathbf{v} = d\mathbf{q}/dt$:

$$\tilde{\sigma}_{jj'} = 2\eta\left(e_{jj'} - \frac{1}{3}\delta_{jj'}\,\mathrm{Tr(e)}\right) + 3\zeta\left(\frac{1}{3}\delta_{jj'}\,\mathrm{Tr(e)}\right), \tag{8.52a}$$

where $e_{jj'}$ are the elements of the symmetrized strain derivative tensor:

$$e_{jj'} \equiv \frac{ds_{jj'}}{dt} = \frac{1}{2}\left(\frac{\partial v_j}{\partial r_{j'}} + \frac{\partial v_{j'}}{\partial r_j}\right). \tag{8.52b}$$

Experiments confirm that Eq. (8.52) gives a good description of the viscosity effects in a broad range of isotropic fluids. The coefficient η is called either the *shear viscosity,* the *dynamic viscosity,* or just *viscosity*, while ζ is called the *second* (or *bulk*) viscosity.

In the most frequent case of a virtually incompressible fluid, $\mathrm{Tr(e)} = d[\mathrm{Tr(s)}]/dt = (dV/dt)/V = 0$, so that the term proportional to ζ vanishes, and η is the only important viscosity parameter[28]. Table 8.1 shows the approximate values of the viscosity, together with the mass density ρ, for several representative fluids. One can see that η may vary in very broad limits; the extreme cases of liquids are glasses

[28] Probably the most important effect we miss by neglecting ζ is the attenuation of the (longitudinal) acoustic waves, into which the second viscosity makes a major (and in some cases, the main) contribution—whose (rather straightforward) analysis is left for a reader's exercise.

Table 8.1. Important parameters of several representative fluids (approximate values).

Fluid (all at 300 K, until indicated otherwise)	η (mPa s)	ρ (kg m^{-3})
Glasses	10^{21}–10^{24}	2200–2500
Earth magmas (at 800–1400 K)	10^{4}–10^{14}	2200–2800
Machine oils (SAE 10–40 W)	65–320	900
Water	0.89	1000
Mercury	1.53	13530
Liquid helium 4 (at 4.2 K, 10^5 Pa)	0.019	130
Air (at 10^5 Pa)	0.018	1.3

(which, somewhat counter-intuitively, are not stable solids even at room temperature, but rather may 'flow', although extremely slowly, until they eventually crystallize) and liquid helium (whose viscosity is of the order of that of gases, despite its much larger density[29]).

Incorporating the additional components of $\sigma_{jj'}$ to the equation (8.23) of fluid motion, just as it was done in the derivation of Eq. (7.107) of the elasticity theory, and with the account of Eq. (8.19), we arrive at the famous *Navier–Stokes* equation[30]:

$$\rho\frac{\partial \mathbf{v}}{\partial t} + \rho(\mathbf{v}\cdot\nabla)\mathbf{v} = -\nabla\mathcal{P} + \mathbf{f} + \eta\nabla^2\mathbf{v} + \left(\zeta + \frac{\eta}{3}\right)\nabla(\nabla\cdot\mathbf{v}). \tag{8.53}$$

The apparent simplicity of this equation should not mask the enormous range of phenomena, notably including turbulence (see the next section), that are described by it, and the complexity of its solutions even for some simple geometries. In most problems that are interesting in practice, the only option is to use numerical methods, but due to the large number of parameters (ρ, η, ζ, plus geometrical parameters of the involved bodies, plus the distribution of bulk forces **f**, plus boundary conditions), this approach is strongly plagued by the 'curse of dimensionality' that was discussed at the end of section 5.8.

Let us see how the Navier–Stokes equation works, using several simple examples. As the simplest case, let us consider the so-called *Couette flow* of an incompressible fluid layer between two wide, horizontal plates (figure 8.10), caused by their mutual sliding with a constant relative velocity $\mathbf{v}_0$. Let us assume a *laminar* (vorticity-free) fluid flow. (As will be discussed in the next section, this assumption is only valid within certain limits.) Then we may use the evident symmetry of the problem, to take, in the reference frame shown in figure 8.10, $\mathbf{v} = \mathbf{n}_z v(y)$. Let the bulk forces be vertical, $\mathbf{f} = \mathbf{n}_y f$, so they do not give an additional drive to fluid flow. Then for the

[29] Actually, at even lower temperatures (for He 4, at $T < T_\lambda \approx 2.17$ K), helium becomes a *superfluid*, i.e. loses its viscosity completely, as result of Bose–Einstein condensation—see, e.g. *Part SM* section 3.4.

[30] Named after C-L Navier (1785–1836) who suggested the equation, and G Stokes (1819–1903) who demonstrated its relevance by solving the equation for several key situations.

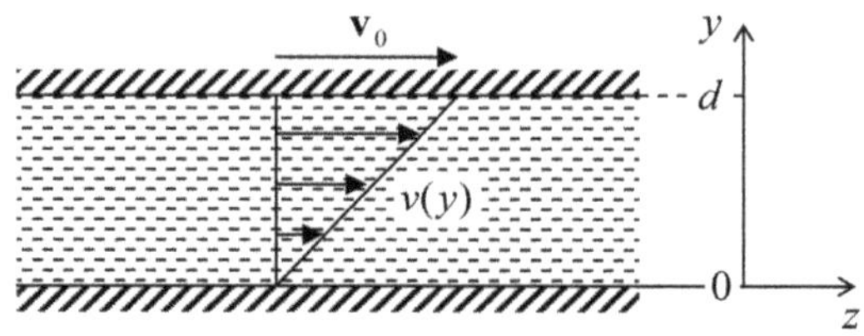

Figure 8.10. The simplest problem of the viscous fluid flow.

stationary flow ($\partial\mathbf{v}/\partial t = 0$), the vertical, y-component of the Navier–Stokes equation is reduced to the static Pascal equation (8.6), showing that the pressure distribution is not affected by the plate (and fluid) motion. In the horizontal, z-component of the equation, only one term, $\nabla^2 v$, survives, so that for the only Cartesian component of the fluid's velocity we obtain the 1D Laplace equation

$$\frac{d^2v}{dy^2} = 0. \qquad (8.54)$$

In contrast to the ideal fluid (see, e.g. Figure 8.8b), the relative velocity of a viscous fluid and a solid wall it flows by should approach zero at the wall[31], so that Eq. (8.54) should be solved with boundary conditions

$$v = \begin{cases} 0, & \text{at} \quad y = 0, \\ v_0, & \text{at} \quad y = d. \end{cases} \qquad (8.55)$$

Using the evident solution of this boundary problem, $v(y) = (y/d)v_0$, illustrated by arrows in figure 8.10, we can now calculate the horizontal *drag force* acting on a unit area of each plate. For the bottom plate,

$$\frac{F_z}{A_y} = \sigma_{zy}|_{y=0} = \eta \frac{\partial v}{\partial y}|_{y=0} = \eta \frac{v_0}{d}. \qquad (8.56)$$

(For the top plate, the derivative $\partial v/\partial y$ has the same value, but the sign of dA_y has to be changed to reflect the direction of the outer normal to the solid surface, so that we obtain a similar force but with the negative sign.) The well-known result (8.56) is often used in undergraduate courses for a definition of the dynamic viscosity, and indeed shows its physical meaning very well[32].

As the next, slightly less trivial example let us consider the so-called *Poiseuille problem*[33]: finding the relation between the constant external pressure gradient

[31] This is essentially an additional experimental fact, but may be understood as follows. A solid may be considered as an ultimate case of a fluid (with infinite viscosity), and the tangential component of the velocity should be a continuous an interface between two viscous fluids, in order to avoid infinite stress—see Eq. (8.52).

[32] The very notion of viscosity η was introduced (by nobody else as the same Sir Isaac Newton) via an equality equivalent to Eq. (8.56), so that any effect resulting in a drag force proportional to velocity is frequently called the *Newtonian viscosity*.

[33] It was solved theoretically by G Stokes in 1845 in order to explain Eq. (8.62), which had been formulated by G Hagen and J Poiseuille in 1839–40 on the basis of their experimental results.

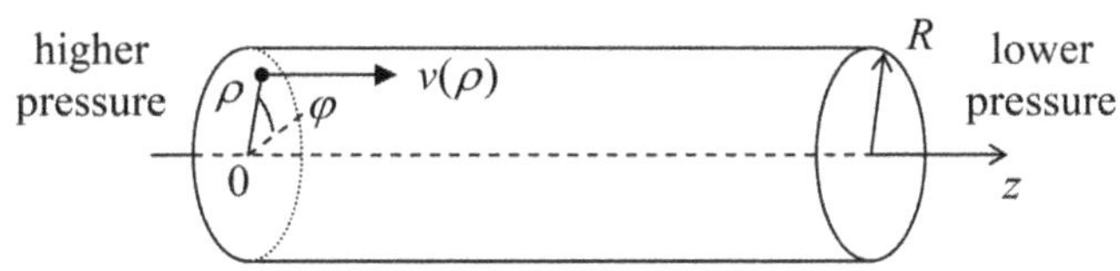

Figure 8.11. The Poiseuille problem.

$\chi \equiv -\partial\mathcal{P}/\partial z$ applied along a round pipe with internal radius R (figure 8.11), and the so-called *discharge Q*—defined as the mass of fluid flowing through the pipe's cross-section in unit time. Again assuming a laminar flow, we can involve the problem's uniformity along the z-axis and its axial symmetry to infer that $\mathbf{v} = \mathbf{n}_z v(\rho)$, and $\mathcal{P} = -\chi z + f(\rho, \phi) + \text{const}$ (where $\boldsymbol{\rho} = \{\rho, \phi\}$ is again the 2D radius-vector rather than the fluid density), so that the Navier–Stokes equation (8.53) for an incompressible fluid (with $\nabla\cdot\mathbf{v} = 0$) is reduced to a 2D Poisson equation

$$\eta\nabla_2^2 v = -\chi. \tag{8.57}$$

After spelling out the 2D Laplace operator in polar coordinates for our axially symmetric case $\partial/\partial\phi = 0$, Eq. (8.57) becomes a simple ordinary differential equation,

$$\eta\frac{1}{\rho}\frac{d}{d\rho}\left(\rho\frac{dv}{d\rho}\right) = -\chi, \tag{8.58}$$

which has to be solved at the segment $0 \leqslant \rho \leqslant R$, with the following boundary conditions:

$$\begin{aligned} v = 0, &\quad \text{at } \rho = R, \\ \frac{dv}{d\rho} = 0, &\quad \text{at } \rho = 0. \end{aligned} \tag{8.59}$$

(The latter condition is required by the axial symmetry.) A straightforward double integration yields:

$$v = \frac{\chi}{4\eta}(R^2 - \rho^2), \tag{8.60}$$

so that the (easy) integration of the mass flow density over the cross-section of the pipe,

$$Q \equiv \int_A \rho v d^2r = 2\pi\rho\frac{\chi}{4\eta}\int_0^R (R^2 - \rho'^2)\rho' d\rho', \tag{8.61}$$

immediately gives us the so-called *Poiseuille* (or 'Hagen–Poiseuille') *law* for the fluid discharge:

$$Q = \frac{\pi}{8}\rho\frac{\chi}{\eta}R^4, \tag{8.62}$$

where (sorry!) ρ is the mass density again.

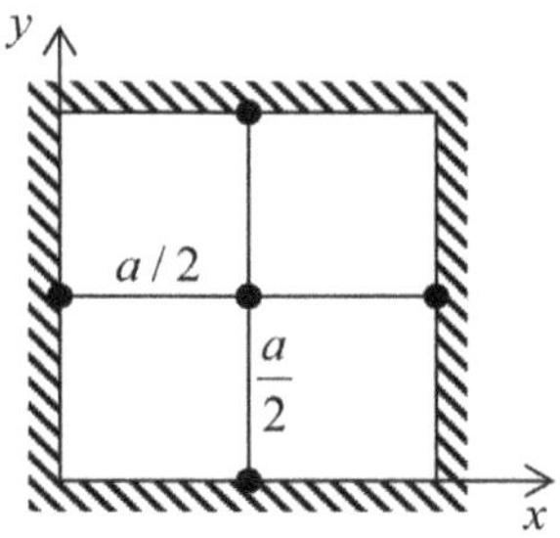

Figure 8.12. Application of the finite-difference method with a very coarse mesh (with step $h = a/2$) to the problem of viscous fluid flow in a pipe with a square cross-section.

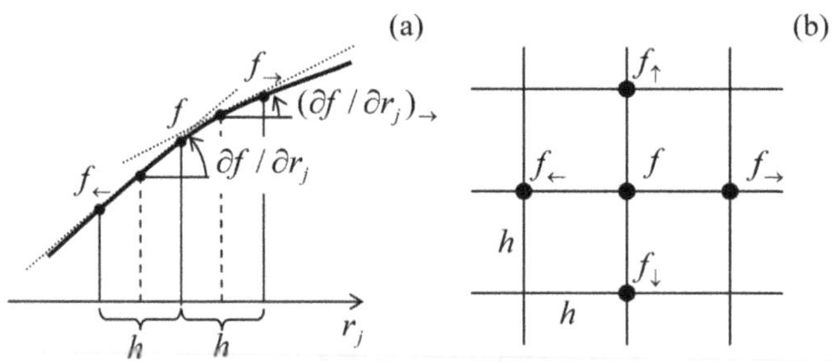

Figure 8.13. Idea of the finite-difference method in (a) one and (b) two dimensions.

Of course, the 2D Poisson equation (8.57) is not so readily solvable for each cross-section shape. For example, consider a very simple, square-shaped cross-section with side a (figure 8.12). For this case, it is natural to use the Cartesian coordinates, so that Eq. (8.57) becomes

$$\frac{\partial^2 v}{\partial x^2} + \frac{\partial^2 v}{\partial y^2} = -\frac{\chi}{\eta} = \text{const}, \qquad \text{for } 0 \leqslant x, y \leqslant a, \tag{8.63}$$

and has to be solved with boundary conditions

$$v = 0, \qquad \text{at } x, y = 0, a. \tag{8.64}$$

For this boundary problem, analytical methods such as the variable separation give answers in the form of the infinite sums (series)[34], which ultimately require computers anyway, for their plotting and comprehension. Let me use this pretext to discuss how explicitly numerical methods may be used for such problems—or for any partial differential equations involving the Laplace operator. The simplest of them is the *finite-difference* method[35] in which the function to be calculated, $f(r_1, r_2, \ldots)$, is represented by its values in discrete points of a rectangular grid (frequently called the *mesh*) of the corresponding dimensionality (figure 8.13).

[34] See, e.g. *Part EM* section 2.5.

[35] For more details see, e.g. [4].

In section 5.7, we have already discussed how to use such a grid to approximate the first derivative—see Eq. (5.98). Its extension to the second derivative is straightforward—see figure 8.13a[36]:

$$\frac{\partial^2 f}{\partial r_j^2} = \frac{\partial}{\partial r_j}\left(\frac{\partial f}{\partial r_j}\right) \approx \frac{1}{h}\left(\frac{\partial f}{\partial r_{j\rightarrow}} - \frac{\partial f}{\partial r_{j\leftarrow}}\right) \approx \frac{1}{h}\left[\frac{f_\rightarrow - f}{h} - \frac{f - f_\leftarrow}{h}\right]$$
$$= \frac{f_\rightarrow + f_\leftarrow - 2f}{h^2}. \tag{8.65}$$

The relative error of this approximation is of the order of $h^2\partial^4 f/\partial r_j^4$, quite acceptable in many cases. As a result, the left-hand side of Eq. (8.63), treated on a square mesh with step h (figure 8.13b), may be approximated with the so-called *five-point scheme*:

$$\frac{\partial^2 v}{\partial x^2} + \frac{\partial^2 v}{\partial y^2} \approx \frac{v_\rightarrow + v_\leftarrow - 2v}{h^2} + \frac{v_\uparrow + v_\downarrow - 2v}{h^2} = \frac{v_\rightarrow + v_\leftarrow + v_\uparrow + v_\downarrow - 4v}{h^2}. \tag{8.66}$$

(The generalization to the *seven-point scheme*, appropriate for 3D problems, is straightforward.)

Let us apply this scheme to the pipe with the square cross-section, using an extremely coarse mesh with step $h = a/2$ (figure 8.12). In this case the fluid velocity v should equal zero at the walls, i.e. in all points of the five-point scheme, except for the central point (in which the velocity is evidently the largest), so that Eqs. (8.63) and (8.66) yield[37]

$$\frac{0 + 0 + 0 + 0 - 4v_{\max}}{(a/2)^2} \approx -\frac{\chi}{\eta}, \qquad \text{i.e. } v_{\max} \approx \frac{1}{16}\frac{\chi a^2}{\eta}. \tag{8.67}$$

This result for the maximal velocity is only ~20% different from the exact value. Using a slightly finer mesh with $h = a/4$, which gives a readily solvable system of three linear equations for three different velocity values (the exercise left for the reader), brings us within a couple percent from the exact result. This shows that such 'numerical' methods may be practically more efficient than the 'analytical' ones, even if the only available tool is a calculator app on your smartphone rather than an advanced computer.

Of course, many practical problems of fluid dynamics do require high-performance computing, particularly in conditions of turbulence (see the next section) with its complex, irregular spatial–temporal structure. In these conditions, the finite-difference approach discussed above may become unsatisfactory, because it implies the same accuracy of the derivative approximation through the whole volume of

[36] As a reminder, in the beginning of section 6.4 we have already discussed the reciprocal transition—from a similar sum to the second derivative in the continuous limit ($h \to 0$).

[37] Note that the value (8.67) of $v_{\max}$ is exactly the same as given by the analytical formula (8.60) for the round cross-section with the radius $R = a/2$. This is not an occasional coincidence. The velocity distribution given by Eq. (8.60) is a quadratic function of both x and y. For such functions, with all derivatives higher than $\partial^2 f/\partial r_j^2$ equal to zero, Eq. (8.66) is exact rather than approximate.

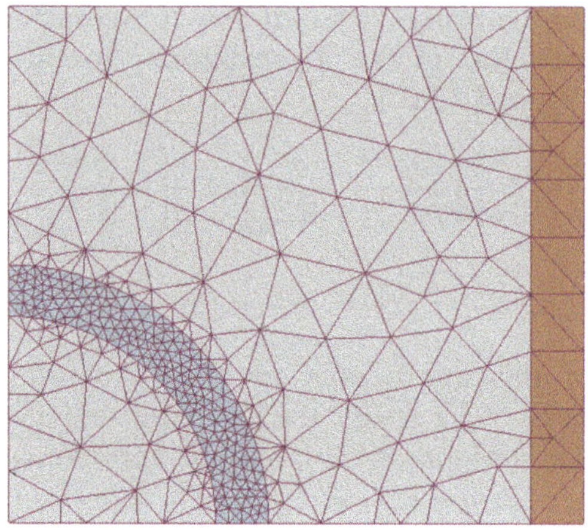

Figure 8.14. A typical finite-element mesh generated automatically for a system of complex geometry—in this case, one round cylindrical shell inside another one, with mutually perpendicular axes. (Adapted from the original by I Zureks https://commons.wikimedia.org/w/index.php?curid=2358783, under the CC license BY-SA 3.0.)

interest. A more powerful (but also much more complex for implementation) approach is the *finite-element method* in which the discrete point mesh is based on triangles with uneven sides, and is (in most cases, automatically) generated in accordance with the system geometry, giving many more mesh points at the location of highest gradients of the calculated function (figure 8.14), and hence better calculation accuracy for the same total number of points. Unfortunately I do not have time to go into the details of that method, so the interested reader is referred to the special literature on this subject[38].

Before proceeding to our next topic, let me mention one more important problem that is analytically solvable using the Navier–Stokes equation: a slow motion of a solid sphere of radius R, with a constant velocity $\mathbf{v}_0$, through an incompressible viscous fluid—or equivalently, a slow flow of the fluid (uniform at large distances) around an immobile sphere. Indeed, in the limit $v \to 0$, the second term on the left-hand side of Eq. (8.53) is negligible (just as at the surface wave analysis in section 8.3), and the equation takes the form

$$-\nabla\mathcal{P} + \eta\nabla^2\mathbf{v} = 0, \tag{8.68}$$

which should be complemented with the incompressibility condition $\nabla\cdot\mathbf{v} = 0$ and boundary conditions

$$\begin{aligned} \mathbf{v} &= 0, \text{ at } r = R, \\ \mathbf{v} &\to \mathbf{v}_0, \text{ at } r \to \infty. \end{aligned} \tag{8.69}$$

[38] I can recommend, e.g. [5] or [6].

In spherical coordinates, with the polar axis directed along the vector $\mathbf{v}_0$, this boundary problem has axial symmetry (so that $\partial\mathbf{v}/\partial\phi = 0$ and $v_\phi = 0$), and allows the following analytical solution:

$$v_r = v_0 \cos\theta\left(1 - \frac{3R}{2r} + \frac{R^3}{2r^2}\right), \qquad v_\theta = -v_0 \sin\theta\left(1 - \frac{3R}{4r} + \frac{R^3}{4r^2}\right). \qquad (8.70)$$

Calculating the pressure distribution from Eq. (8.68), and integrating it over the surface of the sphere, it is now straightforward to obtain the famous *Stokes formula* for the drag force acting on the sphere:

$$F = 6\pi\eta R v_0. \qquad (8.71)$$

Historically, this formula has played an important role in the first precise (better than 1%) calculation of the fundamental electric charge e by R Millikan and H Fletcher from their famous oil drop experiments in 1909–13. For what follows, it is convenient to recast this result into the following form:

$$C_d = \frac{24}{Re}, \qquad (8.72)$$

where C_d is the *drag coefficient* defined as

$$C_d \equiv \frac{F}{\rho v_0^2 A/2}, \qquad (8.73)$$

with $A \equiv \pi R^2$ being the sphere's cross-section 'as seen by the incident fluid flow', and Re is the so-called *Reynolds number*[39]. In the general case, the number is defined as

$$\mathrm{Re} \equiv \frac{\rho v l}{\eta}, \qquad (8.74)$$

where l is the linear-size scale of the problem, and v is its velocity scale. (In the particular case of Eq. (8.72) for the sphere, l is identified with the sphere's diameter $D = 2R$, and v with v_0.) The physical sense of these two definitions will be discussed in the next section.

8.6 Turbulence

As figure 8.15 shows, the Stokes' result (8.71)–(8.72) is only valid at $Re \ll 1$, while for larger values of the Reynolds number, i.e. at higher velocities v_0, the drag force is larger. This fact is not quite surprising, because at the derivation of the Stokes' result, the nonlinear term $(\mathbf{v}\cdot\nabla)\mathbf{v}$ in the Navier–Stokes equation (8.53), which scales as v^2, was neglected in comparison to the linear terms, scaling as v.

What is more surprising is that the function $C_d(Re)$ exhibits such a complicated behavior over many orders of the velocity's magnitude, giving a hint that the fluid

[39] This notion was introduced in 1851 by the same G Stokes, but eventually named after O Reynolds who popularized it three decades later.

flow at large Reynolds numbers should be very complicated. Indeed, the reason for this complexity is a gradual development of very intricate, time-dependent fluid patterns, called *turbulence*, rich with vortices—for an example see figure 8.16. These vortices are particularly pronounced in the region behind the moving body (the so-called *wake*), while the region before the body remains almost unperturbed. As figure 8.15 indicates, the turbulence exhibits rather different behaviors at various velocities (i.e. values of *Re*), and sometimes changes rather abruptly—see, for example, the significant drag drops at $Re \approx 5 \times 10^5$.

In order to understand the conditions of this phenomenon, let us estimate the scale of various terms in the Navier–Stokes equation (8.53) for the generic case of a body with characteristic size l moving in an otherwise static, incompressible fluid, with velocity v. In this case the characteristic time scale of possible non-stationary phenomena is given by the ratio l/v,[40] so that we arrive at the following estimates:

$$\begin{array}{lcccc} \text{Equation term:} & \rho\dfrac{\partial \mathbf{v}}{\partial t} & \rho(\mathbf{v}.\ \nabla)\mathbf{v} & \mathbf{f} & \eta\nabla^2\mathbf{v} \\ \text{Order of magnitude:} & \rho\dfrac{v^2}{l} & \rho\dfrac{v^2}{l} & \rho g & \eta\dfrac{v}{l^2} \end{array} \tag{8.75}$$

(I have skipped the term $\nabla\mathcal{P}$ because, as we saw in the previous section, in typical fluid flow problems it balances the viscosity term and hence is of the same order of magnitude.) This table shows that the relative importance of the terms may be characterized by two dimensionless ratios[41].

The first of them is the so-called *Froude number*

$$F \equiv \frac{\rho v^2/l}{\rho g} = \frac{v^2}{l\, g}, \tag{8.76}$$

which characterizes the relative importance of the gravity—or, upon an appropriate modification, of other bulk forces. In most practical problems (with the important exception of surface waves, see section 8.4) $F \gg 1$, so that the gravity effects may be neglected.

Much more important is another ratio, the Reynolds number (8.74), which may be rewritten as

[40] The time scale of phenomena in non-autonomous systems may be different from l/v; for example, for forced oscillations of a fluid flow with frequency ω, it is given by the oscillation period $\mathcal{T} \equiv 2\pi/\omega$. For such problems, the ratio $S \equiv (l/v)/\mathcal{T}$ serves as another, independent dimensionless constant, commonly called either the *Strouhal number* or the *reduced frequency*.

[41] For substantially compressible fluids (e.g. gases), the most important additional dimensionless parameter is the *Mach number*, $M \equiv v/v_l$, where $v_l = (K/\rho)^{1/2}$ is the velocity of the longitudinal sound—which is, as we already know, the only wave mode possible in an infinite fluid. Particularly significant in practice are *supersonic effects* (including the shock wave in the form of the famous *Mach cone* with half-angle $\theta_M = \sin^{-1} M^{-1}$) that arise at $M > 1$. For a more thorough discussion of these issues, I have to refer the reader to more specialized texts—e.g. chapter 9 of [2], or chapter 15 in [8], which is generally a good book on the subject. Another popular, rather simple, textbook is [9].

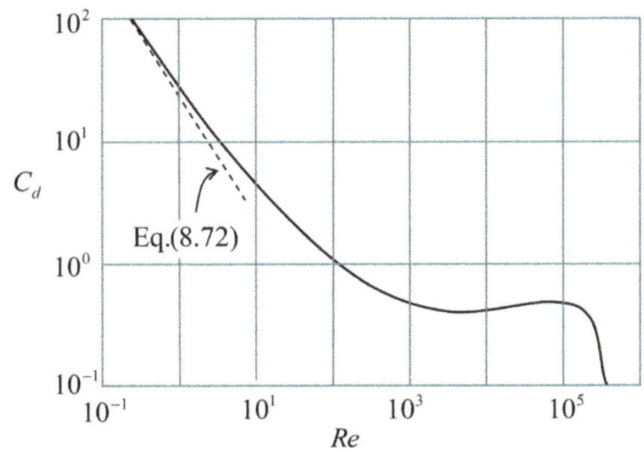

Figure 8.15. The drag coefficient C_d for a sphere in an incompressible fluid, as a function of the Reynolds number (a log–log plot). Experimental data from [7].

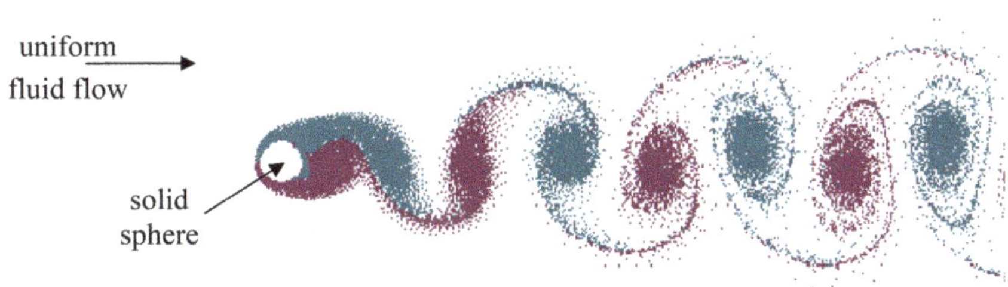

Figure 8.16. A snapshot of the turbulent tail (*wake*) behind a sphere moving in a fluid with a high Reynolds number, showing the so-called *von Kármán vortex street*. Adapted from the original (actually, a very nice animation www.mcef.ep.usp.br/staff/jmeneg/cesareo/vort2.gif) by Cesareo de La Rosa Siquiera, as a copy-right-free material, available at https://commons.wikimedia.org/w/index.php?curid=87351.

$$Re \equiv \frac{\rho v l}{\eta} = \frac{\rho v^2/l}{\eta v/l^2}, \tag{8.77}$$

and hence is a measure of the relative importance of the fluid particle's inertia in comparison with the viscosity effects[42]. So again, it is natural that for a sphere, the role of the vorticity-creating term $(\mathbf{v}\cdot\nabla)\mathbf{v}$ becomes noticeable already at $Re \sim 1$—see figure 8.15. What is very counter-intuitive is the onset of turbulence in systems where the laminar (turbulence-free) flow is formally an exact solution to the Navier–Stokes equation for any Re. For example, at $Re > Re_t \approx 2100$ (with $l \equiv 2R$ and $v \equiv v_{\max}$) the laminar flow in a round pipe, described by Eq. (8.60), becomes unstable, and the resulting turbulence decreases the fluid discharge Q in comparison with the Poiseuille law (8.62). Even more strikingly, the critical value of Re is rather insensitive to the pipe wall roughness, and does not go away even in the limit of perfectly smooth walls.

[42] Note that the 'dynamic' viscosity η participates in this number (and many other problems of fluid dynamics) only in the combination η/ρ, which thereby has deserved the special name of *kinematic viscosity*.

Since $Re \gg 1$ in many real-life situations[43], turbulence is very important in practice. However, despite nearly a century of intensive research, there is no general, quantitative analytical theory of this phenomenon[44] and most results are still obtained either by rather approximate analytical treatments, or by the numerical solution of the Navier–Stokes equations using the approaches discussed in the previous section, or in experiments (e.g. on scaled models[45] in *wind tunnels*). Only certain general, semi-quantitative features may be readily understood from simple arguments.

For example, figure 8.15 shows that within a very broad range of Reynolds numbers, from $\sim 10^2$ to $\sim 3 \times 10^5$, the C_d of a sphere is of the order of 1. Moreover, for a flat, thin, round disk, perpendicular to the incident flow, C_d is very close to 1.1 for any $Re > 10^3$. The approximate equality $C_d \approx 1$, i.e. the drag force $F \approx \rho v_0^2 A/2$, may be understood (in the picture where the object is moved by an external force F with the velocity v_0 through a fluid which was initially at rest) as the equality of the force's power Fv_0 and the fluid's kinetic energy $(\rho v_0^2/2)V$ created in volume $V = v_0 A$ in unit time. This relation would be exact if the object gave velocity v_0 to each and every fluid particle its cross-section runs into, for example by dragging all such particles behind itself. In reality, much of this kinetic energy goes into vortices, where the particle velocity may differ from v_0, so that the equality $C_d \approx 1$ is only approximate.

Unfortunately, due to time/space restrictions, for a more detailed discussion of these results I have to refer the reader to more special literature[46], and will conclude this chapter with a brief discussion of just one issue: can the turbulence be 'explained by a singe mechanism'? (In other words, can it be reduced, at least on a semi-quantitative level, to a set of simpler phenomena that are commonly considered 'well understood'?) Apparently the answer in *no*[47], although nonlinear dynamics of simpler systems may provide some useful insights.

In the middle of the last century, the most popular qualitative explanation of turbulence had been the formation of an 'energy cascade' that would transfer the energy from the regular fluid flow to a hierarchy of vortices of various size. With our background, it is easier to retell that story in the time-domain language (with velocity v serving as the conversion factor), using the fact that in a rotating vortex. each component of the particle's radius-vector oscillates in time, so that to some extent the vortex plays the role of an oscillatory motion mode. Let us consider the

[43] Indeed, the values of η and ρ for water listed in table 8.1 imply that even for a few-meter-sized object (such as a human body or a small boat), $Re > 1000$ at any speed above just ~ 1 mm s^{-1}.

[44] A rare exception is the relatively recent theoretical result by S Orszag (1971) for the turbulence threshold in a flow of an incompressible fluid through a gap of thickness t between two parallel plane walls (see figure 8.10): $Re_t \approx 5772$ (for $l = t/2$, $v = v_{max}$). However, this result does not predict the turbulence patterns at $Re > Re_t$.

[45] The crucial condition of correct modeling is the equality of the Reynolds numbers (8.74) (and if relevant, also of the Froude numbers and/or the Mach numbers) of the object of interest and its model.

[46] See, e.g. [10].

[47] The following famous quote is attributed to W Heisenberg on his deathbed: 'When I meet God, I will ask him two questions: Why relativity? And why turbulence? I think he will have an answer for the first question.' Although probably inaccurate, this story reflects rather well the understandable frustration of the fundamental physics community, renown for their reductionist mentality, with the enormous complexity of phenomena which obey simple (e.g. Navier–Stokes) equations.

passage of a solid body between two, initially close, small parts of fluid. The body pushes them apart, but after its passage these partial volumes are free to return to their initial positions. However, the dominance of inertia effects in motion with $Re \gg 1$ means that the volumes continue to 'oscillate' for a while about those equilibrium positions. (Since elementary volumes of an incompressible fluid cannot merge, these oscillations take the form of rotating vortices—see figure 8.16 again.)

Now, from section 5.8 we know that intensive oscillations in a system with a quadratic nonlinearity, in this case provided by the convective term $(\mathbf{v}\cdot\nabla)\mathbf{v}$, are equivalent, for small perturbations, to the oscillation of the system parameters at the corresponding frequency. On the other hand, as was briefly discussed in section 6.7, in a system with two oscillatory degrees of freedom, a periodic parameter change with frequency ω_p may lead to the non-degenerate parametric excitation ('down-conversion') of oscillations with frequencies $\omega_{1,2}$ satisfying the relation $\omega_1 + \omega_2 = \omega_p$. Moreover, the spectrum of oscillations in such system also has higher combinational frequencies such as $(\omega_p + \omega_1)$, thus pushing the oscillation energy up the frequency scale ('up-conversion'). In the presence of other oscillatory modes, these oscillations may in turn produce, via the same nonlinearity, even higher frequencies, etc. In a fluid, the spectrum of these 'oscillatory modes' (actually, vortex structures) is essentially continuous, so that the above arguments make very plausible a sequential transfer of energy to a broad spectrum of oscillatory modes—whose frequency spectrum is limited from above by the energy dissipation due to the fluid's viscosity. When excited, these modes interact (in particular, mutually phase-lock) via the system's nonlinearity, creating the complex motion we call turbulence.

Although they do not have much quantitative predictive power, such hand-waving explanations, which are essentially based on the excitation of a *large* number of effective degrees of freedom, dominated turbulence reviews until the mid-1960s. At that point, the discovery (or rather re-discovery) of quasi-random motions in classical dynamic systems with just *a few* degrees of freedom altered the discussion substantially. Since this phenomenon, called *deterministic chaos*, extends well beyond fluid dynamics, and I will devote to it the separate (albeit short) following chapter, and at its end will briefly return to the discussion of turbulence.

8.7 Problems

Problem 8.1. Find the first-order correction to the Pascal law, Eq. (8.6), due to small compressibility of a liquid, and evaluate this correction for the water at the bottom of the Earth's oceans.

Problem 8.2. Find the stationary shape of the open surface of an incompressible, heavy fluid in a container rotated about its vertical axis with a constant angular velocity ω—see the figure below.

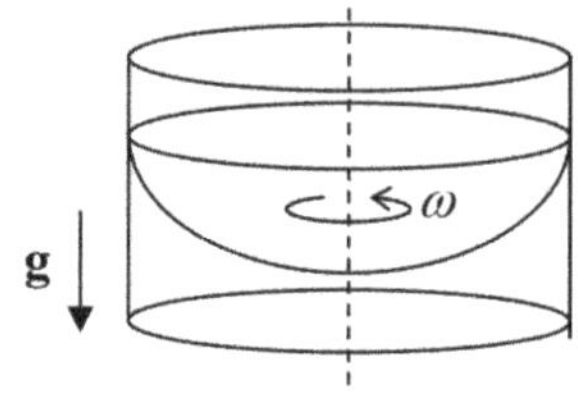

Problem 8.3.* Use two different approaches to calculate the stationary shape of the surface of an incompressible fluid of density ρ near a vertical plane wall, in a uniform gravity field—see the figure below. In particular, find the height h of the liquid's rise at the wall surface as a function of the contact angle θ_c.

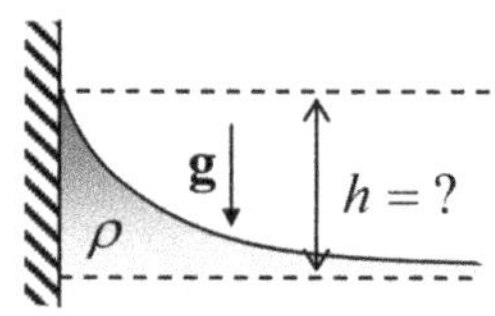

Problem 8.4.* A soap film with surface tension γ is stretched between two similar, coaxial, thin, round rings of radius R, separated by distance d—see the figure below. Neglecting gravity, calculate the equilibrium shape of the film, and the force needed to keep the rings at the fixed distance.

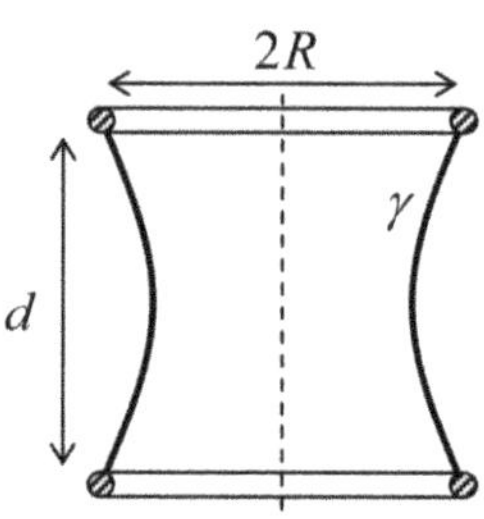

Problem 8.5. A solid sphere of radius R is kept in a steady, vorticity-free flow of an ideal incompressible fluid, with velocity v_0. Find the spatial distribution of velocity and pressure, and in particular their extremal values. Compare the results with those obtained in section 8.4 for a round cylinder.

Problem 8.6. A small source, located at distance d from a plane wall of a container filled with an ideal, incompressible fluid of density ρ, injects additional fluid isotropically, at a constant mass current ('discharge') $Q \equiv dM/dt$—see the figure below. Calculate the fluid's velocity distribution, and its pressure on the wall, created by the flow.

Hint: Recall the charge image method in electrostatics[48] and contemplate its possible analog.

[48] See, e.g. *Part EM* sections 2.9, 3.3, and 4.3.

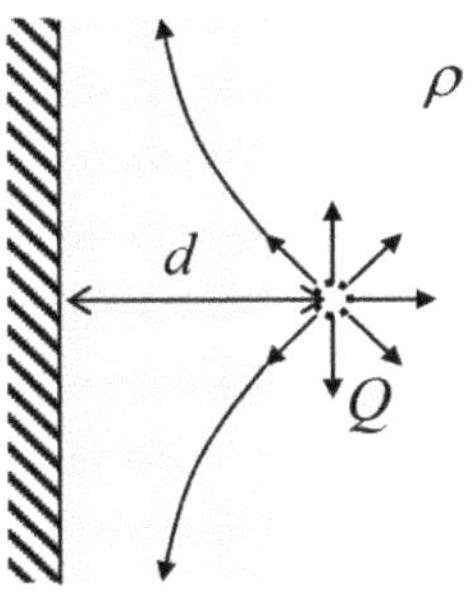

Problem 8.7. Calculate the average kinetic, potential, and full energies (per unit area) of a traveling sinusoidal wave, of a small amplitude q_A, on the horizontal surface on an ideal, incompressible, deep fluid of density ρ, in a uniform gravity field $\mathbf{g}$.

Problem 8.8. Calculate the average power (per unit width of wave's front) carried by the surface wave discussed in the previous problem, and relate the result to the wave's energy.

Problem 8.9. Derive Eq. (8.48) for surface waves on a finite-thickness layer of a heavy liquid.

Problem 8.10. Derive Eq. (8.50) for the capillary waves ('ripples').

Problem 8.11.* Derive a 2D differential equation describing the propagation of relatively long ($\lambda \gg h$) waves on the surface of a broad, plane layer of thickness h, of an ideal, incompressible fluid, and use it to calculate the longest standing wave modes and frequencies in a layer covering a spherical planet of radius $R \gg h$.

Hint: The second task requires some familiarity with the basic properties of spherical harmonics[49].

Problem 8.12. Calculate the velocity distribution and the dispersion relation of the waves propagating along the horizontal interface of two ideal, incompressible fluids of different densities.

Problem 8.13. Use the finite-difference approximation for the Laplace operator, with the mesh step $h = a/4$, to find the maximum velocity and total mass flow Q of a viscous, incompressible fluid through a long pipe with a square-shaped cross-section of side a. Compare the results with those described in section 8.5 for the same problem with the mesh step $h = a/2$ and for a pipe with a circular cross-section of the same area.

Problem 8.14. A layer, of thickness h, of a heavy, viscous, incompressible fluid flows down a long and wide incline plane, under its own weight—see the figure below. Find the stationary velocity distribution profile, and the total fluid discharge (per unit width).

[49] See, e.g. *Part EM* section 2.8 and/or *Part QM* section 3.6.

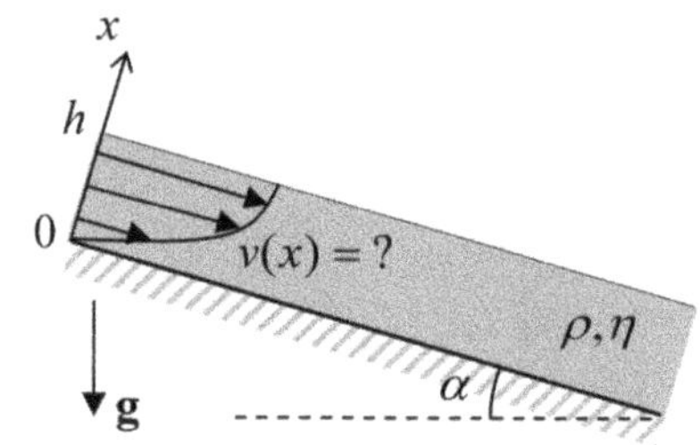

Problem 8.15. Calculate the drag torque exerted on a unit length of a solid round cylinder of radius R that rotates about its axis, with angular velocity ω, inside an incompressible fluid with viscosity η, kept static far from the cylinder.

Problem 8.16. Calculate the tangential force (per unit area) exerted by an incompressible fluid, with density ρ and viscosity η, on a broad solid plane placed over its surface and forced to oscillate, along the surface, with amplitude a and frequency ω.

Problem 8.17. A massive barge, with a flat bottom of area A, floats in shallow water, with a clearance $h \ll A^{1/2}$—see the figure below. Analyze the time dependence of the barge's velocity $V(t)$, and the water velocity profile, after the barge's engine has been turned off. Discuss the limits of large and small values of the dimensionless parameter $M/\rho Ah$.

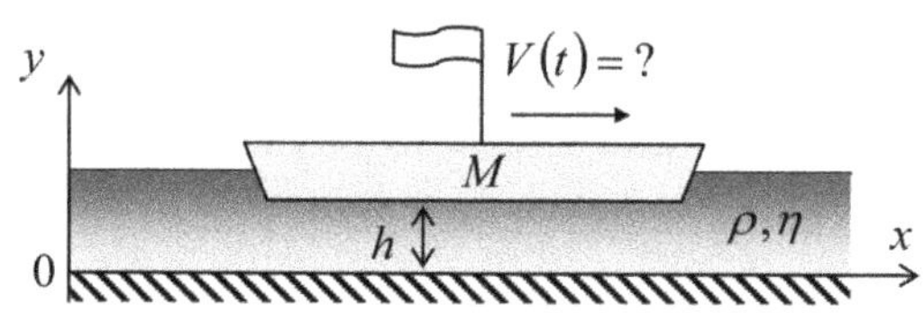

Problem 8.18.* Derive a general expression for mechanical energy loss rate in a viscous incompressible fluid that obeys the Navier–Stokes equation, and use this expression to calculate the attenuation coefficient of the surface waves, assuming that the viscosity is small. (Quantify this condition.)

Problem 8.19. Use the Navier–Stokes equation to calculate the coefficient of attenuation of a plane, sinusoidal acoustic wave.

References

[1] do Camo M P 2016 *Differential Geometry of Curves and Surfaces* 2nd edn (New York: Dover)

[2] Landau L and Lifshitz E 1987 *Fluid Mechanics* 2nd edn (Oxford: Butterworth-Heinemann)

[3] Abbott B *et al* 2016 *Phys. Rev. Lett.* **116** 061102

[4] Leveque R J 2007 *Finite Difference Methods for Ordinary and Partial Differential Equations* (Philadelphia, PA: SIAM)

[5] Johnson C 2009 *Numerical Solution of Partial Differential Equations by the Finite Element Method* (New York: Dover)

[6] Hughes T 2000 *The Finite Element Method* (New York: Dover)
[7] Eisner F 1931 Das Widerstandsproblem *Proc. 3rd Int. Congress on Applied Mechanics (Stockholm)*
[8] Cohen I and Kundu P 2007 *Fluid Mechanics* 4th edn (New York: Academic)
[9] Granger R *Fluid Mechanics* (New York: Dover) 195
[10] Davidson P A 2004 *Turbulence* (Oxford: Oxford University Press)

Chapter 9

Deterministic chaos

This chapter gives a very brief review of chaotic phenomena in deterministic maps and dynamic systems with and without dissipation, and an even shorter discussion of the possible role of chaos in the fluid turbulence.

9.1 Chaos in maps

The possibility of quasi-random dynamics of deterministic systems with a few degrees of freedom (now called *deterministic chaos*—or just 'chaos') had been noticed before the twentieth century[1], but became broadly recognized only after the publication of a 1963 paper by E Lorenz. In that work, he examined numerical solutions of the following system of three nonlinear, ordinary differential equations,

$$\begin{aligned}\dot{q}_1 &= a_1(q_2 - q_1),\\ \dot{q}_2 &= a_2q_1 - q_2 - q_1q_3,\\ \dot{q}_3 &= q_1q_2 - a_3q_3,\end{aligned} \qquad (9.1)$$

as a rudimentary model of heat transfer through a horizontal layer of fluid separating two solid plates. (Experiments show that if the bottom plate is kept hotter than the top one, the liquid may exhibit turbulent convection.) He has found that within a certain range of constants $a_{1,2,3}$, the solutions of Eq. (9.1) follow complex, unpredictable, non-repeating trajectories in the 3D q-space. Moreover, the resulting functions $q_j(t)$ (where $j = 1, 2, 3$) are so sensitive to initial conditions $q_j(0)$ that at sufficiently large times t, solutions corresponding to slightly different initial conditions become completely different.

[1] It may be traced back at least to (apparently, independent) works by J Poincaré in 1892 and by J Hadamard in 1898. Citing Poincaré, '...it may happen that small differences in the initial conditions produce very great ones in the final phenomena. [...] Prediction becomes impossible.'

doi:10.1088/978-0-7503-1398-8ch9

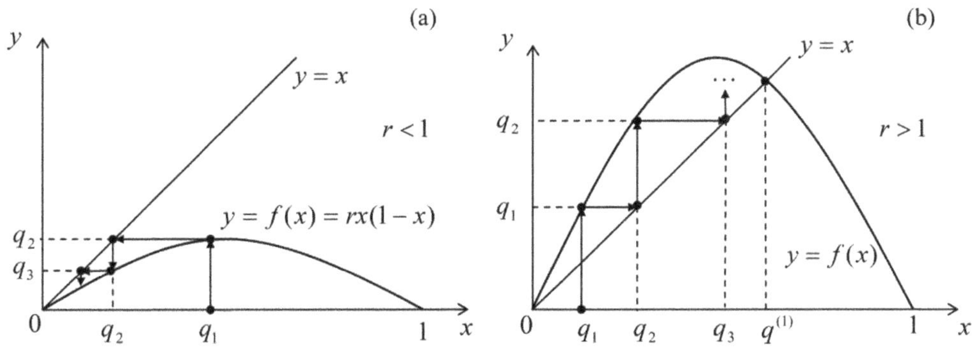

Figure 9.1. Graphical analysis of the logistic map for: (a) $r < 1$ and (b) $r > 1$.

Very soon it was realized that such behavior is typical for even simpler mathematical objects called *maps*, so I will start my discussion of chaos from these objects. A 1D map is essentially a rule for finding the next number q_{n+1} of a discrete sequence numbered by the integer index n, in the simplest case using only its last known value q_n. The most famous example is the so-called *logistic map*[2]:

$$q_{n+1} = f(q_n) \equiv rq_n(1 - q_n). \tag{9.2}$$

The basic properties of this map may be understood using the (hopefully, self-explanatory) graphical representation shown in figure 9.1[3]. One can readily see that at $r < 1$ (figure 9.1a) the logistic map sequence rapidly converges to the trivial fixed point $q^{(0)} = 0$, because each next value of q is less than the previous one. However, if r is increased above 1 (as in the example shown in figure 9.1b), the fixed point $q^{(0)}$ becomes unstable. Indeed, at $q_n \ll 1$, the map yields $q_{n+1} = rq_n$, so that at $r > 1$, the values q_n grow with each iteration. Instead of the unstable point $q^{(0)} = 0$, in the range $1 < r < r_1$, where $r_1 \equiv 3$, the map has a stable fixed point, $q^{(1)}$, that may be found by plugging this value into both sides of Eq. (9.2):

$$q^{(1)} = rq^{(1)}(1 - q^{(1)}), \tag{9.3}$$

giving $q^{(1)} = (1 - 1/r)$—see the left branch in figure 9.2.

At $r > r_1 = 3$, the plot gets thicker: here the fixed point $q^{(1)}$ also becomes unstable. To prove that, let us take $q_n = q^{(1)} + \tilde{q}_n$, assume that the deviation $\tilde{q}_n$ from the fixed point $q^{(1)}$ is small, and linearize the map (9.2) in $\tilde{q}_n$—just as we repeatedly did for differential equations earlier in this course. The result is

$$\tilde{q}_{n+1} = \left.\frac{df}{dq}\right|_{q=q^{(1)}} \tilde{q}_n = r(1 - 2q^{(1)})\tilde{q}_n = (2 - r)\tilde{q}_n. \tag{9.4}$$

[2] Its chaotic properties were first discussed in 1976 by R May, although the map itself is one of the simple ecological models repeatedly discussed earlier, and may be traced back at least to the 1838 work by P Verhulst.

[3] Since the maximum value of the function $f(q)$, achieved at $q = 1/2$, equals $r/4$, the mapping may be limited to segment $x = [0, 1]$ if the parameter r is between 0 and 4. Since all interesting properties of the map, including chaos, may be found within these limits, I will discuss only this range of r.

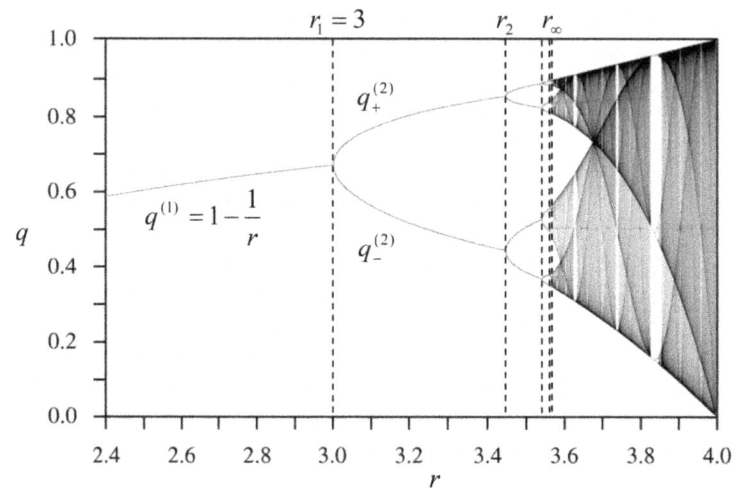

Figure 9.2. The fixed points and chaotic regions of the logistic map. Adapted, under the CCO 1.0 Universal Public Domain Dedication, from the original by Jordan Pierce, available at http://en.wikipedia.org/wiki/Logistic_map. (A very nice live simulation of the map is also available on the same website.)

It shows that at $0 < 2 - r < 1$, i.e. $1 < r < 2$, the deviations $\tilde{q}_n$ decrease monotonically. At $-1 < 2 - r < 0$, i.e. in the range $2 < r < 3$, the deviations' signs alternate, but their magnitudes still decrease (as in a stable focus—see section 5.6). However, at $-1 < 2 - r$, i.e. $r > r_1 = 3$, the deviations are growing in magnitude while still changing sign at each step. Since the map (9.2) has no other fixed points, this means that at $n \to \infty$, the values q_n do not converge to one point; rather, within the range $r_1 < r < r_2$, they approach a *limit cycle* of alternation of two points, $q_+^{(2)}$ and $q_-^{(2)}$, which satisfy the following system of algebraic equations:

$$q_+^{(2)} = f\left(q_-^{(2)}\right), \quad q_-^{(2)} = f\left(q_+^{(2)}\right). \tag{9.5}$$

(These points are also plotted in figure 9.2, as functions of parameter r.) What has happened at point r_1 is called the *period-doubling bifurcation*.

The story repeats at $r = r_2 = 1 + \sqrt{6} \approx 3.45$, where the system goes from the two-point limit cycle to a four-point cycle, then at $r = r_3 \approx 3.54$, where the limit cycle becomes composed of eight alternating points, etc. Most remarkably, the period-doubling bifurcation points r_n, as the number of points in the limit cycle doubles from 2^{n-1} points to 2^n points, become closer and closer. Numerical calculations show that these points obey the following asymptotic behavior:

$$r_n \to r_\infty - \frac{C}{\delta^n}, \quad \text{where } r_\infty = 3.5699\ldots, \ \delta = 4.6692\ldots \tag{9.6}$$

This parameter δ is called the *Feigenbaum constant*; for other maps, and some dynamic systems (see the next section), period-doubling sequences follow a similar law, but with different values of δ.

More important for us, however, is what happens at $r > r_\infty$. Numerous numerical experiments, repeated with increasing precision[4], have confirmed that here the system is fully disordered, with no reproducible limit cycle, although (as figure 9.2 shows) at $r \approx r_\infty$, all sequential values q_n are still confined to a few narrow regions[5]. However, as parameter r is increased well beyond r_∞, these regions broaden and merge. This the so-called *deep chaos*, with no apparent order at all[6].

The most important feature of the chaos (in this and any other system) is the *exponential divergence of trajectories*. For a 1D map, this means that even if the initial conditions q_1 in two map implementations differ by a very small amount Δq_1, the difference Δq_n between the corresponding sequences q_n is growing, on the average, exponentially with n. Such exponents may be used to characterize chaos. Indeed, an evident generalization of Eq. (9.4) to an arbitrary point q_n is

$$\Delta q_{n+1} = e_n \Delta q_n, \qquad e_n = \left.\frac{df}{dq}\right|_{q=q_n}. \tag{9.7}$$

Let us assume that Δq_1 is so small that N first values q_n are relatively close to each other. Then using Eq. (9.7) iteratively for these steps, we obtain

$$\Delta q_N = \Delta q_1 \prod_{n=1}^{N} e_n, \quad \text{so that} \quad \ln\left|\frac{\Delta q_N}{\Delta q_1}\right| = \sum_{n=1}^{N} \ln|e_n|. \tag{9.8}$$

Numerical experiments show that in most chaotic regimes, at $N \to \infty$ such a sum fluctuates about an average, which grows as λN, with parameter

$$\lambda = \lim_{\Delta q_1 \to 0} \lim_{N \to \infty} \frac{1}{N} \sum_{n=1}^{N} \ln|e_n|, \tag{9.9}$$

called the *Lyapunov exponent*[7], being independent on the initial conditions. The bottom panel in figure 9.3 shows λ as a function of the parameter r for the logistic map (9.2). (Its top panel shows the same data as figure 9.2 and is reproduced here for the sake of comparison.)

Note that at $r < r_\infty$, λ is negative, indicating the sequence's stability, apart from points $r_1, r_2, \ldots$ where λ would become positive if the limit-cycle change had not brought it back into the negative territory. However, at $r > r_\infty$, the exponent

---

[4] The reader should remember that just as the usual ('nature') experiments, numerical experiments also have a limited accuracy, due to unavoidable rounding errors.

[5] The geometry of these regions is essentially *fractal*, i.e. has a dimensionality intermediate between 0 (which any final set of geometric points would have) and 1 (pertinent to a 1D continuum). An extensive discussion of fractal geometries, and their relation to the deterministic chaos may be found, e.g. in the book [1].

[6] This does not mean that the chaos development is a monotonic function of r. As figure 9.2 shows, within certain intervals of this parameter, the chaotic behavior suddenly disappears, being replaced, typically, with a few-point limit cycle, just to resume on the other side of the interval. Sometimes (but not always!) the 'route to chaos' on the borders of these intervals follows the same Feigenbaum sequence of period-doubling bifurcations.

[7] After A Lyapunov (1857–1918), famous for his studies of the stability of dynamic systems.

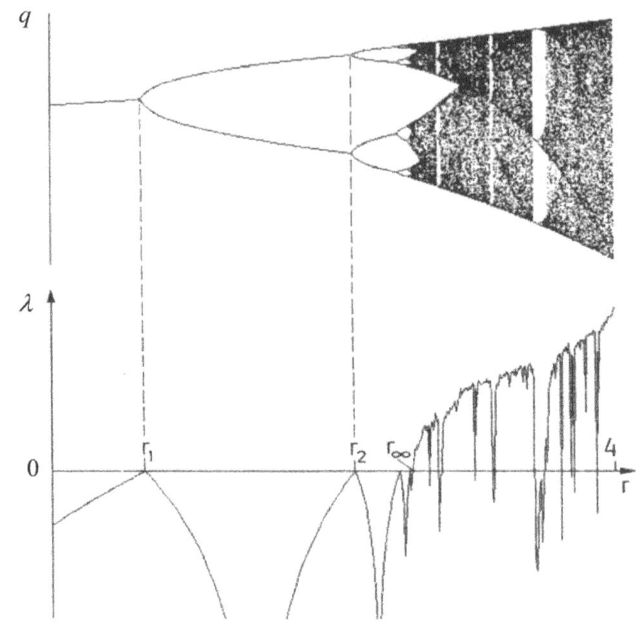

Figure 9.3. The Lyapunov exponent for the logistic map. Adapted with permission from [2]. Copyright 2005 Wiley-VCH Verlag GmbH & Co. KGaA.

becomes positive, returning to negative values only in limited intervals of stable limit cycles. It is evident that in numerical experiments (which dominate the studies of the deterministic chaos) the Lyapunov exponent may be used as a good measure of the chaos' 'depth'[8].

Despite all the abundance of results published for particular maps[9], and several interesting general observations (such as the existence of the Feigenbaum bifurcation sequences), to the best of my knowledge, nobody can yet predict patterns such as those shown in figures 9.2 and 9.3, from just looking at the map rule itself, i.e. without carrying out actual numerical experiments. Unfortunately the understanding of deterministic chaos in other systems is not much better.

9.2 Chaos in dynamic systems

Proceeding to the discussion of chaos in dynamic systems, it is more natural with our background to illustrate this discussion not with the Lorenz' system (9.1), but with the system of equations describing a dissipative pendulum driven by a sinusoidal external force, which was repeatedly discussed in chapter 5. Introducing two new variables, the normalized momentum $p \equiv (dq/dt)/\omega_0$, and the external force's full

[8] N-dimension maps, which relate N-dimensional vectors rather than scalars, may be characterized by N Lyapunov exponents rather than one. In order to have chaotic behavior, it is sufficient for just one of them to become positive. For such systems, another measure of chaos, the *Kolmogorov entropy*, may be more relevant. This measure and its relation to the Lyapunov exponents are discussed, e.g. in *Part SM* section 2.2.

[9] See, e.g. chapters 2–4 in [2] or chapters 8–9 in [3].

phase $\psi \equiv \omega t$, we may rewrite Eq. (5.42), describing the pendulum, in a form similar to Eq. (9.1), i.e. as a system of three first-order ordinary differential equations:

$$\begin{aligned} \dot{q} &= \omega_0 p, \\ \dot{p} &= -\omega_0 \sin q - 2\delta p + (f_0/\omega_0)\cos\psi, \\ \dot{\psi} &= \omega. \end{aligned} \tag{9.10}$$

Figure 9.4 shows several results of a numerical solution of Eq. (9.10)[10]. In all cases, the parameters δ, ω_0, and f_0 are fixed, while the external frequency ω is gradually changed. For the case shown in the top panel, the system still tends to a stable periodic solution, with a very low contents of higher harmonics. If the external force frequency is reduced by a just few percent, the third subharmonic may be excited. (This effect has already been discussed in section 5.8—see, e.g. figure 5.15.) The next panel shows that just a very small further reduction of the frequency ω leads to a new tripling of the period, i.e. the generation of a complex waveform with the ninth subharmonic. Finally (see the bottom panel of figure 9.4), even a minor further change of parameters leads to oscillations without any visible period, e.g. to chaos.

In order to trace this transition, a direct inspection of the oscillation waveforms $q(t)$ is not very convenient, and trajectories on the phase plane $[q, p]$ also become messy if plotted for many periods of the external frequency. In situations like this, the Poincaré (or 'stroboscopic') plane, already discussed in section 5.6, is much more useful. As a reminder, this is essentially just the phase plane $[q, p]$, but with the points highlighted only once a period, e.g. at $\psi = 2\pi n$, with $n = 1, 2, \ldots$. On this plane, periodic oscillations of frequency are presented just as one fixed point—see, e.g. the top panel in the right-hand column of figure 9.4. The beginning of the third subharmonic generation, shown on the next panel, means the oscillation period tripling and is reflected on the Poincaré plane as splitting of the fixed point into three. It is evident that this transition is similar to the period-doubling bifurcation in the logistic map, in addition to the fact (already discussed in section 5.8) that in systems with an asymmetric nonlinearity, such as the pendulum (9.10), the third subharmonic is easier to excite. From this point, the ninth harmonic generation (shown on the third panel of figure 9.4), i.e. one more splitting of the points on the Poincaré plane, may be understood as one more step on the Feigenbaum-like route to chaos—see the bottom panel of the figure.

So, the transition to chaos in dynamic systems may be at least qualitatively similar to that in 1D maps, with a law similar to Eq. (9.6) for the critical values of some parameter of the system (in figure 9.4, frequency ω), though with a system-specific value of the exponent δ. Moreover, it is evident that we can always consider the first two differential equations of the system (9.10) as a 2D map that relates the

[10] In the actual simulation, a small term εq, with $\varepsilon \ll 1$, has been added to the left-hand side of this equation. This term slightly tames the trend of the solution to spread along the q-axis, and makes the presentation of results easier, without affecting the system's dynamics too much.

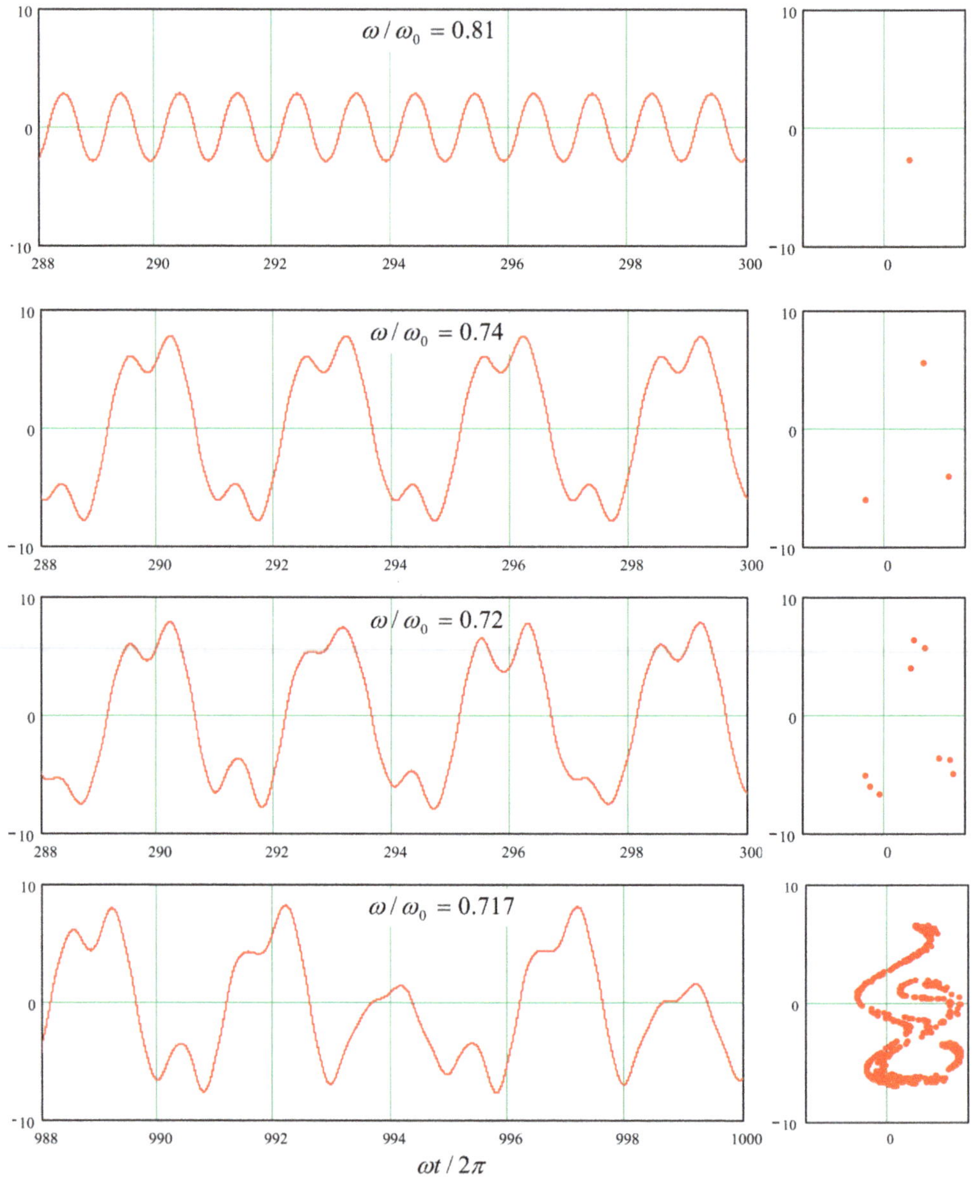

Figure 9.4. Oscillations in a pendulum with weak damping, $\delta/\omega_0 = 0.1$, driven by a sinusoidal external force with a fixed effective amplitude $f_0/\omega_0^2 = 1$, and several close values of the frequency ω (listed on the panels). Left column: the oscillation waveforms $q(t)$ recorded after certain initial transient intervals. Right column: representations of the same processes on the Poincaré plane of variables $[q, p]$, with the axis q turned vertically, for the convenience of comparison with the left panel.

vector $\{q_{n+1}, p_{n+1}\}$ of the coordinate and velocity, measured at $\psi = 2\pi(n + 1)$, with the previous value $\{q_n, p_n\}$ of that vector (reached at $\psi = 2\pi n$).

Unfortunately this similarity also implies that chaos in dynamical systems is at least as complex, and is as little understood, as in maps. For example, figure 9.5

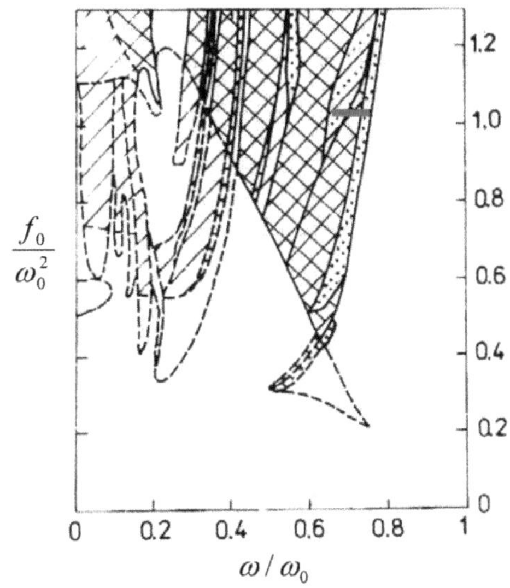

Figure 9.5. The phase diagram of an externally driven pendulum with weak damping ($\delta/\omega_0 = 0.1$). Regions of oscillations with the basic period are not shaded; the notation for other regions is as follows. Doted: subharmonic generation; cross-hatched: chaos; hatched: either chaos or the basic period (depending on the initial conditions); hatch-dotted: either the basic period or subharmonics. Solid lines show the boundaries of single-regime regions, while dashed lines are the boundaries of the regions where several types of motion are possible. (Figure courtesy of V Kornev.)

shows (part of) the phase diagram of the externally driven pendulum, with the red bar marking the route to chaos traced in figure 9.4, and shading/hatching styles marking different oscillation regimes. One can see that the pattern is at least as complex as that shown in figures 9.2 and 9.3, and, apart from a few features[11], is equally unpredictable from the form of the equation.

Are there any valuable general results concerning the chaos in dynamic systems? The most important (although almost self-evident) result is that this phenomenon is impossible in any system described by one or two first-order differential equations with time-independent right-hand sides. Indeed, let us start with a single equation

$$\dot{q} = f(q), \tag{9.11}$$

where $f(q)$ is any single-valued function. This equation may be directly integrated to give

$$t = \int^{q} \frac{dq'}{f(q')} + \text{const}, \tag{9.12}$$

showing that the relation between q and t is unique and hence does not leave any place for chaos.

[11] In some cases, it is possible to predict a parameter region where chaos *cannot* happen, due to the lack of any instability-amplification mechanism. Unfortunately, typically the analytically predicted boundaries of such a region form a rather loose envelope of the actual (numerically simulated) chaotic regions.

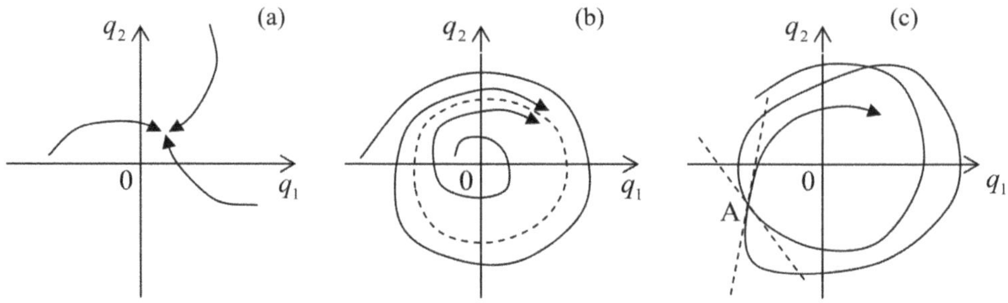

Figure 9.6. Attractors in dynamical systems: (a) a fixed point, (b) a limit cycle, and (c) a strange attractor.

Next, let us explore the system of two such equations:

$$\begin{aligned} \dot{q}_1 &= f_1(q_1, q_2), \\ \dot{q}_2 &= f_2(q_1, q_2). \end{aligned} \tag{9.13}$$

Consider its phase plane shown schematically in figure 9.6. In a 'usual' system, the trajectories approach either some fixed point (figure 9.6a) describing static equilibrium, or a limit cycle (figure 9.6b) describing periodic oscillations. (Both notions are united by the term *attractor*, because they 'attract' trajectories launched from various initial conditions.) However, phase-plane trajectories of a chaotic system of equations that describe real physical variables (which cannot tend to infinity), should be confined to a limited phase-plane area, and simultaneously cannot start repeating each other. (This topology is frequently called the *strange attractor*.) For that, the 2D trajectories need to cross—see, e.g. point A in figure 9.6c.

However, in the case described by Eq. (9.13), such a crossing is clearly impossible, because according to these equations, the tangent of a phase-plane trajectory is a unique function of the coordinates $\{q_1, q_2\}$:

$$\frac{dq_1}{dq_2} = \frac{f_1(q_1, q_2)}{f_2(q_1, q_2)}. \tag{9.14}$$

Thus, in this case the deterministic chaos is impossible[12]. It becomes, however, readily possible if the right-hand sides of a system similar to Eq. (9.13) depend either on other variables of the system or on time. For example, if we consider the first two differential equations of system (9.10), in the case $f_0 = 0$ they have the structure of the system (9.13) and hence the chaos is impossible, even at $\delta < 0$ when (as we know from section 5.4) the system allows self-excitation of oscillations, leading to a limit-cycle attractor. However, if $f_0 \neq 0$, this argument does not work any longer and (as we have already seen) the system may have a strange attractor—which is, for dynamic systems, a synonym for the deterministic chaos.

[12] A mathematically strict formulation of this statement is called the *Poincaré–Bendixon theorem*, which was proved by I Bendixon as early as 1901.

Thus, chaos is only possible in autonomous dynamic systems described by three or more differential equations of the first order[13].

9.3 Chaos in Hamiltonian systems

The last conclusion is of course valid for Hamiltonian systems, which are just a particular type of dynamic systems. However, one may wonder whether these systems—which feature at least one first integral of motion, H = const, and hence are more 'ordered' than the systems discussed above—can exhibit chaos at all. The answer is *yes*, because such systems still can have mechanisms for an exponential growth of a small initial perturbation.

As the simplest way to show it, let us consider the so-called *mathematical billiard*, i.e. system with a ballistic particle (a 'ball') moving freely by inertia on a horizontal plane surface ('table') limited by rigid impenetrable walls. In this idealized model of the usual game of billiards, the ball's velocity $\mathbf{v}$ is conserved when it moves on the table, and when it runs into a wall the ball is elastically reflected from it as from a mirror[14], with the reversal of the sign of the normal velocity v_n and the conservation of the tangential velocity v_τ, and hence without any loss of its kinetic (and hence the full) energy

$$E = H = T = \frac{m}{2}v^2 = \frac{m}{2}(v_n^2 + v_\tau^2). \tag{9.15}$$

This model, while being a legitimate 2D dynamic system[15], allows geometric analyses for several simple table shapes. The simplest case is a rectangular billiard table of area $a \times b$ (figure 9.7), whose analysis may be readily carried out just by the replacement of each *ball* reflection event with the mirror reflection of the *table* in that wall—see dashed lines in panel (a).

Such analysis (left for the reader's pleasure) shows that if the tangent of the ball launching angle φ is commensurate with the side-length ratio,

$$\tan\varphi = \pm\frac{m}{n}\frac{b}{a}, \tag{9.16}$$

where n and m are non-negative integers without common integer multipliers, the ball returns exactly to the launch point O, after bouncing m times from each wall of length a, and n times from each wall of length b. (The red lines in figure 9.7a show an example of such a trajectory for $n = m = 1$, while blue lines, for $m = 3$, $n = 1$.) The larger is the sum $(m + n)$, the more complex is such closed trajectory—'orbit'.

[13] Since a typical dynamic system with one degree of freedom is described by two such equations, the number of first-order equations describing a dynamic system is sometimes called the number of *half-degrees of freedom*. This notion is very useful and popular in statistical mechanics—see, e.g. *Part SM* section 2.2.

[14] A more scientific-sounding name for such a reflection is *specular* (from the Latin word 'speculum' meaning a metallic mirror).

[15] Indeed, it is fully described by the following Lagrangian function: $L = mv^2/2 - U(\boldsymbol{\rho})$, with $U(\boldsymbol{\rho}) = 0$ for 2D radius-vectors $\boldsymbol{\rho}$ belonging to the table area and $U(\boldsymbol{\rho}) = +\infty$ outside the area.

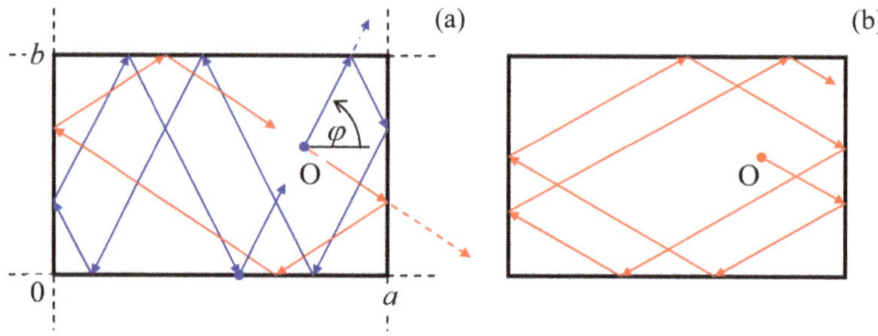

Figure 9.7. Ball motion on a rectangular billiard at (a) a commensurate and (b) an incommensurate launch angle.

Finally, if $(n + m) \to \infty$, i.e. tan φ and b/a are incommensurate (meaning that their ratio is an irrational number), the trajectory covers all the table area, and the ball never returns exactly into the launch point. Still, this is not real chaos. Indeed, a small shift of the launch point O shifts all the trajectory fragments by the same displacement. Moreover, at any time t, each of the Cartesian components $v_j(t)$ of the ball's velocity (with coordinate axes parallel to the table sides) may take only two values, $\pm v_j(0)$, and hence may vary only as much as the initial velocity is being changed.

In 1963, Ya Sinai showed that the situation changes completely if an additional wall, in the shape of a circle, is inserted into the rectangular billiard (figure 9.8). For most initial conditions, the ball's trajectory eventually runs into the circle (see the red line on panel (a) as an example), and the further trajectory becomes essentially chaotic. Indeed, let us consider the ball's reflection from the circle-shaped wall—figure 9.8b. Due to the conservation of the tangential velocity, and the sign change of the normal velocity component, the reflection obeys the reflection law (cf figure 7.12 and its discussion): $\theta_r = \theta_i$. Figure 9.8b shows that as the result, a small difference $\delta\varphi$ between the angles of two close trajectories (as measured in the lab system) doubles by magnitude at each reflection from the curved wall. This means that the small deviation grows along the ball trajectory as

$$|\delta\varphi(N)| \sim |\delta\varphi(0)| \times 2^N = |\delta\varphi(0)|e^{N \ln 2}, \tag{9.17}$$

where N is the number of reflections from the convex wall[16]. As we already know, such exponential divergence of trajectories, with a positive Lyapunov exponent, is the main feature of the deterministic chaos[17].

---

[16] Superficially, Eq. (9.17) is also valid for a plane wall but, as was discussed above, a billiard with such walls features a full correlation between sequential reflections, so that the angle φ always returns to its initial value. In a Sinai billiard, such correlation disappears. Because of that, concave walls may also make a billiard chaotic. A famous example is the *stadium billiard*, suggested by L Bunimovich, with two straight, parallel walls connecting two semi-circular, concave walls. Another example that allows a straightforward analysis is the *Hadamard billiard*: an infinite (or rectangular) table with a non-horizontal surface of negative curvature.

[17] Curved-wall billiards are also a convenient platform for studies of quantum properties of classically chaotic systems (for a conceptual discussion, see *Part QM* section 3.5), in particular the features called 'quantum scars'—see, e.g. the spectacular numerical simulation results by E J Heller [4].

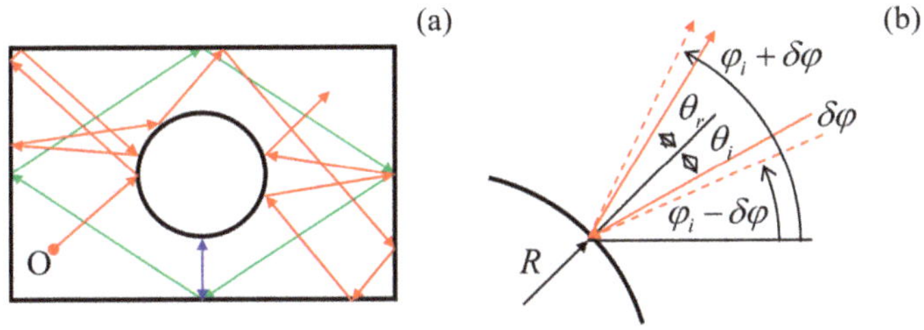

Figure 9.8. (a) Motion on a Sinai billiard table and (b) the mechanism of the exponential divergence of close trajectories.

The most important new feature of the dynamic chaos in Hamiltonian systems is its dependence on initial conditions. (In the systems discussed in the previous two sections, that lack the integrals of motion, the initial conditions are rapidly 'forgotten' and the chaos is usually characterized after an initial transient period—see, e.g. Figure 9.4.) Indeed, even a Sinai billiard allows periodic motion along closed orbits at certain initial conditions—see the blue and green lines in figure 9.8a as examples. Thus the chaos 'depth' in such systems may be characterized by the 'fraction'[18] of the phase space of initial parameters (for a 2D billiard, of the 3D space of initial values of x, y, and φ) resulting in chaotic trajectories.

This conclusion is also valid for the Hamiltonian systems that are met in experiments more frequently than billiards, for example, coupled nonlinear oscillators without damping. Perhaps, the earliest and the most popular example is the so-called *Hénon–Heiles* system[19], which may be descrbed by the following Lagrangian function:

$$L = \frac{m_1}{2}\left(\dot{q}_1^2 - \omega_1^2 q_1^2\right) + \frac{m_2}{2}\left(\dot{q}_2^2 - \omega_2^2 q_2^2\right) - \varepsilon\left(q_1^2 - \frac{1}{3}q_2^2\right)q_2. \tag{9.18}$$

It is straightworward to use Eq. (2.19) to derive the corresponding Lagrange equations of motion,

$$\begin{aligned} m_1\left(\ddot{q}_1 + \omega_1^2 q_1\right) &= -2\varepsilon q_1 q_2, \\ m_2\left(\ddot{q}_2 + \omega_2^2 q_2\right) &= -\varepsilon\left(q_1^2 - q_2^2\right), \end{aligned} \tag{9.19}$$

and find its first integral of motion (physically, the energy conservation law):

$$H = E = \frac{m_1}{2}\left(\dot{q}_1^2 + \omega_1^2 q_1^2\right) + \frac{m_2}{2}\left(\dot{q}_2^2 + \omega_2^2 q_2^2\right) + \varepsilon\left(q_1^2 - \frac{1}{3}q_2^2\right)q_2 = \text{const.} \tag{9.20}$$

[18] Actually, quantitative characterization of the fraction is not trivial, because it may have fractal dimensionality. Unfortunately, due to lack of time I have to refer the reader interested in this issue to special literature, e.g. the monograph by B Mandelbrot [5] and references therein.

[19] It was first studied in 1964 by M Hénon and C Heiles as a simple model of star rotation about a gallactic center. Most studies of this equation have been carried out for the following particular case: $m_2 = 2m_1$, $m_1\omega_1^2 = m_2\omega_2^2$. In this case, introducing new variables $x \equiv \varepsilon q_1$, $y \equiv \varepsilon q_2$, and $\tau \equiv \omega_1 t$, it is possible to rewrite Eqs. (9.18)–(9.20) in parameter-free forms. All the results shown in figure 9.9 below are for this case.

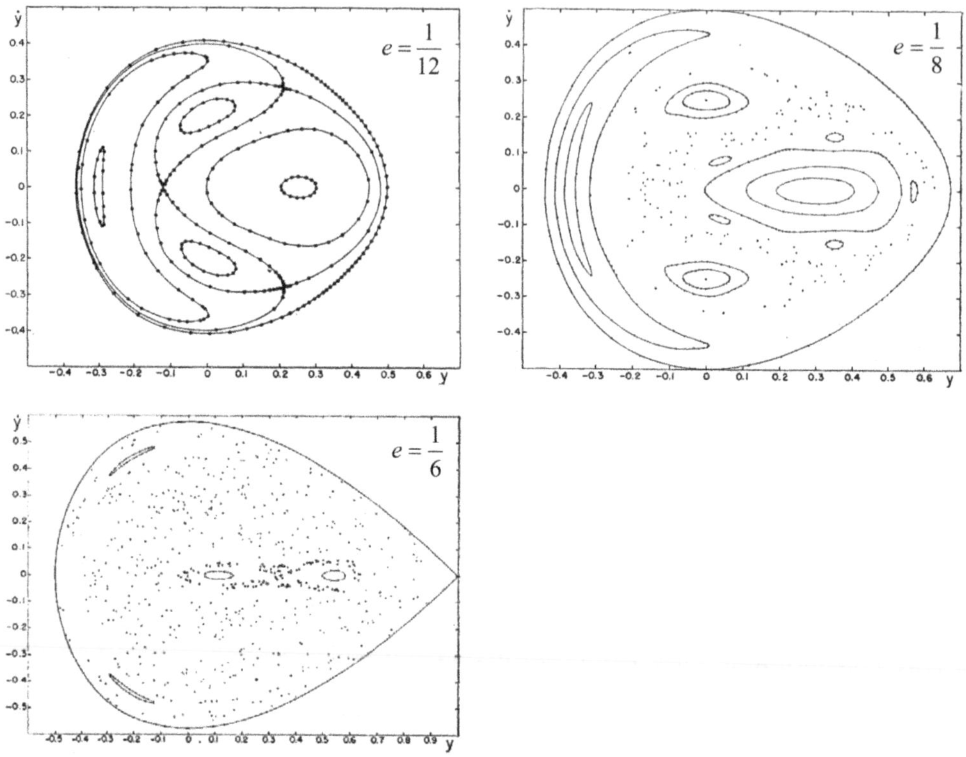

Figure 9.9. Poincaré planes of the Hénon–Heiles system (9.19), in notation $y \equiv \varepsilon q_2$, for three values of the dimensionless energy $e \equiv E/E_0$, with $E_0 \equiv m_1\omega_1^2/\varepsilon^2$. Adapted with permission from [7]. Copyright 1964 Americal Astronomical Society.

In the conext of our discussions in chapters 5 and 6, Eq. (9.19) may be readily interpreted as those describing two oscillators, with small-oscillation eigenfrequencies ω_1 and ω_2, coupled only as described by the quadratic terms on the right-hand sides of the equations. This means that as the oscillation amplitudes $A_{1,2}$, and hence the total energy E of the system, tend to zero, the oscillator subsystems are virtually independent, each performing sinusoidal oscillations at its own frequency. This observation suggests a convenient way to depict the system motion[20]. Let us consider a Poincaré plane for one of the oscillators (say, with the coordinate q_2), similar to that discussed in section 9.2 above, with the only difference being that (because of the absence of an explicit function of time in system's equations), the trajectory on the phase plane $[q_2, \dot{q}_2]$ is highlighted at the moments when $q_1 = 0$.

Let us start from the limit $A_{1,2}\to 0$, when oscillations of q_2 are virtually sinusoidal. As we already know (see figure 5.9 and its discussion), if the representation point

[20] Generally, the system has a trajectory in 4D space, e.g. that of coordinates $q_{1,2}$ and their time derivatives, although the first integral of motion (9.20) means that for each fixed energy E, the motion is limited to a 3D subspace. Still, this is one dimension too many for a convenient representation of the motion.

highlighting was perfectly synchronous with the frequency ω_2 of the oscillations, there would be only one point on the Poincaré plane—see, e.g. the top-right panel in figure 9.4. However, at the q_1-initiated highlighting there is no such synchronism, so at each period a different point of the elliptical (at the proper scaling of the velocity, circular) trajectory is highlighted, so that the resulting points, for certain initial conditions, reside on a circle of radius A_2. If we now vary the initial conditions—i.e. redistribute the initial energy between the oscillators, but keep the total energy E constant—on the Poincaré plane we obtain a set of ellipses.

Now, if the initial energy is increased, the nonlinear interaction of the oscillations start to deform these ellipses, causing also their crossings—see, e.g. the top-left panel of figure 9.9. Still, below a certain threshold value of E, all Poincaré points belonging to a certain initial condition sit on a single closed contour. Moreover, these contours may be calculated approximately, but with a pretty good accuracy, using a straighforward generalization of the method discussed in section 5.2[21].

However, starting from some value of energy, certain initial conditions lead to sequences of points scattered over parts of the Poincaré plane, with non-zero area—see the top right panel of figure 9.9. This means that the corresponding oscillations $q_2(t)$ do not repeat from one (quasi-)period to the next one—cf figure 9.4 for the dissipative, forced pendulum. This is chaos[22]. However, some other initial conditions still lead to closed contours. This feature is similar to Sinai billiards and is typical for Hamiltonian systems. As the energy is increased, larger and larger parts of the Poincaré plane belong to the chaotic motion, signifying deeper and deeper chaos—see the bottom panel of figure 9.9.

9.4 Chaos and turbulence

This extremely short section consists of essentially just one statement, extending the discussion in section 8.5. The (re-)discovery of the deterministic chaos in systems with just a few degrees of freedom in the 1960s changed the tone of the debates concerning turbulence origins, very considerably. At first, an extreme point of view that equated the notions of chaos and turbulence became the debate's favorite[23]. However, after some initial excitement, significant evidence for the Landau-style mechanisms involving many degrees of freedom were rediscovered and could not be ignored any longer. To the best knowledge of the author, who is a very distant albeit interested observer of that field, most experimental and numerical-simulation data carry features of both mechanisms, so the debate continues[24]. Due to the age differential, most readers of these notes have a much better chance than the author to see where this discussion will end[25].

---

[21] See, e.g. [6].

[22] This fact complies with the necessary condition of chaos, discussed at the end of section 9.2, because Eqs. (9.19) may be rewritten as a system of *four* differential equations of the first order.

[23] An important milestone on that road was the work by S Newhouse *et al* [8], who proved the existence of a strange attractor in a rather abstract model of fluid flow.

[24] See, e.g. [9].

[25] The reader interested in deterministic chaos as such may also like to have a look at a very popular book by S Strogatz [10].

9.5 Problems

Problem 9.1. Generalize the reasoning of section 9.1 to an arbitrary 1D map $q_{n+1} = f(q_n)$, with function $f(q)$ differentiable at all points of interest. In particular, derive the condition of stability of an N-point limit cycle $q^{(1)} \to q^{(2)} \to \ldots \to q^{(N)} \to q^{(1)}$.

Problem 9.2. Use the stability condition, derived in the previous problem, to analyze the possibility of deterministic chaos in the so-called *tent map*, with

$$f(q) = \begin{cases} rq, & \text{for } 0 \leqslant q \leqslant 1/2, \\ r(1-q), & \text{for } 1/2 \leqslant q \leqslant 1, \end{cases} \quad \text{with } 0 \leqslant r \leqslant 2.$$

Problem 9.3. A dynamic system is described by the following system of differential equations:

$$\dot{q}_1 = -q_1 + a_1 q_2^3,$$
$$\dot{q}_2 = a_2 q_2 - a_3 q_2^3 + a_4 q_2 \left(1 - q_1^2\right).$$

Can it exhibit chaos at some set of constant parameters a_1–a_4?

Problem 9.4. A periodic function of time has been added to the right-hand side of the first equation of the system considered in the previous problem. Is the deterministic chaos possible now?

References

[1] Mandelbrot B B 1983 *The Fractal Geometry of Nature* (San Francisco, CA: Freeman)
[2] Schuster H and Just W 2005 *Deterministic Chaos* 4th edn (New York: Wiley)
[3] Thompson J and Stewart H 2002 *Nonlinear Dynamics and Chaos* 2nd edn (New York: Wiley)
[4] Heller E J 1984 *Phys. Rev. Lett.* **53** 1515
[5] Mandelbrot B 1983 *The Fractal Geometry of Nature* (San Francisco, CA: Freeman)
[6] Berry M V 1978 *AIP Conf. Proc.* **46** 16–120
[7] Hénon M and Heiles C 1964 *Astron. J.* **69** 73
[8] Newhouse S *et al* 1978 *Comm. Math. Phys.* **64** 35
[9] Frisch U 1996 *Turbulence: The Legacy of A N Kolmogorov* (Cambridge: Cambridge University Press)
[10] Strogatz S 2001 *Nonlinear Dynamics and Chaos* (Westview)

IOP Publishing

Classical Mechanics
Lecture notes
Konstantin K Likharev

Chapter 10

A bit more of analytical mechanics

This concluding chapter reviews two alternative approaches to analytical mechanics, whose main advantage is a closer parallel to quantum mechanics in general and to its quasi-classical (WKB) approximation in particular. One of them, the Hamiltonian formalism, may also be used to derive an important asymptotic result, the adiabatic invariance, for classical systems with slowly changing parameters.

10.1 Hamilton equations

Throughout this course we have seen how analytical mechanics, in its Lagrangian form, is invaluable for solving various particular problems of classical mechanics. Now let us discuss several alternative formulations[1] that may not be much more useful for this purpose, but shed light on possible extensions of classical mechanics, most importantly to quantum mechanics.

As was already discussed in section 2.3, the partial derivative $p_j \equiv \partial L/\partial \dot{q}_j$ participating in the Lagrange equations (2.19)

$$\frac{d}{dt}\frac{\partial L}{\partial \dot{q}_j} - \frac{\partial L}{\partial q_j} = 0, \qquad (10.1)$$

may be considered as the generalized momentum corresponding to the generalized coordinate q_j, and the full set of these momenta may be used to define the Hamiltonian function (2.32):

$$H \equiv \sum_j p_j \dot{q}_j - L. \qquad (10.2)$$

[1] Due to not only W Hamilton (1805–65), but also C Jacobi (1804–51).

doi:10.1088/978-0-7503-1398-8ch10

Now let us rewrite the full differential of this function[2] in the following form:

$$dH = d\left(\sum_j p_j \dot{q}_j - L\right) = \sum_j \left[d(p_j)\dot{q}_j + p_j d(\dot{q}_j)\right] - dL$$
$$= \sum_j \left[d(p_j)\dot{q}_j + p_j d(\dot{q}_j)\right] - \left[\frac{\partial L}{\partial t}dt + \sum_j \left(\frac{\partial L}{\partial q_j}d(q_j) + \frac{\partial L}{\partial \dot{q}_j}d(\dot{q}_j)\right)\right]. \quad (10.3)$$

According to the definition (2.31) of the generalized momentum, the second terms of each sum over j cancel each other, while according to the Lagrange equation (10.1), the derivative $\partial L/\partial q_j$ is just $\dot{p}_j$, so that

$$dH = \frac{\partial L}{\partial t}dt + \sum_j \left(\dot{q}_j dp_j - \dot{p}_j dq_j\right). \quad (10.4)$$

So far, this is just a universal identity. Now comes the main trick of Hamilton's approach: let us consider H a function of the following independent arguments: time t, the generalized coordinates q_j, and the generalized momenta p_j (rather than generalized velocities). With this commitment, the general rule of differentiation of a function of several arguments gives

$$dH = \frac{\partial H}{\partial t}dt + \sum_j \left(\frac{\partial H}{\partial q_j}dq_j + \frac{\partial H}{\partial p_j}dp_j\right), \quad (10.5)$$

where dt, dq_j, and dp_j are independent differentials. Since Eq. (10.5) should be valid for any choice of these argument differentials, it should hold in particular if the differentials correspond to the real law of motion, for which Eq. (10.4) is valid as well. The comparison of Eqs. (10.4) and (10.5) gives us three relations:

$$\frac{\partial H}{\partial t} = -\frac{\partial L}{\partial t}, \quad (10.6)$$

$$\dot{q}_j = \frac{\partial H}{\partial p_j},$$
$$\dot{p}_j = -\frac{\partial H}{\partial q_j}. \quad (10.7)$$

Comparing the first of them with Eq. (2.35), we see that

$$\frac{dH}{dt} = \frac{\partial H}{\partial t}, \quad (10.8)$$

[2] Actually, this differential was already spelled out (partly and implicitly) in section 2.3—see Eqs (2.33)–(2.35).

meaning that function $H(t, q_j, p_j)$ can change in time only via its explicit dependence on t. Eqs. (10.7) are even more substantial: provided that such function $H(t, q_j, p_j)$ has been calculated, they give us two first-order differential equations (called the *Hamilton equations*) for the time evolution of the generalized coordinate and generalized momentum of each degree of freedom of the system[3].

Let us have a look at these equations for the simplest case of a system with one degree of freedom, with the Lagrangian function (3.3):

$$L = \frac{m_{\text{ef}}}{2}\dot{q}^2 - U_{\text{ef}}(q, t). \tag{10.9}$$

In this case, $p \equiv \partial L/\partial \dot{q} = m_{\text{ef}}\dot{q}$, and $H \equiv p\dot{q} - L = m_{\text{ef}}\dot{q}^2/2 + U_{\text{ef}}(q, t)$. In order to honor our new commitment, we need to express the Hamiltonian function explicitly via t, q, and p (rather than $\dot{q}$). From the above expression for p, we immediately have $\dot{q} = p/m_{\text{ef}}$; plugging this back to (10.9), we get

$$H = \frac{p^2}{2m_{\text{ef}}} + U_{\text{ef}}(q, t). \tag{10.10}$$

Now we can spell out Eqs. (10.7) for this particular case:

$$\dot{q} \equiv \frac{\partial H}{\partial p} = \frac{p}{m_{\text{ef}}}, \tag{10.11}$$

$$\dot{p} \equiv -\frac{\partial H}{\partial q} = -\frac{\partial U_{\text{ef}}}{\partial q}. \tag{10.12}$$

While the first of these equations just repeats the definition of the generalized momentum corresponding to coordinate q, the second one gives the equation of momentum change. Differentiating Eq. (10.11) over time, and plugging Eq. (10.12) into the result, we obtain:

$$\ddot{q} = \frac{\dot{p}}{m_{\text{ef}}} = -\frac{1}{m_{\text{ef}}}\frac{\partial U_{\text{ef}}}{\partial q}. \tag{10.13}$$

So, we have returned to the same equation (3.4) that was derived from the Lagrangian approach[4].

Thus, the Hamiltonian formalism does not give much new for the solution of this problem—and indeed most problems of classical mechanics. (This is why its discussion had been postponed until the very end of this course.) Moreover, since the Hamiltonian function $H(t, q_j, p_j)$ does not include generalized velocities explicitly, the phenomenological introduction of dissipation in this approach is less straightforward than in the Lagrangian equations, whose precursor form (2.17)

[3] Of course, the right-hand side of each Eq. (10.7) can generally include coordinates and momenta of other degrees of freedom as well, so that the equations of motion for different j are generally coupled.

[4] The reader is highly encouraged to perform a similar check for a few more problems, for example those listed at the end of the chapter, to obtain a better feeling of the Hamiltonian formalism.

is valid for dissipative forces as well. However, the Hamilton equations (10.7), which treat the generalized coordinates and momenta in a manifestly symmetric way, are heuristically fruitful—in addition to being aesthetically appealing. This is particularly true in the cases where these arguments participate in H in a similar way. For example, for the very important case of a dissipation-free linear ('harmonic') oscillator, for which $U_{\text{ef}} = \kappa_{\text{ef}} q^2/2$, Eq. (10.10) gives the famous symmetric form

$$H = \frac{p^2}{2m_{\text{ef}}} + \frac{\kappa_{\text{ef}} x^2}{2} \equiv \frac{p^2}{2m_{\text{ef}}} + \frac{m_{\text{ef}} \omega_0^2 x^2}{2}, \quad \text{where } \omega_0^2 \equiv \frac{\kappa_{\text{ef}}}{m_{\text{ef}}}. \tag{10.14}$$

The Hamilton equations (10.7) for this system preserve this symmetry, especially evident if we introduce the normalized momentum $\wp \equiv p/m_{\text{ef}}\omega_0$ (implicitly used in sections 5.6 and 9.2):

$$\frac{dq}{dt} = \omega_0 \wp, \qquad \frac{d\wp}{dt} = -\omega_0 q. \tag{10.15}$$

More practically, the Hamilton approach gives additional tools for the search for the integrals of motion. In order to see that, let us consider the full time derivative of an arbitrary function $f(t, q_j, p_j)$:

$$\frac{df}{dt} = \frac{\partial f}{\partial t} + \sum_j \left(\frac{\partial f}{\partial q_j} \dot{q}_j + \frac{\partial f}{\partial p_j} \dot{p}_j \right). \tag{10.16}$$

Plugging in $\dot{q}_j$ and $\dot{p}_j$ from the Hamilton equations (10.7), we obtain

$$\frac{df}{dt} = \frac{\partial f}{\partial t} + \sum_j \left(\frac{\partial H}{\partial p_j} \frac{\partial f}{\partial q_j} - \frac{\partial H}{\partial q_j} \frac{\partial f}{\partial p_j} \right) = \frac{\partial f}{\partial t} + \{H, f\}. \tag{10.17}$$

The last term on the right-hand side of this expression is the so-called *Poisson bracket*[5], and is defined for two arbitrary functions $f(t, q_j, p_j)$ and $g(t, q_j, p_j)$ as

$$\{g, f\} = \sum_j \left(\frac{\partial g}{\partial p_j} \frac{\partial f}{\partial q_j} - \frac{\partial f}{\partial p_j} \frac{\partial g}{\partial p_j} \right). \tag{10.18}$$

From this definition, one can readily verify that apart from the evident relations $\{f, f\} = 0$ and $\{f, g\} = -\{g, f\}$, the Poisson brackets obey the following important *Jacobi identity*:

$$\{f, \{g, h\}\} + \{g, \{h, f\}\} + \{h, \{f, g\}\} = 0. \tag{10.19}$$

[5] Named after S Poisson—of the Poisson equation and the Poisson statistical distribution fame.

Now let us use these relations for a search for integrals of motion. First, Eq. (10.17) shows that if a function f does not depend on time explicitly, and

$$\{H, f\} = 0, \tag{10.20}$$

then $df/dt = 0$, i.e. the function is an integral of motion. Moreover, it turns out that if we already know two integrals of motion, say f and g, then the following function,

$$F \equiv \{f, g\}, \tag{10.21}$$

is also an integral of motion—the so-called *Poisson theorem*. In order to prove it, we may use the Jacobi identity (10.19) with $h = H$. Now using Eq. (10.17) to express the Poisson brackets $\{g, H\}$, $\{H, g\}$, and $\{H,\{f, g\}\} = \{H, F\}$ via the full and partial time derivatives of functions f, g, and F, we obtain

$$\left\{f, \frac{\partial g}{\partial t} - \frac{dg}{dt}\right\} + \left\{g, \frac{df}{dt} - \frac{\partial f}{\partial t}\right\} + \frac{dF}{dt} - \frac{\partial F}{\partial t} = 0, \tag{10.22}$$

so that if f and g are indeed integrals of motion, i.e. $df/dt = dg/dt = 0$, then

$$\frac{dF}{dt} = \frac{\partial F}{\partial t} + \left\{g, \frac{\partial f}{\partial t}\right\} - \left\{f, \frac{\partial g}{\partial t}\right\} = \frac{\partial F}{\partial t} - \left[\left\{\frac{\partial f}{\partial t}, g\right\} + \left\{f, \frac{\partial g}{\partial t}\right\}\right]. \tag{10.23}$$

Plugging Eq. (10.21) into the first term of the right-hand side of this relation, and differentiating it by parts, we obtain $dF/dt = 0$, i.e. F is indeed an integral of motion as well.

Finally, one more important role of the Hamilton formalism is that it allows one to trace the close connection between classical and quantum mechanics. Indeed, using Eq. (10.18) to calculate the Poisson brackets of the generalized coordinates and momenta, we readily obtain

$$\{q_j, q_{j'}\} = 0, \quad \{p_j, p_{j'}\} = 0, \quad \{q_j, p_{j'}\} = -\delta_{jj'}. \tag{10.24}$$

In quantum mechanics, the operators of these quantities ('observables') obey commutation relations[6]

$$\left[\hat{q}_j, \hat{q}_{j'}\right] = 0, \quad \left[\hat{p}_j, \hat{p}_{j'}\right] = 0, \quad \left[\hat{q}_j, \hat{p}_{j'}\right] = i\hbar\delta_{jj'}, \tag{10.25}$$

where the definition of the commutator, $[\hat{g}, \hat{f}\,] \equiv \hat{g}\hat{f} - \hat{f}\hat{g}$, is to a certain extent[7] similar to that (10.18) of the Poisson bracket. We see that the classical relations (10.24) are similar to the quantum-mechanical relations (10.25) if the following parallel has been made:

$$\{g, f\} \leftrightarrow \frac{i}{\hbar}[\hat{g}, \hat{f}\,]. \tag{10.26}$$

---

[6] See, e.g. *Part QM* section 2.1.

[7] There is of course a conceptual difference between the 'usual' products of the function derivatives participating in the Poisson brackets, and the operator 'products' (meaning their sequential action on a state vector) forming the commutator.

This analogy extends well beyond Eqs. (10.24)–(10.25). For example, making the replacement Eq. (10.26) in Eq. (10.17), we obtain

$$\frac{d\hat{f}}{dt} = \frac{\partial \hat{f}}{\partial t} + \frac{i}{\hbar}[\hat{H}, \hat{f}\,], \quad \text{i.e.} \quad i\hbar\frac{d\hat{f}}{dt} = i\hbar\frac{\partial \hat{f}}{\partial t} + [\hat{f}, \hat{H}], \tag{10.27}$$

which is the correct equation of operator evolution in the Heisenberg picture of quantum mechanics[8].

The parallel (10.26) may give an important clue in the search for the proper quantum-mechanical operator of a given observable—which is not always elementary.

10.2 Adiabatic invariance

One more application of the Hamiltonian formalism in classical mechanics is the solution of the following problem[9]. Earlier in the course, we already studied some effects of the time variation of parameters of a single oscillator (section 5.5) and coupled oscillators (section 6.5). However, those discussions were focused on the case when the parameter variation speed is comparable with the own oscillation frequency (or frequencies) of the system. Another practically important case is when some system's parameter (let us call it λ) is changed much more slowly (*adiabatically*[10]),

$$\left|\frac{\dot{\lambda}}{\lambda}\right| \ll \frac{1}{\mathcal{T}}, \tag{10.28}$$

where $\mathcal{T}$ is a typical time period of oscillations in the system. Let us consider a 1D system whose Hamiltonian $H(q, p, \lambda)$ depends on time only via such a slow evolution of the parameter $\lambda = \lambda(t)$, and whose initial energy restricts the system's motion to a finite coordinate interval—see, e.g. figure 3.2c.

Then, as we know from section 3.3, if the parameter λ is constant, the system performs a periodic (though not necessarily sinusoidal) motion back and forth on axis q or, in a different language, along a closed trajectory on the phase plane $[q, p]$—see figure 10.1[11]. According to Eq. (10.8), in this case H is constant on the trajectory. (In order to distinguish this particular *value* from the Hamiltonian *function* as such, I will call it E, assuming that this constant coincides with the full mechanical energy E, as it does for the function (10.10), although this assumption is not necessary for the calculation made below.)

---

[8] See, e.g. *Part QM* section 4.6.

[9] Various aspects of this problem and its quantum-mechanical extension were first discussed by L Le Cornu (1895), Lord Rayleigh (1902), H Lorentz (1911), P Ehrenfest (1916), and M Born and V Fock (1928).

[10] This term is also used in the thermodynamics and statistical mechanics, where it implies not only a slow parameter variation, but also thermal insulation of the system—see, e.g. *Part SM* section 1.3. Evidently, the latter condition is irrelevant in our current context.

[11] In section 5.6, we discussed this plane for the particular case of sinusoidal oscillations—see figure 5.9

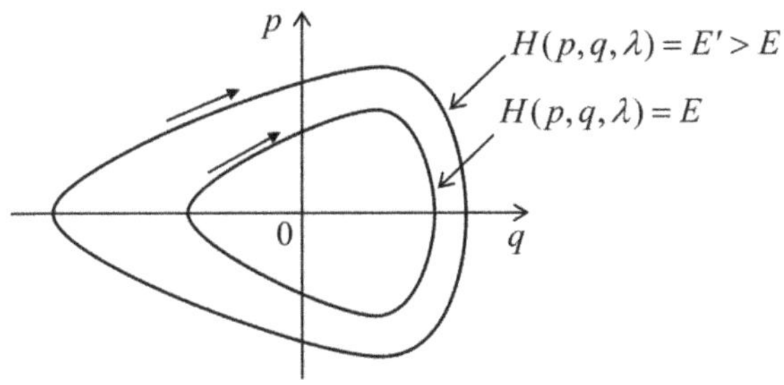

Figure 10.1. Phase-plane representation of periodic oscillations of a 1D Hamiltonian system, for two values of energy (schematically).

The oscillation period $\mathcal{T}$ may be calculated as a contour integral along this closed trajectory:

$$\mathcal{T} \equiv \int_0^{\mathcal{T}} dt = \oint \frac{dt}{dq} dq = \oint \frac{1}{\dot{q}} dq. \tag{10.29}$$

Using the first of the Hamilton Eqs. (10.7), we may represent this integral as

$$\mathcal{T} = \oint \frac{1}{\partial H / \partial p} dq. \tag{10.30}$$

At each given point q, $H = E$ is a function of p alone, so that we may flip the partial derivative in the denominator just as the full derivative, and rewrite Eq. (10.30) as

$$\mathcal{T} = \oint \frac{\partial p}{\partial E} dq. \tag{10.31}$$

For the particular Hamiltonian function (10.10), this relation is immediately reduced to Eq. (3.27), now in the form of a contour integral:

$$\mathcal{T} = \left(\frac{m_{\text{ef}}}{2}\right)^{1/2} \oint \frac{1}{[E - U_{\text{ef}}(q)]^{1/2}} dq. \tag{10.32}$$

Superficially, it looks like these formulas may also be used to find the motion period change when the parameter λ is being changed adiabatically, for example, by plugging known functions $m_{\text{ef}}(\lambda)$ and $U_{\text{ef}}(q, \lambda)$ into (10.32). However, there is no guarantee that the energy E in that integral would stay constant as the parameter changes, and indeed we will see below that this is not necessarily the case. Even more interestingly, in the most important case of the harmonic oscillator ($U_{\text{ef}} = \kappa_{\text{ef}} q^2/2$), whose oscillation period $\mathcal{T}$ does not depend on E (see Eq. (3.29) and its discussion), its variation in the adiabatic limit (10.28) may be readily predicted: $\mathcal{T}(\lambda) = 2\pi/\omega_0(\lambda) = 2\pi[m_{\text{ef}}(\lambda)/\kappa_{\text{ef}}(\lambda)]^{1/2}$, but the dependence of the oscillation energy E (and hence the oscillation amplitude) on λ is not immediately obvious.

In order to address this issue, let us use Eq. (10.8) (with $E = H$) to represent the energy change with $\lambda(t)$, i.e. in time, as

$$\frac{dE}{dt} = \frac{\partial H}{\partial t} = \frac{\partial H}{\partial \lambda} \frac{d\lambda}{dt}. \tag{10.33}$$

Since we are interested in a very slow (adiabatic) time evolution of energy, we can average Eq. (10.33) over fast oscillations in the system, for example over one oscillation period $\mathcal{T}$, treating $d\lambda/dt$ as a constant during this averaging[12]. The averaging yields

$$\frac{d\bar{E}}{dt} \approx \frac{d\lambda}{dt}\frac{\partial \bar{H}}{\partial \lambda} \equiv \frac{d\lambda}{dt}\frac{1}{\mathcal{T}}\int_0^{\mathcal{T}} \frac{\partial H}{\partial \lambda} dt. \tag{10.34}$$

Transforming this time integral into a contour integral, just as we did at the transition from Eq. (10.29) to Eq. (10.30), and then using Eq. (10.31) for $\mathcal{T}$, we obtain

$$\frac{d\bar{E}}{dt} = \frac{d\lambda}{dt}\frac{\oint \frac{\partial H/\partial \lambda}{\partial H/\partial p} dq}{\oint \frac{\partial p}{\partial E} dq}. \tag{10.35}$$

At each point q of the contour, H is a function of not only λ, but also of p, which may be also λ-dependent, so that if E is fixed, the partial differentiation of the relation $E = H$ over λ yields

$$\frac{\partial H}{\partial \lambda} + \frac{\partial H}{\partial p}\frac{\partial p}{\partial \lambda} = 0, \quad \text{i.e.} \quad \frac{\partial H/\partial \lambda}{\partial H/\partial p} = -\frac{\partial p}{\partial \lambda}. \tag{10.36}$$

Plugging the last relation into Eq. (10.35), we obtain

$$\frac{d\bar{E}}{dt} = -\frac{d\lambda}{dt}\frac{\oint \frac{\partial p}{\partial \lambda} dq}{\oint \frac{\partial p}{\partial E} dq}. \tag{10.37}$$

Since the left-hand side of Eq. (10.37), and the derivative $d\lambda/dt$ do not depend on q, we may move them into the integrals over q as constants, and rewrite Eq. (10.37) as

$$\oint \left(\frac{\partial p}{\partial E}\frac{d\bar{E}}{dt} + \frac{\partial p}{\partial \lambda}\frac{d\lambda}{dt} \right) dq = 0. \tag{10.38}$$

Now let us consider the following integral over the same phase plane contour,

$$J \equiv \frac{1}{2\pi}\oint p dq, \tag{10.39}$$

called the *action variable*. Just to understand its physical sense, let us calculate J for a harmonic oscillator (10.14). As we know very well from chapter 5, for such an oscillator $q = A\cos\Psi$, $p = -m_{ef}\omega_0 A \sin \Psi$ (with $\Psi = \omega_0 t$ + const), so that J may be

[12] This is the most critical point of this proof, because at any non-vanishing rate of parameter change the oscillations are, strictly speaking, non-periodic. Because of the approximate nature of this conjecture (which is very close to the assumptions made at the derivation of the reduced equations in section 5.3), new, stricter (but also much more cumbersome) proofs of the final Eq. (10.42) are still being offered in the literature—see, e.g. [1] and/or [2].

easily expressed either via the oscillations' amplitude A, or via their energy $E = H = m_{ef}\omega_0{}^2A^2/2$:

$$\begin{aligned} J &= \frac{1}{2\pi}\oint pdp \\ &= \frac{1}{2\pi}\int_{\Psi=0}^{\Psi=2\pi}(-m_{ef}\omega_0 A\sin\Psi)d(A\cos\Psi) = \frac{1}{2\pi}\frac{m_{ef}\omega_0}{2}A^2 = \frac{E}{\omega_0}. \end{aligned} \tag{10.40}$$

Returning to a general system with an adiabatically changed parameter λ, let us use the definition of J (10.39) to calculate its time derivative, again taking into account that at each point q of the trajectory, p is a function of E and λ:

$$\frac{dJ}{dt} = \frac{1}{2\pi}\oint\frac{dp}{dt}dq = \frac{1}{2\pi}\oint\left(\frac{\partial p}{\partial E}\frac{dE}{dt} + \frac{\partial p}{\partial\lambda}\frac{d\lambda}{dt}\right)dq. \tag{10.41}$$

Within the accuracy of our approximation, in which the contour integrals (10.38) and (10.41) are calculated along a closed trajectory, the factor dE/dt is indistinguishable from its time average, and these integrals coincide, so that the result Eq. (10.38) is applicable to Eq. (10.41) as well. Hence, we have finally arrived at a very important result—at a slow parameter variation, $dJ/dt = 0$, i.e. the action variable remains constant:

$$J = \text{const}. \tag{10.42}$$

This is the famous *adiabatic invariance*[13]. In particular, according to Eq. (10.40), in a harmonic oscillator the energy of oscillation changes proportionately to its own (slowly changed) frequency.

Before moving on, let me briefly note that the adiabatic invariance is not the only application of the action variable J. Since the initial choice of generalized coordinates and velocities (and hence the generalized momenta) in analytical mechanics is arbitrary (see section 2.1), it is almost evident that J may be taken for a new generalized momentum corresponding to a certain new generalized coordinate Θ,[14] and that the pair $\{J, \Theta\}$ should satisfy the Hamilton equations (10.7), in particular,

$$\frac{d\Theta}{dt} = \frac{\partial H}{\partial J}. \tag{10.43}$$

Following the commitment of section 10.1 (made there for the 'old' arguments q_j, p_j), before the differentiation on the right-hand side in (10.43), H should be expressed as a function (in addition to t) of the 'new' arguments J and Θ. For time-independent

[13] For certain particular oscillators, e.g. a point pendulum, Eq. (10.42) may be also proved directly—an exercise highly recommended to the reader.

[14] This, again, is a plausible argument but not a strict proof. Indeed, although according to its definition (10.39), J is nothing more than a sum of several (formally, an infinite number of) values of momentum p, they are not independent, but have to be selected on the same closed trajectory on the phase-plane. For more mathematical rigor, the reader is referred to section 45 of [3], which discusses the general rules of the so-called *canonical transformations* from one set of Hamiltonian arguments to another—say from $\{p, q\}$ to $\{J, \Theta\}$.

Hamiltonian systems, H is uniquely defined by J—see, e.g. Eq. (10.40). Hence in this case the right-hand side of Eq. (10.43) does not depend on either t or Θ, so that according to that equation, Θ (called the *angle variable*) is a linear function of time:

$$\Theta = \frac{\partial H}{\partial J}t + \text{const.} \tag{10.44}$$

For a harmonic oscillator, according to Eq. (10.40), the derivative $\partial H/\partial J = \partial E/\partial J$ is just $\omega_0 = 2\pi/\mathcal{T}$, so that $\Theta = \omega_0 t + \text{const}$, i.e. it is just the full phase Ψ that has been used repeatedly in this volume—in particular in chapter 5. It may be shown that a more general form of this relation,

$$\frac{\partial H}{\partial J} = \frac{2\pi}{\mathcal{T}}, \tag{10.45}$$

is valid for an arbitrary system described by (10.10). Thus, Eq. (10.44) becomes

$$\Theta = 2\pi\frac{t}{\mathcal{T}} + \text{const.} \tag{10.46}$$

This means that for an arbitrary (nonlinear) 1D oscillator, the angle variable Θ is a convenient generalization of the full phase Ψ. For this reason, variables J and Θ present a convenient tool for discussion of certain fine points of the dynamics of strongly nonlinear oscillators—for whose discussion I, unfortunately, do not have time[15].

10.3 The Hamilton principle

Now let me show that the Lagrangian equations of motion, that were derived in section 2.1 from Newton's laws, may be also obtained from the so-called *Hamilton principle*[16], namely the condition of a minimum (or rather an extremum) of the integral called *action*:

$$S \equiv \int_{t_{\text{ini}}}^{t_{\text{fin}}} L dt, \tag{10.47}$$

where t_{ini} and t_{fin} are, respectively, the initial and final moments of time, at which moments all generalized coordinates and velocities are considered fixed (not varied)—see figure 10.2.

The proof of that statement (in the realm of classical mechanics) is rather simple. Considering, similarly to section 2.1, a possible virtual variation of the motion described by infinitesimal deviations $\{\delta q_j(t), \delta\dot{q}_j(t)\}$ from the real motion, the necessary condition for S to be minimal is

$$\delta S \equiv \int_{t_{\text{ini}}}^{t_{\text{fin}}} \delta L dt = 0, \tag{10.48}$$

[15] The interested reader may see, for example, chapter 6 in [4].

[16] It is also called the 'principle of least action', or the 'principle of stationary action'. (These names may be more fair in the context of a long history of the development of the principle, starting from its simpler particular forms, which also includes the names P de Fermat, P Maupertuis, L Euler, and J-L Lagrange.)

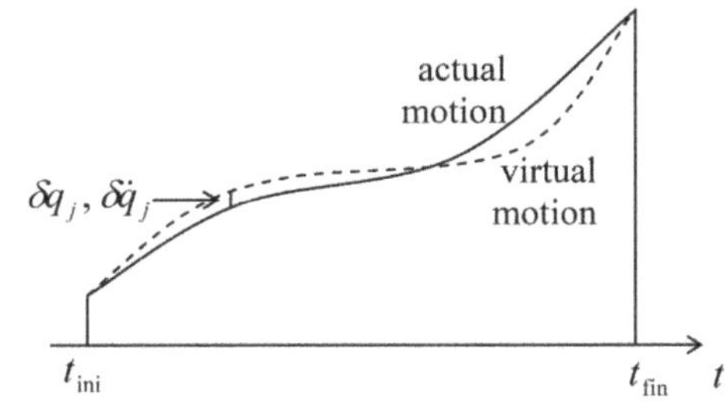

Figure 10.2. Deriving the Hamilton principle.

where δS and δL are the variations of the action and the Lagrange function, corresponding to the set $\{\delta q_j(t),\ \delta \dot{q}_j(t)\}$. As has been already discussed in section 2.1, we can use the operation of variation just as the usual differentiation (but at fixed time, see figure 2.2), swapping these two operations if needed—see figure 2.3 and its discussion. Thus, we may write

$$\delta L = \sum_j \left(\frac{\partial L}{\partial q_j}\delta q_j + \frac{\partial L}{\partial \dot{q}_j}\delta \dot{q}_j \right) = \sum_j \frac{\partial L}{\partial q_j}\delta q_j + \sum_j \frac{\partial L}{\partial \dot{q}_j}\frac{d}{dt}\delta q_j. \tag{10.49}$$

After plugging the last expression into Eq. (10.48), we can integrate the second term by parts:

$$\begin{aligned} \delta S &= \int_{t_{\text{ini}}}^{t_{\text{fin}}} \sum_j \frac{\partial L}{\partial q_j}\delta q_j dt + \sum_j \int_{t_{\text{ini}}}^{t_{\text{fin}}} \frac{\partial L}{\partial \dot{q}_j}\frac{d}{dt}\delta q_j dt. \\ &= \int_{t_{\text{ini}}}^{t_{\text{fin}}} \sum_j \frac{\partial L}{\partial q_j}\delta q_j dt + \sum_j \left[\frac{\partial L}{\partial \dot{q}_j}\delta q_j \right]_{t_{\text{ini}}}^{t_{\text{fin}}} - \sum_j \int_{t_{\text{ini}}}^{t_{\text{fin}}} \delta q_j d\left(\frac{\partial L}{\partial \dot{q}_j} \right) = 0. \end{aligned} \tag{10.50}$$

Since the generalized coordinates in the initial and final points are considered fixed (not affected by the variation), all $\delta q_j(t_{\text{ini}})$ and $\delta q_j(t_{\text{fin}})$ vanish, so that the second term on the right-hand side of Eq. (10.50) vanishes as well. Now multiplying and dividing the last term of that part by dt, we finally obtain

$$\begin{aligned} \delta S &= \int_{t_{\text{ini}}}^{t_{\text{fin}}} \sum_j \frac{\partial L}{\partial q_j}\delta q_j dt - \sum_j \int_{t_{\text{ini}}}^{t_{\text{fin}}} \delta q_j \frac{d}{dt}\left(\frac{\partial L}{\partial \dot{q}_j} \right) dt \\ &= -\int_{t_{\text{ini}}}^{t_{\text{fin}}} \sum_j \left[\frac{d}{dt}\left(\frac{\partial L}{\partial \dot{q}_j} \right) - \frac{\partial L}{\partial q_j} \right] \delta q_j dt = 0. \end{aligned} \tag{10.51}$$

This relation should hold for an arbitrary set of functions $\delta q_j(t)$, and for any time interval, so that it is only possible if the expressions in square brackets equal zero for all j, giving us the set of the Lagrange equations (2.19). So, the Hamilton principle indeed gives the Lagrange equations of motion.

It is fascinating to see how the Hamilton principle works for particular cases. As a very simple example, let us consider the usual 1D linear oscillator, with the Lagrangian function used so many times in this course:

$$L = \frac{m}{2}\dot{q}^2 - \frac{m\omega_0^2}{2}q^2. \tag{10.52}$$

As we know very well, the Lagrange equations of motion for this L are exactly satisfied by any sinusoidal function of frequency ω_0, in particular by a symmetric function of time

$$q_e(t) = A\cos\omega_0 t, \quad \text{so that } \dot{q}_e(t) = -A\omega_0\sin\omega_0 t. \tag{10.53}$$

On a limited time interval, say $0 \leqslant \omega_0 t \leqslant +\pi/2$, this function is rather smooth, and may be evidently well approximated by other simple, reasonably selected functions of time, for example

$$q_a(t) = A(1 - \lambda t^2), \quad \text{so that } \dot{q}_a(t) = -2A\lambda t, \tag{10.54}$$

provided that the parameter λ is also selected in a reasonable way. Let us take $\lambda = (\pi/2\omega_0)^2$, so that the approximate function $q_a(t)$ coincides with the exact function $q_e(t)$ at both ends of our time interval (figure 10.3):

$$q_a(t_{\text{ini}}) = q_a(t_{\text{ini}}) = A, \quad q_a(t_{\text{fin}}) = q_a(t_{\text{fin}}) = 0, \\ \text{where } t_{\text{ini}} \equiv 0, \quad t_{\text{fin}} \equiv \frac{\pi}{2\omega_0}, \tag{10.55}$$

and check which of them the Hamilton principle 'prefers', i.e. which would give the least action. An elementary calculation of the values of action (10.47) corresponding to these two functions yields

$$S_e = \left(\frac{\pi}{8} - \frac{\pi}{8}\right)m\omega_0 A^2 = 0, \\ S_a = \left(\frac{4}{3\pi} - \frac{2\pi}{15}\right)m\omega_0 A^2 \approx (0.4244 - 0.4189)m\omega_0 A^2 > 0, \tag{10.56}$$

with the first terms in all the parentheses coming from the time integrals of the kinetic energy and the second terms from those of the potential energy.

This result shows, first, that the exact function of time, for which these two contributions *exactly* cancel[17], is indeed 'preferable' for minimizing the action. Second, in the approximate function, the two contributions to action are rather close to the exact ones, and hence *almost* cancel each other, signaling that this approximation is very reasonable. It is evident that in some cases when the exact

[17] Such cancellation, i.e. the equality $S = 0$, is of course not the general requirement; it is specific only for this particular example.

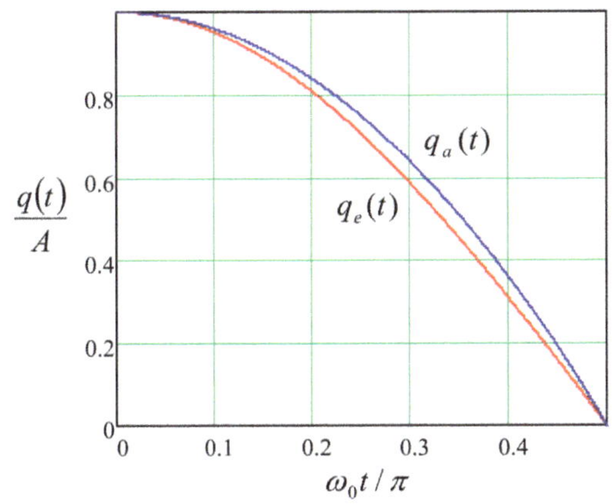

Figure 10.3. Plots of the functions $q(t)$ given by (10.53) and (10.54).

analytical solution of the equations of motion cannot be found, the minimization of S by adjusting one or more free parameters, incorporated into a 'trial' function, may be used to find a reasonable approximation for the actual law of motion[18].

It is very useful to make the notion of action S, defined by Eq. (10.47), more transparent by calculating it for the simple case of a single particle moving in a potential field that conserves its energy $E = T + U$. In this case the Lagrangian function $L = T - U$ may be represented as

$$L = T - U = 2T - (T + U) = 2T - E = mv^2 - E, \tag{10.57}$$

with a time-independent E, so that

$$S = \int L dt = \int mv^2 dt - Et + \text{const.} \tag{10.58}$$

Representing the expression under the remaining integral as $m\mathbf{v} \cdot \mathbf{v}dt = \mathbf{p} \cdot (d\mathbf{r}/dt)dt = \mathbf{p} \cdot d\mathbf{r}$, we finally obtain

$$S = \int \mathbf{p} \cdot d\mathbf{r} - Et + \text{const} = S_0 - Et + \text{const}, \tag{10.59}$$

where the time-independent integral

$$S_0 \equiv \int \mathbf{p} \cdot d\mathbf{r} \tag{10.60}$$

is frequently called the *abbreviated action*.

This expression may be used to establish one more important connection between the classical and quantum mechanics, now in its Schrödinger picture. Indeed, in the

[18] This is essentially a classical analog of the variational method used in quantum mechanics—see, e.g. *Part QM* section 2.9.

quasi-classical (WKB) approximation of that picture[19] a particle of a fixed energy is described by a de Broglie wave

$$\Psi(\mathbf{r}, t) \propto \exp\left\{i\left(\int \mathbf{k} \cdot d\mathbf{r} - \omega t + \text{const}\right)\right\}, \tag{10.61}$$

where the wavevector **k** is proportional to the particle's momentum, and frequency ω, to its energy:

$$\mathbf{k} = \frac{\mathbf{p}}{\hbar}, \quad \omega = \frac{E}{\hbar}. \tag{10.62}$$

Plugging these expressions into Eq. (10.61) and comparing the result with Eq. (10.59), we see that the WKB wavefunction may be represented as

$$\Psi \propto \exp\{iS/\hbar\}. \tag{10.63}$$

Hence the Hamilton's principle (10.48) means that the total phase of the quasi-classical wavefunction should be minimal along the particle's real trajectory. But this is exactly the so-called *eikonal minimum principle* well known from optics (although valid for any other waves as well), where it serves to define the ray paths in the geometric optics limit—similar to the WKB approximation condition. Thus, the ratio $S/\hbar$ may be considered just as the eikonal, i.e. the total phase accumulation, of the de Broglie waves[20].

Now, comparing Eq. (10.60) and (10.39), we see that the action variable J is just the change of the abbreviated action S_0 along a single phase-plane contour (divided by 2π). This means, in particular, that in the WKB approximation, J is the number of de Broglie waves along the classical trajectory of a particle, i.e. an integer value of the corresponding quantum number. If system's parameters are changed slowly, the quantum number has to stay integer, and hence J cannot change, giving a quantum-mechanical interpretation of the adiabatic invariance. It is really fascinating that a fact of classical mechanics may be 'derived' (or at least understood) more easily from the quantum mechanics' standpoint[21].

10.4 The Hamilton–Jacobi equation

The action S, defined by Eq. (10.47), may be used for one more formulation of classical mechanics. For that, we need one more, different commitment: S needs to be considered as a function of the following independent arguments: the final time point t_{fin} (which I will, for brevity, denote as t in this section), and the set of generalized coordinates (but not of the generalized velocities!) at that point:

$$S \equiv \int_{t_{\text{ini}}}^{t} L dt = S[t, q_j(t)]. \tag{10.64}$$

[19] See, e.g. *Part QM* section 3.1.

[20] Indeed, Eq. (10.63) was the starting point for R Feynman's development of his path-integral formulation of quantum mechanics—see, e.g. *Part QM* section 5.3.

[21] As a reminder, we have run into a similarly pleasant surprise at our discussion of the non-degenerate parametric excitation in section 6.7.

Let us calculate a variation of this (from the variational point of view, new!) function, resulting from an arbitrary combination of variations of the final values $q_j(t)$ of the coordinates, while keeping t fixed. Formally this may be done by repeating the variation calculations described by Eqs. (10.49)–(10.51), expect that now the variations δq_j at the finite point (t) do not necessarily equal zero. As a result, we obtain

$$\delta S = \sum_j \frac{\partial L}{\partial \dot{q}_j} \delta q_j|_t - \int_{t_{\text{ini}}}^{t} dt \sum_j \left[\frac{d}{dt}\left(\frac{\partial L}{\partial \dot{q}_j} \right) - \frac{\partial L}{\partial q_j} \right] \delta q_j. \tag{10.65}$$

For the motion along the real trajectory, i.e. satisfying the Lagrange equations the second term of this expression equals zero. Hence Eq. (10.65) shows that, for (any) fixed time t,

$$\frac{\partial S}{\partial q_j} = \frac{\partial L}{\partial \dot{q}_j}. \tag{10.66}$$

But the last derivative is nothing else than the generalized momentum p_j—see Eq. (2.31), so that

$$\frac{\partial S}{\partial q_j} = p_j. \tag{10.67}$$

(As a reminder, both sides of this relation refer to the final moment t of the trajectory.) As a result, the full derivative of the action $S[t, q_j(t)]$ over time takes the form

$$\frac{dS}{dt} = \frac{\partial S}{\partial t} + \sum_j \frac{\partial S}{\partial q_j} q_j = \frac{\partial S}{\partial t} + \sum_j p_j \dot{q}_j. \tag{10.68}$$

Now, by the very definition (10.64), the full derivative dS/dt is nothing more that the Lagrangian function L, so that Eq. (10.68) yields

$$\frac{\partial S}{\partial t} = L - \sum_j p_j \dot{q}_j. \tag{10.69}$$

However, according to the definition Eq. (10.2) of the Hamiltonian function H, the right-hand side of Eq. (10.68) is just $(-H)$, so that we obtain an extremely simply-looking *Hamilton–Jacobi equation*

$$\frac{\partial S}{\partial t} = -H. \tag{10.70}$$

This simplicity is, however, rather deceiving, because in order to use this equation for the calculation of function $S(t, q_j)$ for any particular problem, the Hamiltonian function has to be first expressed as a function of time t, generalized coordinates q_j, and the generalized momenta p_j (which may be, according to Eq. (10.67), represented just as derivatives $\partial S/\partial q_j$). Let us see how this procedure works for the simplest case of a 1D system with the Hamiltonian function given by Eq. (10.10). In this case, the only generalized momentum is $p = \partial S/\partial q$, so that

$$H = \frac{p^2}{2m_{\text{ef}}} + U_{\text{ef}}(q,\, t) = \frac{1}{2m_{\text{ef}}}\left(\frac{\partial S}{\partial q}\right)^2 + U_{\text{ef}}(q,\, t), \tag{10.71}$$

and the Eq. (10.70) is reduced to a partial differential equation,

$$\frac{\partial S}{\partial t} + \frac{1}{2m_{\text{ef}}}\left(\frac{\partial S}{\partial q}\right)^2 + U_{\text{ef}}(q,\, t) = 0. \tag{10.72}$$

Its solution may be readily found in the particular case of time-independent potential energy $U_{\text{ef}} = U_{\text{ef}}(q)$. In this case, Eq. (10.72) is evidently satisfied by the following variable-separated solution:

$$S(t,\, q) = S_0(q) + \text{const} \times t. \tag{10.73}$$

Plugging this solution into Eq. (10.72), we see that since the sum of two last terms on the left-hand side of that equation represent the full mechanical energy E, the constant in Eq. (10.73) is nothing but $(-E)$. Thus for the function $S_0(q)$ we obtain an ordinary differential equation

$$-H + \frac{1}{2m_{\text{ef}}}\left(\frac{dS_0}{dq}\right)^2 + U_{\text{ef}}(q) = 0. \tag{10.74}$$

Integrating it, we obtain

$$S_0 = \int \{2m_{\text{ef}}[E - U_{\text{ef}}(q)]\}^{1/2} dq + \text{const}, \tag{10.75}$$

so that, finally, the action is equal to

$$S = \int \{2m_{\text{ef}}[E - U_{\text{ef}}(q)]\}^{1/2} dq - Et + \text{const}. \tag{10.76}$$

For the case of 1D motion of a single 1D particle, i.e. for $q = x$, $m_{\text{ef}} = m$, $U_{\text{ef}}(q) = U(x)$, this solution is just the 1D case of the more general relations (10.59)–(10.60), which were obtained above in a much simpler way. (In particular, S_0 is just the abbreviated action.)

This particular case illustrates that the Hamilton–Jacobi equation is not the most efficient way for solution of most practical problems. However, it may be rather useful for studies of certain mathematical aspects of dynamics[22]. Moreover, in the early 1950s this approach was extended to a completely different field—the optimal control theory, in which the role of action S is played by the so-called *cost function*—a certain functional of a dynamic system, that should be minimized by an optimal

[22] See, e.g. chapters 6–9 in [5].

choice of a *control signal*—a function of time that affects the system's dynamics. From the point of view of this mathematical theory, Eq. (10.70) is a particular case of a more general *Hamilton–Jacobi–Bellman* equation[23].

10.5 Problems

In each of problems 10.1–10.3, for the given system:

(i) derive the Hamilton equations of motion, and

(ii) check whether the equations are equivalent to those derived from the Lagrangian formalism.

Problem 10.1. Our 'testbed' system: a bead on a ring, being rotated with a fixed angular velocity ω about its vertical diameter—see figure 2.1, reproduced below.

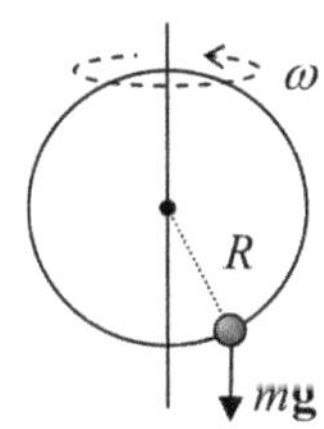

Problem 10.2. The system considered in problem 2.3: a pendulum hanging from a horizontal support whose motion law $x_0(t)$ is fixed—see the figure below. (No vertical plane constraint.)

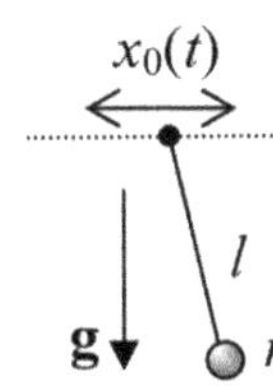

Problem 10.3. The system considered in problem 2.8: a block of mass m that can slide, without friction, along the inclined surface of a heavy wedge (mass m'). The wedge is free to move, also without friction, along a horizontal surface—see figure below. (Both motions are within the vertical plane containing the steepest slope line.)

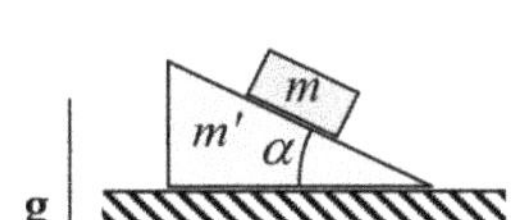

[23] See, e.g. [6]. The reader should not be intimidated by the unnatural term 'dynamic programming', which was invented by the founding father of this field, R Bellman, to lure government bureaucrats into funding his research, deemed too theoretical at that time. (Currently, it has a broad range of important applications.)

Problem 10.4. Find and solve the equations of motion of a particle with the following Hamiltonian function:

$$H = \frac{1}{2m}(\mathbf{p} + a\mathbf{r})^2,$$

where a is a constant scalar.

Problem 10.5. Let L be the Lagrangian function and H the Hamiltonian function of the same system. Which three of the following four statements,

$$\text{(i)}\ \frac{dL}{dt} = 0, \quad \text{(ii)}\ \frac{\partial L}{\partial t} = 0, \quad \text{(iii)}\ \frac{dH}{dt} = 0, \quad \text{(iv)}\ \frac{\partial H}{\partial t} = 0,$$

are equivalent? Give an example of when the three equalities hold, but the fourth one does not.

Problem 10.6. Calculate the Poisson brackets of Cartesian components of the angular momentum **L** of a particle moving in a central force field and its Hamiltonian function H, and discuss the most evident implication of the result.

Problem 10.7. After small oscillations have been initiated in the point pendulum shown in the figure below, the string is being pulled up slowly, so that the pendulum's length l is being reduced. Neglecting dissipation:

(i) prove by a direct calculation that the oscillation energy is indeed changing proportionately to the oscillation frequency, as follows from the constancy of the corresponding adiabatic invariant (10.40), and

(ii) find the l-dependence of the amplitudes of the angular and linear deviations from the equilibrium.

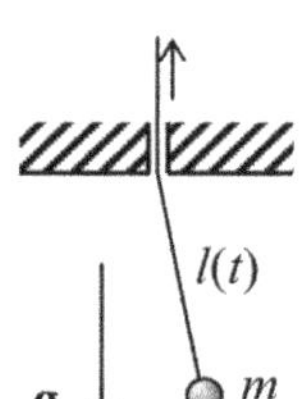

Problem 10.8. The mass m of a small body that performs 1D oscillations in the potential well $U(x) = ax^{2n}$, with $n > 0$, is being changed slowly. Calculate the oscillation energy E as a function of m.

Problem 10.9. A stiff ball is bouncing vertically from the floor of an elevator whose upward acceleration changes very slowly. Neglecting the energy dissipation, calculate how much the bounce height h changes during the acceleration's increase from 0 to g. Is your result valid for an equal but abrupt increase of the elevator's acceleration?

References

[1] Wells C and Siklos S 2007 *Eur. J. Phys.* **28** 105
[2] Lobo A *et al* 2012 *Eur. J. Phys.* **33** 1063
[3] Landau L and Lifshitz E 1976 *Mechanics* 3rd edn (Oxford: Butterworth-Heinemann)
[4] Jose J and Saletan E 1998 *Classical Dynamics* (Cambridge: Cambridge University Press)
[5] Percival I C and Richards D 1983 *Introduction to Dynamics* (Cambridge: Cambridge University Press)
[6] Bertsekas T P 2005 *Dynamic Programming and Optimal Control* vol 1 (Aetna Scientific)
Bertsekas T P 2007 *Dynamic Programming and Optimal Control* vol 2 (Aetna Scientific)

IOP Publishing

Classical Mechanics
Lecture notes
Konstantin K Likharev

Appendix A

Selected mathematical formulas

This appendix lists selected mathematical formulas that are used in this series, but not always remembered by students (and some instructors:-).

A.1 Constants

A Euclidean circle's *length-to-diameter ratio*:

$$\pi = 3.141\ 592\ 653\ \ldots; \qquad \pi^{1/2} \approx 1.77. \tag{A.1}$$

Natural logarithm base:

$$e \equiv \lim_{n\to\infty}\left(1 + \frac{1}{n}\right)^n = 2.718\ 281\ 828\ \ldots; \tag{A.2a}$$

from that value, the logarithm base conversion factors are as follows:

$$\frac{\ln x}{\log_{10} x} = \ln 10 \approx 2.303, \qquad \frac{\log_{10} x}{\ln x} = \frac{1}{\ln 10} \approx 0.434. \tag{A.2b}$$

The *Euler* (or 'Euler–Mascheroni') *constant*:

$$\gamma \equiv \lim_{n\to\infty}\left(1 + \frac{1}{2} + \frac{1}{3} + \ldots \frac{1}{n} - \ln n\right) = 0.577\ 156\ 649\ 0\ \ldots; \quad e^{\gamma} \approx 1.781. \tag{A.3}$$

A.2 Combinatorics, sums, and series

(i) *Combinatorics*

- The number of different *permutations*, i.e. *ordered* sequences of k elements selected from a set of n distinct elements ($n \geqslant k$), is

$$^{n}P_k \equiv n \cdot (n-1) \cdot \ldots \cdot (n-k+1) = \frac{n!}{(n-k)!}; \tag{A.4a}$$

in particular, the number of different permutations of *all* elements of the set $(n = k)$ is

$$^kP_k = k \cdot (k-1) \cdot \ldots \cdot 2 \cdot 1 = k!. \tag{A.4b}$$

- The number of different *combinations*, i.e. *unordered* sequences of k elements from a set of $n \geqslant k$ distinct elements, is equal to the binomial coefficient

$$^nC_k \equiv \binom{n}{k} \equiv \frac{^nP_k}{^kP_k} = \frac{n!}{k!(n-k)!}. \tag{A.5}$$

  In an alternative, very popular 'ball/box language', ^nC_k is the number of different ways to put in a box, in an arbitrary order, k balls selected from n distinct balls.
- A generalization of the binomial coefficient notion is the multinomial coefficient,

$$^nC_{k_1,k_2,\ldots k_l} \equiv \frac{n!}{k_1!k_2!\ldots k_l!}, \qquad \text{with } n = \sum_{j=1}^{l} k_j, \tag{A.6}$$

which, in the standard mathematical language, is a number of different permutations in a multiset of l distinct element types from an n-element set which contains k_j ($j = 1, 2, \ldots l$) elements of each type. In the 'ball/box language', the coefficient (A.6) is the number of different ways to distribute n balls between l different boxes, each time keeping the number (k_j) of balls in the jth box fixed, but ignoring their order inside the box. The binomial coefficient ^nC_k (A.5), is a particular case of the multinomial coefficient (A.6) for $l = 2$—counting the explicit box as the first, and the remaining space as the second box, so that if $k_1 \equiv k$, then $k_2 = n - k$.
- One more important combinatorial quantity is the number $M_n^{(k)}$ of ways to place n *indistinguishable* balls into k distinct boxes. It may be readily calculated from Eq. (A.5) as the number of different ways to select $(k-1)$ partitions between the boxes in an imagined linear row of $(k - 1 + n)$ 'objects' (balls in the boxes *and* partitions between them):

$$M_n^{(k)} = {}^{n-1+k}C_{k-1} \equiv \frac{(k-1+n)!}{(k-1)!n!}. \tag{A.7}$$

(ii) *Sums and series*

- *Arithmetic progression*:

$$r + 2r + \ldots + nr \equiv \sum_{k=1}^{n} kr = \frac{n(r+nr)}{2}; \tag{A.8a}$$

in particular, at $r = 1$ it is reduced to the sum of n first natural numbers:

$$1 + 2 + \ldots + n \equiv \sum_{k=1}^{n} k = \frac{n(n+1)}{2}. \tag{A.8b}$$

- Sums of squares and cubes of n first natural numbers:

$$1^2 + 2^2 + \ldots + n^2 \equiv \sum_{k=1}^{n} k^2 = \frac{n(n+1)(2n+1)}{6}; \tag{A.9a}$$

$$1^3 + 2^3 + \ldots + n^3 \equiv \sum_{k=1}^{n} k^3 = \frac{n^2(n+1)^2}{4}. \tag{A.9b}$$

- The *Riemann zeta function*:

$$\zeta(s) \equiv 1 + \frac{1}{2^s} + \frac{1}{3^s} + \ldots = \sum_{k=1}^{\infty} \frac{1}{k^s}; \tag{A.10a}$$

the particular values frequently met in applications are

$$\begin{aligned} &\zeta\left(\frac{3}{2}\right) \approx 2.612, \ \zeta(2) = \frac{\pi^2}{6}, \ \zeta\left(\frac{5}{2}\right) \approx 1.341, \\ &\zeta(3) \approx 1.202, \ \zeta(4) = \frac{\pi^4}{90}, \ \zeta(5) \approx 1.037. \end{aligned} \tag{A.10b}$$

Finite geometric progression (for real $\lambda \neq 1$):

$$1 + \lambda + \lambda^2 + \ldots + \lambda^{n-1} \equiv \sum_{k=0}^{n-1} \lambda^k = \frac{1 - \lambda^n}{1 - \lambda}; \tag{A.11a}$$

in particular, if $\lambda^2 < 1$, the progression has a finite limit at $n \to \infty$ (called the *geometric series*):

$$\lim_{n\to\infty} \sum_{k=0}^{n-1} \lambda^k = \sum_{k=0}^{\infty} \lambda^k = \frac{1}{1 - \lambda}. \tag{A.11b}$$

- *Binomial sum* (or the 'binomial theorem'):

$$(1 + a)^n = \sum_{k=0}^{n} {}^nC_k a^k, \tag{A.12}$$

where ${}^nC_k$ are the binomial coefficients defined by Eq. (A.5).

- The *Stirling formula*:

$$\lim_{n\to\infty} \ln(n!) = n(\ln n - 1) + \frac{1}{2}\ln(2\pi n) + \frac{1}{12n} - \frac{1}{360n^3} + \ldots; \quad \text{(A.13)}$$

for most applications in physics, the first term[1] is sufficient.

- The *Taylor* (or 'Taylor–Maclaurin') *series*: for any infinitely differentiable function $f(x)$,

$$\begin{aligned}\lim_{\tilde{x}\to 0} f(x+\tilde{x}) + f(x) + \frac{df}{dx}(x)\tilde{x} + \frac{1}{2!}\frac{d^2f}{dx^2}(x)\tilde{x}^2 + \ldots \\ = \sum_{k=0}^{\infty}\frac{1}{k!}\frac{d^kf}{dx^k}(x)\tilde{x}^k;\end{aligned} \quad \text{(A.14}a\text{)}$$

note that for many functions this series converges only within a limited, sometimes small range of deviations $\tilde{x}$. For a function of several arguments, $f(x_1,x_2,\ldots,x_N)$, the first terms of the Taylor series are

$$\begin{aligned}\lim_{\tilde{x}_k\to 0} f(x_1+\tilde{x}_1+x_2+\tilde{x}_2\ldots) = f|_{\tilde{x}_k=0} + \sum_{k=1}^{N}\frac{\partial f}{\partial x_k}|_{\tilde{x}_k=0}\tilde{x}_k \\ + \frac{1}{2!}\sum_{k,k'=1}^{N}\frac{\partial^2 f}{\partial x_k \partial x_{k'}}|_{\tilde{x}_k,\tilde{x}_{k'}=0}\tilde{x}_k\tilde{x}_{k'} + \ldots\end{aligned} \quad \text{(A.14}b\text{)}$$

- The *Euler–Maclaurin formula*, valid for any infinitely differentiable function $f(x)$:

$$\begin{aligned}\sum_{n=0}^{\infty} f(n) = \int_0^{\infty} f(x)dx - \frac{1}{2}f(0) - \frac{1}{6}\cdot\frac{1}{2!}\frac{df}{dx}(0) \\ + \frac{1}{30}\cdot\frac{1}{4!}\frac{d^3f}{dx^3}(0) + \ldots;\end{aligned} \quad \text{(A.15}a\text{)}$$

the coefficients participating in this formula are the so-called *Bernoulli numbers*[2]:

$$\begin{aligned}B_1 = \frac{1}{2},\ B_2 = \frac{1}{6},\ B_3 = 0,\ B_4 = \frac{1}{30},\ B_5 = 0, \\ B_6 = \frac{1}{42},\ B_7 = 0,\ B_8 = \frac{1}{30},\ \ldots\end{aligned} \quad \text{(A.15}b\text{)}$$

---

[1] Actually, this leading term was derived by A de Moivre in 1733, before J Stirling's work, but nevertheless is also commonly called the Stirling approximation.

[2] Note that definitions of B_k (or rather their signs and indices) differ even in the most popular handbooks.

A.3 Basic trigonometric functions

- Trigonometric functions of the sum and the difference of two arguments[3]:

$$\cos(a \pm b) = \cos a \cos b \mp \sin a \sin b, \tag{A.16a}$$

$$\sin(a \pm b) = \sin a \cos b \pm \cos a \sin b. \tag{A.16b}$$

- Sums of two functions of arbitrary arguments:

$$\cos a + \cos b = 2\cos\frac{a+b}{2}\cos\frac{b-a}{2}, \tag{A.17a}$$

$$\cos a - \cos b = 2\sin\frac{a+b}{2}\sin\frac{b-a}{2}, \tag{A.17b}$$

$$\sin a \pm \sin b = 2\sin\frac{a\pm b}{2}\cos\frac{\pm b-a}{2}. \tag{A.17c}$$

- Trigonometric function products:

$$2\cos a\cos b = \cos(a+b) + \cos(a-b), \tag{A.18a}$$

$$2\sin a\cos b = \sin(a+b) + \sin(a-b), \tag{A.18b}$$

$$2\sin a\sin b = \cos(a-b) - \cos(a+b). \tag{A.18c}$$

For the particular case of equal arguments, $b = a$, these three formulas yield the following expressions for the squares of trigonometric functions, and their product:

$$\cos^2 a = \frac{1}{2}(1+\cos 2a),\ \sin a\cos a = \frac{1}{2}\sin 2a, \quad \sin^2 a = \frac{1}{2}(1-\cos 2a). \tag{A.18d}$$

- Cubes of trigonometric functions:

$$\cos^3 a = \frac{3}{4}\cos a + \frac{1}{4}\cos 3a, \qquad \sin^3 a = \frac{3}{4}\sin a - \frac{1}{4}\sin 3a. \tag{A.19}$$

[3] I am confident that the reader is quite capable of deriving relations (A.16) by representing the exponent in the elementary relation $e^{i(a\pm b)} = e^{ia}e^{\pm ib}$ as a sum of its real and imaginary parts, Eqs. (A.18) directly from Eqs. (A.16), and Eqs. (A.17) from Eqs. (A.18), by variable replacement; however, I am still providing these formulas to save his or her time. (Quite a few formulas below are included because of the same reason.)

- Trigonometric functions of a complex argument:

$$\begin{aligned}\sin(a+ib) &= \sin a\cosh b + i\cos a\sinh b,\\ \cos(a+ib) &= \cos a\cosh b - i\sin a\sinh b.\end{aligned} \tag{A.20}$$

- Sums of trigonometric functions of n equidistant arguments:

$$\sum_{k=1}^{n}\left\{\begin{matrix}\sin\\ \cos\end{matrix}\right\}k\xi = \left\{\begin{matrix}\sin\\ \cos\end{matrix}\right\}\left(\frac{n+1}{2}\xi\right)\sin\left(\frac{n}{2}\xi\right)\Big/\sin\left(\frac{\xi}{2}\right). \tag{A.21}$$

A.4 General differentiation

- Full differential of a product of two functions:

$$d(fg) = (df)g + f(dg). \tag{A.22}$$

- Full differential of a function of several independent arguments, $f(x_1, x_2, \ldots, x_n)$:

$$df = \sum_{k=1}^{n}\frac{\partial f}{\partial x_k}dx_k. \tag{A.23}$$

- Curvature of the Cartesian plot of a 1D function $f(x)$:

$$\kappa \equiv \frac{1}{R} = \frac{|\, d^2f/dx^2 \,|}{[1+(df/dx)^2]^{3/2}}. \tag{A.24}$$

A.5 General integration

- Integration *by parts*—immediately follows from Eq. (A.22):

$$\int_{g(A)}^{g(B)} f dg = fg|_A^B - \int_{f(A)}^{f(B)} g df. \tag{A.25}$$

- Numerical (approximate) integration of 1D functions: the simplest *trapezoidal rule*,

$$\begin{aligned}\int_a^b f(x)dx &\approx h\left[f\left(a+\frac{h}{2}\right)+f\left(a+\frac{3h}{2}\right)+\ldots+f\left(b-\frac{h}{2}\right)\right]\\ &= h\sum_{n=1}^{N} f\left(a-\frac{h}{2}+nh\right), \quad \text{where } h \equiv \frac{b-a}{N}.\end{aligned} \tag{A.26}$$

has relatively low accuracy (error of the order of $(h^3/12)d^2f/dx^2$ per step), so that the following *Simpson formula*,

$$\int_a^b f(x)dx \approx \frac{h}{3}[f(a) + 4f(a+h) + 2f(a+2h) + \ldots + 4f(b-h) + f(b)], \quad \text{where } h \equiv \frac{b-a}{2N}, \tag{A.27}$$

whose error per step scales as $(h^5/180)d^4f/dx^4$, is used much more frequently[4].

A.6 A few 1D integrals[5]

(i) *Indefinite integrals*

- Integrals with $(1+\xi^2)^{1/2}$:

$$\int (1+\xi^2)^{1/2} d\xi = \frac{\xi}{2}(1+\xi^2)^{1/2} + \frac{1}{2}\ln|\xi + (1+\xi^2)^{1/2}|, \tag{A.28}$$

$$\int \frac{d\xi}{(1+\xi^2)^{1/2}} = \ln|\xi + (1+\xi^2)^{1/2}|, \tag{A.29a}$$

$$\int \frac{d\xi}{(1+\xi^2)^{3/2}} = \frac{\xi}{(1+\xi^2)^{1/2}}. \tag{A.29b}$$

- Miscellaneous indefinite integrals:

$$\int \frac{d\xi}{\xi(\xi^2 + 2a\xi - 1)^{1/2}} = \arccos\frac{a\xi - 1}{|\xi|(a^2+1)^{1/2}}, \tag{A.30a}$$

$$\int \frac{(\sin\xi - \xi\cos\xi)^2}{\xi^5} d\xi = \frac{2\xi\sin 2\xi + \cos 2\xi - 2\xi^2 - 1}{8\xi^4}, \tag{A.30b}$$

$$\int \frac{d\xi}{a + b\cos\xi} = \frac{2}{(a^2-b^2)^{1/2}} \tan^{-1}\left[\frac{(a-b)}{(a^2-b^2)^{1/2}} \tan\frac{\xi}{2}\right], \quad \text{for } a^2 > b^2. \tag{A.30c}$$

$$\int \frac{d\xi}{1+\xi^2} = \tan^{-1}\xi. \tag{A.30d}$$

[4] Higher-order formulas (e.g. the *Bode rule*), and other guidance including ready-for-use codes for computer calculations may be found, for example, in the popular reference texts by W Press *et al* [1]. In addition, some advanced codes are used as subroutines in the software packages listed in section A.16. In some cases, the Euler–Maclaurin formula (A.15) also may be useful for numerical integration.

[5] A powerful (and free) interactive online tool for working out integrals analytically is available at www.wolframalpha.com/calculators/integral-calculator/.

(ii) *Semi-definite integrals*:

- Integrals with $1/(e^{\xi} \pm 1)$:

$$\int_a^\infty \frac{d\xi}{e^\xi + 1} = \ln(1 + e^{-a}), \tag{A.31a}$$

$$\int_{a>0}^\infty \frac{d\xi}{e^\xi - 1} = \ln \frac{1}{1 - e^{-a}}. \tag{A.31b}$$

(iii) *Definite integrals:*

- Integrals with $1/(1 + \xi^2)$:[6]

$$\int_0^\infty \frac{d\xi}{1 + \xi^2} = \frac{\pi}{2}, \tag{A.32a}$$

$$\int_0^\infty \frac{d\xi}{(1 + \xi^2)^{3/2}} = 1; \tag{A.32b}$$

more generally,

$$\int_0^\infty \frac{d\xi}{(1 + \xi^2)^n} = \frac{\pi}{2}\frac{(2n-3)!!}{(2n-2)!!} \equiv \frac{\pi}{2}\frac{1 \cdot 3 \cdot 5 \ldots (2n-3)}{2 \cdot 4 \cdot 6 \ldots (2n-2)}, \quad \text{for } n = 2, 3, \ldots \tag{A.32c}$$

- Integrals with $(1 - \xi^{2n})^{1/2}$:

$$\int_0^1 \frac{d\xi}{(1 - \xi^{2n})^{1/2}} = \frac{\pi^{1/2}}{2n}\Gamma\left(\frac{1}{2n}\right)\Big/\Gamma\left(\frac{n+1}{2n}\right), \tag{A.33a}$$

$$\int_0^1 (1 - \xi^{2n})^{1/2} d\xi = \frac{\pi^{1/2}}{4n}\Gamma\left(\frac{1}{2n}\right)\Big/\Gamma\left(\frac{3n+1}{2n}\right), \tag{A.33b}$$

where $\Gamma(s)$ is the *gamma-function,* which is most often defined (for $\text{Re}\, s > 0$) by the following integral:

$$\int_0^\infty \xi^{s-1} e^{-\xi} d\xi = \Gamma(s). \tag{A.34a}$$

[6] Eq. (A.32*a*) follows immediately from Eq. (A.30*d*), and Eq. (A.32*b*) from Eq. (A.29*b*)—a couple more examples of the (intentional) redundancy of this list.

The key property of this function is the recurrence relation, valid for any $s \neq 0, -1, -2, \ldots$:

$$\Gamma(s+1) = s\Gamma(s). \tag{A.34b}$$

Since, according to Eq. (A.34*a*), $\Gamma(1) = 1$, Eq. (A.34*b*) for non-negative integers takes the form

$$\Gamma(n+1) = n!, \quad \text{for } n = 0, 1, 2, \ldots \tag{A.34c}$$

(where $0! \equiv 1$). Because of this, for integer $s = n + 1 \geqslant 1$, Eq. (A.34*a*) is reduced to

$$\int_0^\infty \xi^n e^{-\xi} d\xi = n!. \tag{A.34d}$$

Other frequently met values of the gamma-function are those for positive semi-integer arguments:

$$\Gamma\left(\frac{1}{2}\right) = \pi^{1/2}, \quad \Gamma\left(\frac{3}{2}\right) = \frac{1}{2}\pi^{1/2}, \quad \Gamma\left(\frac{5}{2}\right) = \frac{1}{2}\cdot\frac{3}{2}\pi^{1/2}, \quad \Gamma\left(\frac{7}{2}\right) = \frac{1}{2}\cdot\frac{3}{2}\cdot\frac{5}{2}\pi^{1/2}, \ \ldots, \text{etc.} \tag{A.34e}$$

- Integrals with $1/(e^\xi \pm 1)$:

$$\int_0^\infty \frac{\xi^{s-1} d\xi}{e^\xi + 1} = (1 - 2^{1-s})\Gamma(s)\zeta(s), \quad \text{for } s > 0, \tag{A.35a}$$

$$\int_0^\infty \frac{\xi^{s-1} d\xi}{e^\xi - 1} = \Gamma(s)\zeta(s), \quad \text{for } s > 1, \tag{A.35b}$$

where $\zeta(s)$ is the Riemann zeta-function—see Eq. (A.10). Particular cases: for $s = 2n$,

$$\int_0^\infty \frac{\xi^{2n-1} d\xi}{e^\xi + 1} = \frac{2^{2n-1} - 1}{2n}\pi^{2n} B_{2n}, \tag{A.35c}$$

$$\int_0^\infty \frac{\xi^{2n-1} d\xi}{e^\xi - 1} = \frac{(2\pi)^{2n}}{4n} B_{2n}, \tag{A.35d}$$

where B_n are the Bernoulli numbers—see Eq. (A.15). For the particular case $s = 1$ (when Eq. (A.35*a*) yields uncertainty),

$$\int_0^\infty \frac{d\xi}{e^\xi + 1} = \ln 2. \tag{A.35e}$$

- Integrals with $\exp\{-\xi^2\}$:

$$\int_0^\infty \xi^s e^{-\xi^2} d\xi = \frac{1}{2}\Gamma\left(\frac{s+1}{2}\right), \quad \text{for } s > -1; \tag{A.36a}$$

for applications the most important particular even values of s are 0 and 2:

$$\int_0^\infty e^{-\xi^2} d\xi = \frac{1}{2}\Gamma\left(\frac{1}{2}\right) = \frac{\pi^{1/2}}{2}, \tag{A.36b}$$

$$\int_0^\infty \xi^2 e^{-\xi^2} d\xi = \frac{1}{2}\Gamma\left(\frac{3}{2}\right) = \frac{\pi^{1/2}}{4}, \tag{A.36c}$$

although we will also run into the cases $s = 4$ and $s = 6$:

$$\begin{aligned} \int_0^\infty \xi^4 e^{-\xi^2} d\xi &= \frac{1}{2}\Gamma\left(\frac{5}{2}\right) = \frac{3\pi^{1/2}}{8}, \\ \int_0^\infty \xi^6 e^{-\xi^2} d\xi &= \frac{1}{2}\Gamma\left(\frac{7}{2}\right) = \frac{15\pi^{1/2}}{16}; \end{aligned} \tag{A.36d}$$

for odd values $s = 2n + 1$ (with $n = 0, 1, 2, \ldots$), Eq. (A.36*a*) takes a simpler form

$$\int_0^\infty \xi^{2n+1} e^{-\xi^2} d\xi = \frac{1}{2}\Gamma(n+1) = \frac{n!}{2}. \tag{A.36e}$$

- Integrals with cosine and sine functions:

$$\int_0^\infty \cos(\xi^2) d\xi = \int_0^\infty \sin(\xi^2) d\xi = \left(\frac{\pi}{8}\right)^{1/2}, \tag{A.37}$$

$$\int_0^\infty \frac{\cos\xi}{a^2 + \xi^2} d\xi = \frac{\pi}{2a} e^{-a}, \tag{A.38}$$

$$\int_0^\infty \left(\frac{\sin\xi}{\xi}\right)^2 d\xi = \frac{\pi}{2}. \tag{A.39}$$

- Integrals with logarithms:

$$\int_0^1 \ln\frac{a + (1-\xi^2)^{1/2}}{a - (1-\xi^2)^{1/2}} d\xi = \pi[a - (a^2 - 1)^{1/2}], \quad \text{for } a \geqslant 1, \tag{A.40}$$

$$\int_0^1 \ln\frac{1 + (1-\xi)^{1/2}}{\xi^{1/2}} d\xi = 1. \tag{A.41}$$

- Integral representations of the Bessel functions of integer order:

$$J_n(\alpha) = \frac{1}{2\pi}\int_{-\pi}^{+\pi} e^{i(\alpha \sin\xi - n\xi)}d\xi,$$
$$\text{so that } e^{i\alpha \sin\xi} = \sum_{k=-\infty}^{\infty} J_k(\alpha)e^{ik\xi}; \qquad \text{(A.42}a\text{)}$$

$$I_n(\alpha) = \frac{1}{\pi}\int_0^{\pi} e^{\alpha \cos\xi}\cos n\xi d\xi. \qquad \text{(A.42}b\text{)}$$

A.7 3D vector products

(i) *Definitions*:

- *Scalar* ('dot-') *product*:

$$\mathbf{a}\cdot\mathbf{b} = \sum_{j=1}^{3} a_j b_j, \qquad \text{(A.43)}$$

where a_j and b_j are vector components in any orthogonal coordinate system. In particular, vector squared (the same as norm squared):

$$a^2 \equiv \mathbf{a}\cdot\mathbf{a} = \sum_{j=1}^{3} a_j^2 \equiv \|\mathbf{a}\|^2. \qquad \text{(A.44)}$$

- *Vector* ('cross-') *product*:

$$\mathbf{a}\times\mathbf{b} \equiv \mathbf{n}_1(a_2b_3 - a_3b_2) + \mathbf{n}_2(a_3b_1 - a_1b_3) + \mathbf{n}_3(a_1b_2 - a_2b_1) = \begin{vmatrix} \mathbf{n}_1 & \mathbf{n}_2 & \mathbf{n}_2 \\ a_1 & a_2 & a_3 \\ b_1 & b_2 & b_3 \end{vmatrix}, \qquad \text{(A.45)}$$

where $\{\mathbf{n}_j\}$ is the set of mutually perpendicular unit vectors[7] along the corresponding coordinate system axes[8]. In particular, Eq. (A.45) yields

$$\mathbf{a}\times\mathbf{a} = 0. \qquad \text{(A.46)}$$

(ii) *Corollaries* (readily verified by Cartesian components):

- Double vector product (the so-called *bac minus cab* rule):

$$\mathbf{a}\times(\mathbf{b}\times\mathbf{c}) = \mathbf{b}(\mathbf{a}\cdot\mathbf{c}) - \mathbf{c}(\mathbf{a}\cdot\mathbf{b}). \qquad \text{(A.47)}$$

[7] Popular alternative notations for this vector set are $\{\mathbf{e}_j\}$ and $\{\hat{\mathbf{r}}_j\}$.

[8] It is easy to use Eq. (A.45) to check that the direction of the product vector corresponds to the well-known 'right-hand rule' and to the even more convenient *corkscrew rule*: if we rotate a corkscrew (or a screwdriver) from the first operand toward the second one, its body moves in the direction of the product.

- Mixed scalar-vector product (called the *operand rotation rule*):

$$\mathbf{a}\cdot(\mathbf{b}\times\mathbf{c})=\mathbf{b}\cdot(\mathbf{c}\times\mathbf{a})=\mathbf{c}\cdot(\mathbf{a}\times\mathbf{b}).\qquad\text{(A.48)}$$

- Scalar product of vector products:

$$(\mathbf{a}\times\mathbf{b})\cdot(\mathbf{c}\times\mathbf{d})=(\mathbf{a}\cdot\mathbf{c})(\mathbf{b}\cdot\mathbf{d})-(\mathbf{a}\cdot\mathbf{d})(\mathbf{b}\cdot\mathbf{c});\qquad\text{(A.49}a\text{)}$$

in the particular case of two similar operands (say, $\mathbf{a}=\mathbf{c}$ and $\mathbf{b}=\mathbf{d}$), the last formula is reduced to

$$(\mathbf{a}\times\mathbf{b})^2=(ab)^2-(\mathbf{a}\cdot\mathbf{b})^2.\qquad\text{(A.49}b\text{)}$$

A.8 Differentiation in 3D Cartesian coordinates

- Definition of the *del* (or 'nabla') vector-operator ∇:[9]

$$\nabla\equiv\sum_{j=1}^{3}\mathbf{n}_j\frac{\partial}{\partial r_j},\qquad\text{(A.50)}$$

where r_j is a set of linear and orthogonal (called *Cartesian*) coordinates along directions $\mathbf{n}_j$. In accordance with this definition, the operator ∇ acting on a *scalar* function of coordinates, $f(\mathbf{r})$,[10] gives its gradient, i.e. a new *vector*:

$$\nabla f\equiv\sum_{j=1}^{3}\mathbf{n}_j\frac{\partial f}{\partial r_j}\equiv\mathbf{grad}\,f.\qquad\text{(A.51)}$$

- The *scalar product* of the del by a *vector* function of coordinates (a *vector field*),

$$\mathbf{f}(\mathbf{r})\equiv\sum_{j=1}^{3}\mathbf{n}_j f_j(\mathbf{r}),\qquad\text{(A.52)}$$

formally following Eq. (A.43), is a *scalar* function—the *divergence* of the initial function,

$$\nabla\cdot\mathbf{f}\equiv\sum_{j=1}^{3}\frac{\partial f_j}{\partial r_j}\equiv\mathbf{div}\,\mathbf{f},\qquad\text{(A.53)}$$

[9] One can run into the following notation, $\nabla\equiv\partial/\partial\mathbf{r}$, which is convenient is some cases, but may be misleading in quite a few others, so it will be not used in this series.

[10] In this, and the four next sections, all scalar and vector functions are assumed to be differentiable.

while the *vector product* of ∇ and **f**, formed in a formal accordance with Eq. (A.45), is a new vector—the *curl* (in European tradition, called 'rotor' and denoted **rot**) of **f**:

$$\nabla \times \mathbf{f} \equiv \begin{vmatrix} \mathbf{n}_1 & \mathbf{n}_2 & \mathbf{n}_2 \\ \dfrac{\partial}{\partial r_1} & \dfrac{\partial}{\partial r_2} & \dfrac{\partial}{\partial r_3} \\ f_1 & f_2 & f_3 \end{vmatrix} = \mathbf{n}_1\left(\frac{\partial f_3}{\partial r_2} - \frac{\partial f_2}{\partial r_3}\right) + \mathbf{n}_2\left(\frac{\partial f_1}{\partial r_3} - \frac{\partial f_3}{\partial r_1}\right) + \mathbf{n}_3\left(\frac{\partial f_2}{\partial r_1} - \frac{\partial f_1}{\partial r_2}\right) \equiv \mathbf{curl}\,\mathbf{f}. \quad \text{(A.54)}$$

- One more frequently encountered 'product' is $(\mathbf{f}\cdot\nabla)\mathbf{g}$, where **f** and **g** are two arbitrary vector functions of **r**. This product should be also understood in the sense implied by Eq. (A.43), i.e. as a vector whose *j*th Cartesian component is

$$[(\mathbf{f}\cdot\nabla)\mathbf{g}]_j = \sum_{j'=1}^{3} f_{j'} \frac{\partial g_j}{\partial r_{j'}}. \quad \text{(A.55)}$$

A.9 The Laplace operator $\nabla^2 \equiv \nabla \cdot \nabla$

- Expression in Cartesian coordinates—in formal accordance with Eq. (A.44):

$$\nabla^2 = \sum_{j=1}^{3} \frac{\partial^2}{\partial r_j^2}. \quad \text{(A.56)}$$

- According to its definition, the Laplace operator acting on a *scalar* function of coordinates gives a new scalar function:

$$\nabla^2 f \equiv \nabla \cdot (\nabla f) = \mathbf{div}(\mathbf{grad}\, f) = \sum_{j=1}^{3} \frac{\partial^2 f}{\partial r_j^2}. \quad \text{(A.57)}$$

- On the other hand, acting on a *vector* function (A.52), the operator ∇^2 returns another *vector*:

$$\nabla^2 \mathbf{f} = \sum_{j=1}^{3} \mathbf{n}_j \nabla^2 f_j. \quad \text{(A.58)}$$

Note that Eqs. (A.56)–(A.58) are only valid in Cartesian (i.e. orthogonal and linear) coordinates, but generally not in other (even orthogonal) coordinates—see, e.g. Eqs. (A.64) and (A.70) below.

A.10 Operators ∇ and ∇^2 in the most important systems of orthogonal coordinates[11]

(i) *Cylindrical*[12] *coordinates* $\{\rho, \varphi, z\}$ (see the figure below) may be defined by their relations with the Cartesian coordinates:

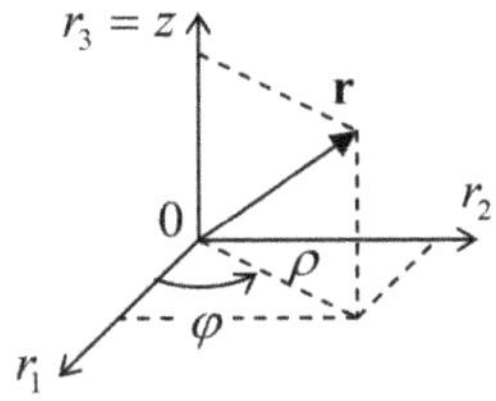

$$r_1 = \rho\cos\varphi,\quad r_2 = \rho\sin\varphi,\quad r_3 = z. \tag{A.59}$$

- Gradient of a scalar function:

$$\nabla f = \mathbf{n}_\rho \frac{\partial f}{\partial \rho} + \mathbf{n}_\varphi \frac{1}{\rho}\frac{\partial f}{\partial \varphi} + \mathbf{n}_z \frac{\partial f}{\partial z}. \tag{A.60}$$

- The Laplace operator of a scalar function:

$$\nabla^2 f = \frac{1}{\rho}\frac{\partial}{\partial \rho}\left(\rho\frac{\partial f}{\partial \rho}\right) + \frac{1}{\rho^2}\frac{\partial^2 f}{\partial \phi^2} + \frac{\partial^2 f}{\partial z^2}. \tag{A.61}$$

- Divergence of a vector function of coordinates ($\mathbf{f} = \mathbf{n}_\rho f_\rho + \mathbf{n}_\varphi f_\varphi + \mathbf{n}_z f_z$):

$$\nabla\cdot\mathbf{f} = \frac{1}{\rho}\frac{\partial(\rho f_\rho)}{\partial \rho} + \frac{1}{\rho}\frac{\partial f_\varphi}{\partial \varphi} + \frac{\partial f_z}{\partial z}. \tag{A.62}$$

- Curl of a vector function:

$$\nabla\times\mathbf{f} = \mathbf{n}_\rho\left(\frac{1}{\rho}\frac{\partial f_z}{\partial \varphi} - \frac{\partial f_\varphi}{\partial z}\right) + \mathbf{n}_\varphi\left(\frac{\partial f_\rho}{\partial z} - \frac{\partial f_z}{\partial \rho}\right) + \mathbf{n}_z\frac{1}{\rho}\left(\frac{\partial(\rho f_\varphi)}{\partial \rho} - \frac{\partial f_\rho}{\partial \varphi}\right). \tag{A.63}$$

- The Laplace operator of a vector function:

$$\nabla^2\mathbf{f} = \mathbf{n}_\rho\left(\nabla^2 f_\rho - \frac{1}{\rho^2}f_\rho - \frac{2}{\rho^2}\frac{\partial f_\varphi}{\partial \varphi}\right) + \mathbf{n}_\varphi\left(\nabla^2 f_\varphi - \frac{1}{\rho^2}f_\varphi + \frac{2}{\rho^2}\frac{\partial f_\rho}{\partial \varphi}\right) + \mathbf{n}_z\nabla^2 f_z. \tag{A.64}$$

[11] Some other orthogonal curvilinear coordinate systems are discussed in *Part EM* section 2.3.

[12] In the 2D geometry with fixed coordinate z, these coordinates are called *polar*.

(ii) *Spherical coordinates* $\{r, \theta, \varphi\}$ (see the figure below) may be defined as:

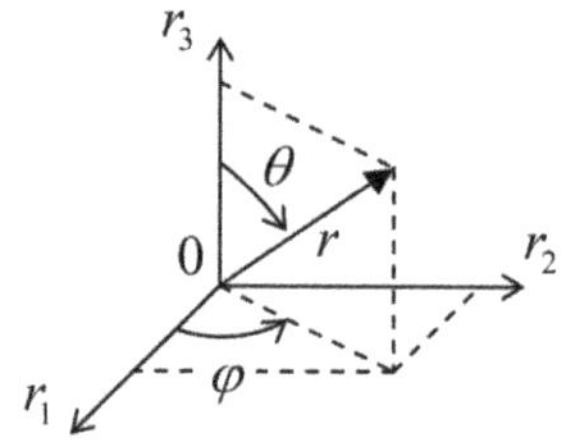

$$r_1 = r\sin\theta\cos\varphi,\quad r_2 = r\sin\theta\sin\varphi,\quad r_3 = r\cos\theta. \tag{A.65}$$

- Gradient of a scalar function:

$$\nabla f = \mathbf{n}_r\frac{\partial f}{\partial r} + \mathbf{n}_\theta\frac{1}{r}\frac{\partial f}{\partial \theta} + \mathbf{n}_\varphi\frac{1}{r\sin\theta}\frac{\partial f}{\partial\varphi}. \tag{A.66}$$

- The Laplace operator of a scalar function:

$$\nabla^2 f = \frac{1}{r^2}\frac{\partial}{\partial r}\left(r^2\frac{\partial f}{\partial r}\right) + \frac{1}{r^2\sin\theta}\frac{\partial}{\partial\theta}\left(\sin\theta\frac{\partial f}{\partial\theta}\right) + \frac{1}{(r\sin\theta)^2}\frac{\partial^2 f}{\partial\varphi^2}. \tag{A.67}$$

- Divergence of a vector function $\mathbf{f} = \mathbf{n}_r f_r + \mathbf{n}_\theta f_\theta + \mathbf{n}_\varphi f_\varphi$:

$$\nabla\cdot\mathbf{f} = \frac{1}{r^2}\frac{\partial(r^2 f_r)}{\partial r} + \frac{1}{r\sin\theta}\frac{\partial(f_\theta\sin\theta)}{\partial\theta} + \frac{1}{r\sin\theta}\frac{\partial f_\varphi}{\partial\varphi}. \tag{A.68}$$

- Curl of a similar vector function:

$$\begin{aligned}\nabla\times\mathbf{f} = {} & \mathbf{n}_r\frac{1}{r\sin\theta}\left(\frac{\partial(f_\varphi\sin\theta)}{\partial\theta} - \frac{\partial f_\theta}{\partial\varphi}\right)\\ & + \mathbf{n}_\theta\frac{1}{r}\left(\frac{1}{\sin\theta}\frac{\partial f_r}{\partial\varphi} - \frac{\partial(rf_\varphi)}{\partial r}\right) + \mathbf{n}_\varphi\frac{1}{r}\left(\frac{\partial(rf_\theta)}{\partial r} - \frac{\partial f_r}{\partial\theta}\right).\end{aligned} \tag{A.69}$$

- The Laplace operator of a vector function:

$$\begin{aligned}\nabla^2\mathbf{f} = {} & \mathbf{n}_r\left(\nabla^2 f_r - \frac{2}{r^2}f_r - \frac{1}{r^2\sin\theta}\frac{\partial}{\partial\theta}(f_\theta\sin\theta) - \frac{2}{r^2\sin\theta}\frac{\partial f_\varphi}{\partial\varphi}\right)\\ & + \mathbf{n}_\theta\left(\nabla^2 f_\theta - \frac{1}{r^2\sin^2\theta}f_\theta + \frac{2}{r^2}\frac{\partial f_r}{\partial\theta} - \frac{2\cos\theta}{r^2\sin^2\theta}\frac{\partial f_\varphi}{\partial\varphi}\right)\\ & + \mathbf{n}_\varphi\left(\nabla^2 f_\varphi - \frac{1}{r^2\sin^2\theta}f_\varphi + \frac{2}{r^2\sin\theta}\frac{\partial f_r}{\partial\varphi} + \frac{2\cos\theta}{r^2\sin^2\theta}\frac{\partial f_\theta}{\partial\varphi}\right).\end{aligned} \tag{A.70}$$

A.11 Products involving vector ∇

(i) *Useful zeros*
- For any scalar function, $f(\mathbf{r})$

$$\nabla \times (\nabla f) \equiv \mathbf{curl}(\mathbf{grad} f) = 0. \tag{A.71}$$

- For any vector function $\mathbf{f}(\mathbf{r})$,

$$\nabla \cdot (\nabla \times \mathbf{f}) \equiv \mathbf{div}(\mathbf{curl}\, \mathbf{f}) = 0. \tag{A.72}$$

(ii) *Laplace operator* expressed via the curl of a curl:

$$\nabla^2 \mathbf{f} = \nabla(\nabla \cdot \mathbf{f}) - \nabla \times (\nabla \times \mathbf{f}). \tag{A.73}$$

(iii) *Spatial differentiation* of a product of a scalar function by a vector function:
- The scalar 3D generalization of Eq. (A.22) is

$$\nabla \cdot (f\mathbf{g}) = (\nabla f) \cdot \mathbf{g} + f(\nabla \cdot \mathbf{g}), \tag{A.74a}$$

and its vector generalization is similar:

$$\nabla \times (f\mathbf{g}) = (\nabla f) \times \mathbf{g} + f(\nabla \times \mathbf{g}). \tag{A.74b}$$

(iv) *3D spatial differentiation* of products of two vector functions:

$$\nabla \times (\mathbf{f} \times \mathbf{g}) = \mathbf{f}(\nabla \cdot \mathbf{g}) - (\mathbf{f} \cdot \nabla)\mathbf{g} - (\nabla \cdot \mathbf{f})\mathbf{g} + (\mathbf{g} \cdot \nabla)\mathbf{f}, \tag{A.75}$$

$$\nabla(\mathbf{f} \cdot \mathbf{g}) = (\mathbf{f} \cdot \nabla)\mathbf{g} + (\mathbf{g} \cdot \nabla)\mathbf{f} + \mathbf{f} \times (\nabla \times \mathbf{g}) + \mathbf{g} \times (\nabla \times \mathbf{f}), \tag{A.76}$$

$$\nabla \cdot (\mathbf{f} \times \mathbf{g}) = \mathbf{g} \cdot (\nabla \times \mathbf{f}) - \mathbf{f} \cdot (\nabla \times \mathbf{g}). \tag{A.77}$$

A.12 Integro-differential relations

(i) For an arbitrary surface S limited by closed contour C:
- The *Stokes theorem*, valid for any differentiable vector field $\mathbf{f}(\mathbf{r})$:

$$\int_S (\nabla \times \mathbf{f}) \cdot d^2\mathbf{r} \equiv \int_S (\nabla \times \mathbf{f})_n d^2 r = \oint_C \mathbf{f} \cdot d\mathbf{r} \equiv \oint_C f_\tau dr, \tag{A.78}$$

where $d^2\mathbf{r} \equiv \mathbf{n} d^2 r$ is the elementary area vector (normal to the surface), and $d\mathbf{r}$ is the elementary contour length vector (tangential to the contour line).

(ii) For an arbitrary volume V limited by closed surface S:

- *Divergence* (or 'Gauss') *theorem*, valid for any differentiable vector field $\mathbf{f}(\mathbf{r})$:

$$\int_V (\nabla \cdot \mathbf{f}) d^3r = \oint_S \mathbf{f} \cdot d^2\mathbf{r} \equiv \oint_S f_n d^2r. \tag{A.79}$$

- *Green's theorem*, valid for two differentiable scalar functions $f(\mathbf{r})$ and $g(\mathbf{r})$:

$$\int_V (f\nabla^2 g - g\nabla^2 f) d^3r = \oint_S (f\nabla g - g\nabla f)_n d^2r. \tag{A.80}$$

- An identity valid for any two scalar functions f and g, and a vector field $\mathbf{j}$ with $\nabla \cdot \mathbf{j} = 0$ (all differentiable):

$$\int_V [f(\mathbf{j} \cdot \nabla g) + g(\mathbf{j} \cdot \nabla f)] d^3r = \oint_S fgj_n \, d^2r. \tag{A.81}$$

A.13 The Kronecker delta and Levi–Civita permutation symbols

The *Kronecker delta symbol* (defined for integer indices):

$$\delta_{jj'} \equiv \begin{cases} 1, & \text{if } j' = j, \\ 0, & \text{otherwise}. \end{cases} \tag{A.82}$$

- The *Levi–Civita permutation symbol* (most frequently used for three integer indices, each taking one of the values 1, 2, or 3):

$$\varepsilon_{jj'j''} \equiv \begin{cases} +1, & \text{if the indices follow in the 'correct'} \\ & \text{order: } 1 \to 2 \to 3 \to 1 \to 2 \ldots, \\ -1, & \text{if the indices follow in the 'incorrect'} \\ & \text{order: } 1 \to 3 \to 2 \to 1 \to 3 \ldots, \\ 0, & \text{if any two indices coincide.} \end{cases} \tag{A.83}$$

- Relation between the Levi–Civita and the Kronecker delta products:

$$\varepsilon_{jj'j''}\varepsilon_{kk'k''} = \sum_{l,l',l''=1}^{3} \begin{vmatrix} \delta_{jl} & \delta_{jl'} & \delta_{jl''} \\ \delta_{j'l} & \delta_{j'l'} & \delta_{j'l''} \\ \delta_{j''l} & \delta_{j''l'} & \delta_{j''l''} \end{vmatrix}; \tag{A.84a}$$

summation of this relation, written for three different values of $j = k$, over these values yields the so-called *contracted epsilon identity*:

$$\sum_{j=1}^{3} \varepsilon_{jj'j''}\varepsilon_{jk'k''} = \delta_{j'k'}\delta_{j''k''} - \delta_{j'k''}\delta_{j''k'}. \tag{A.84b}$$

A.14 Dirac's delta-function, sign function, and theta-function

- Definition of the 1D *delta-function* (for real $a < b$):

$$\int_a^b f(\xi)\delta(\xi)d\xi = \begin{cases} f(0), & \text{if } a < 0 < b, \\ 0, & \text{otherwise,} \end{cases} \tag{A.85}$$

where $f(\xi)$ is any function continuous near $x = 0$. In particular (if $f(x) = 1$ near $x = 0$), the definition yields

$$\int_a^b \delta(x)dx = \begin{cases} 1, & \text{if } a < 0 < b, \\ 0, & \text{otherwise.} \end{cases} \tag{A.86}$$

Relation to the *theta-function* $\theta(\xi)$ and the *sign function* $\text{sgn}(\xi)$

$$\delta(\xi) = \frac{d}{d\xi}\theta(\xi) = \frac{1}{2}\frac{d}{d\xi}\text{sgn}(\xi), \tag{A.87a}$$

where

$$\theta(\xi) \equiv \frac{\text{sgn}(\xi) + 1}{2} = \begin{cases} 0, & \text{if } \xi < 0, \\ 1, & \text{if } \xi > 1, \end{cases}$$
$$\text{sgn}(\xi) \equiv \frac{\xi}{|\xi|} = \begin{cases} -1, & \text{if } \xi < 0, \\ +1, & \text{if } \xi > 1. \end{cases} \tag{A.87b}$$

- An important integral[13]:

$$\int_{-\infty}^{+\infty} e^{is\xi}ds = 2\pi\delta(\xi). \tag{A.88}$$

- 3D generalization of the delta-function of the radius-vector (the 2D generalization is similar):

$$\int_V f(\mathbf{r})\delta(\mathbf{r})d^3r = \begin{cases} f(0), & \text{if } 0 \in V, \\ 0, & \text{otherwise;} \end{cases} \tag{A.89}$$

it may be represented as a product of 1D delta-functions of Cartesian coordinates:

$$\delta(\mathbf{r}) = \delta(r_1)\delta(r_2)\delta(r_3). \tag{A.90}$$

[13] The coefficient in this relation may be readily recalled by considering its left-hand side as the Fourier-integral representation of function $f(s) \equiv 1$, and applying Eq. (A.85) to the reciprocal Fourier transform

$$f(s) \equiv \frac{1}{2\pi}\int_{-\infty}^{+\infty} e^{-is\xi}[2\pi\delta(\xi)]d\xi.$$

A.15 The Cauchy theorem and integral

Let a complex function $f(z)$ be analytic within a part of the complex plane z, that is limited by a closed contour C and includes point z'. Then

$$\oint_C f(z)dz = 0, \tag{A.91}$$

$$\oint_C f(z)\frac{dz}{z-z'} = 2\pi i f(z'). \tag{A.92}$$

The first of these relations is usually called the *Cauchy integral theorem* (or the 'Cauchy–Goursat theorem'), and the second one—the *Cauchy integral* (or the 'Cauchy integral formula').

A.16 Literature

(i) Properties of some *special functions* are briefly discussed at the relevant points of the lecture notes; in alphabetical order:

- Airy functions: *Part QM* section 2.4;
- Bessel functions: *Part EM* section 2.7;
- Fresnel integrals: *Part EM* section 8.6;
- Hermite polynomials: *Part CM* section 2.9;
- Laguerre polynomials (both *simple* and *associated*): *Part QM* section 3.7;
- Legendre polynomials, associated Legendre functions: *Part EM* section 2.8 and *Part QM* section 3.6;
- Spherical harmonics: *Part QM* section 3.6;
- Spherical Bessel functions: *Part QM* sections 3.6 and 3.8.

(ii) For *more formulas*, and their discussion, I can recommend the following handbooks[14]:

- *Handbook of Mathematical Formulas* [2];
- *Tables of Integrals, Series, and Products* [3];
- *Mathematical Handbook for Scientists and Engineers* [4];
- *Integrals and Series* volumes 1 and 2 [5];
- A popular textbook *Mathematical Methods for Physicists* [6] may be also used as a formula manual.

Many formulas are also available from the symbolic calculation modules of the commercially available software packages listed in section (iv) below.

(iii) Probably the most popular collection of *numerical calculation codes* are the twin manuals by W Press *et al* [1]:

- *Numerical Recipes in Fortran 77*;
- *Numerical Recipes* [in C++—KKL].

[14] On a personal note, perhaps 90% of all formula needs throughout my research career were satisfied by a tiny, wonderfully compiled old book [7], used copies of which, rather amazingly, are still available on the Web.

These lecture notes include very brief introductions to numerical methods of differential equation solution:

- ordinary differential equations: *Part CM*, section 5.7;
- partial differential equations: *Part CM* section 8.5 and *Part EM* section 2.11, which include references to literature for further reading.

(iv) The following are the most popular *software packages* for numerical and symbolic calculations, all with plotting capabilities (in alphabetical order):

- Maple (www.maplesoft.com/products/maple/);
- MathCAD (www.ptc.com/engineering-math-software/mathcad/);
- Mathematica (www.wolfram.com/mathematica/);
- MATLAB (www.mathworks.com/products/matlab.html).

References

[1] Press W H *et al* 1992 *Numerical Recipes in Fortran 77* 2nd edn (Cambridge: Cambridge University Press)
Press W H *et al* 2007 *Numerical Recipes* 3rd edn (Cambridge: Cambridge University Press)

[2] Abramowitz M and Stegun I A (ed) 1965 *Handbook of Mathematical Formulas* (New York: Dover) , and numerous later printings. An updated version of this collection is now available online at http://dlmf.nist.gov/.

[3] Gradshteyn I S and Ryzhik I M 1980 *Tables of Integrals, Series, and Products* 5th edn (New York: Academic)

[4] Korn G A and Korn T M 2000 *Mathematical Handbook for Scientists and Engineers* 2nd edn (New York: Academic)

[5] Prudnikov A P *et al* 1986 *Integrals and Series* vol 1 (Boca Raton, FL: CRC Press)
Prudnikov A P *et al* 1986 *Integrals and Series* vol 2 (Boca Raton, FL: CRC Press)

[6] Arfken G *et al* 2012 *Mathematical Methods for Physicists* 7th edn (New York: Academic)

[7] Dwight H B 1961 *Tables of Integrals and Other Mathematical Formulas* 4th edn (London: Macmillan)

Appendix B

Selected physical constants[1]

Table B.1.

Symbol	Quantity	SI value and unit	Gaussian value and unit	Relative rms uncertainty
c	speed of light in free space	$2.997\ 924\ 58 \times 10^{8}$ m s^{-1}	$2.997\ 924\ 58 \times 10^{10}$ cm s^{-1}	0 (defined value)
G	gravitation constant	$6.674\ 1 \times 10^{-11}$ m^{3} kg^{-1} s^{-2}	$6.674\ 1 \times 10^{-8}$ cm ^{3} g^{-1} s^{-2}	$\sim 5 \times 10^{-5}$
$\hbar$	Planck constant	$1.054\ 5718\ 0 \times 10^{-34}$ J s	$1.054\ 571\ 80 \times 10^{-27}$ erg s	$\sim 1 \times 10^{-8}$
e	elementary electric charge	$1.602\ 176\ 2 \times 10^{-19}$ C	$4.803\ 203 \times 10^{-10}$ statcoulomb	$\sim 6 \times 10^{-9}$
m_e	electron's rest mass	$0.910\ 938\ 35 \times 10^{-30}$ kg	$0.910\ 938\ 35 \times 10^{-27}$ G	$\sim 1 \times 10^{-8}$
m_p	proton's rest mass	$1.672\ 621\ 90 \times 10^{-27}$ kg	$1.672\ 621\ 90 \times 10^{-24}$ G	$\sim 1 \times 10^{-8}$
μ_0	magnetic constant	$4\pi \times 10^{-7}$ N A^{-2}	–	0 (defined value)
ε_0	electric constant	$8.854\ 187\ 817 \times 10^{-12}$ F m^{-1}	–	0 (defined value)
k_B	Boltzmann constant	$1.380\ 648 \times 10^{-23}$ J K^{-1}	$1.380\ 650\ 9 \times 10^{-16}$ erg K^{-1}	$\sim 2 \times 10^{-6}$

Comments:

1. The fixed value of c was defined by an international convention in 1983, in order to extend the official definition of a second (as 'the duration of 9 192 631 770 periods of the radiation corresponding to the transition between the two hyperfine levels of the ground state of the cesium-133 atom')

[1] The listed numerical values of the constants are from the most recent (2014) International CODATA recommendation (see, e.g. http://physics.nist.gov/cuu/Constants/index.html), besides a newer result for k_B—see [1].

doi:10.1088/978-0-7503-1398-8ch12

to that of a meter. The values are back-compatible with the legacy definitions of the meter (initially, as 1/40 000 000th of the Earth's meridian length) and the second (for a long time, as $1/(24 \times 60 \times 60) = 1/86\,400$th of the Earth's rotation period), within the experimental errors of those measures.

2. ε_0 and μ_0 are not really the fundamental constants; in the SI system of units one of them (say, μ_0) is selected arbitrarily[2], while the other one is defined via the relation $\varepsilon_0\mu_0 = 1/c^2$.
3. The Boltzmann constant k_B is also not quite fundamental, because its only role is to comply with the independent definition of the kelvin (K), as the temperature unit in which the triple point of water is exactly 273.16 K. If temperature is expressed in energy units k_BT (as is done, for example, in *Part SM* of this lecture note series), this constant disappears altogether.
4. The dimensionless *fine structure* ('Sommerfeld's) *constant* α is numerically the same in any system of units:

$$\alpha \equiv \begin{Bmatrix} e^2/4\pi\varepsilon_0\hbar c & \text{in SI units} \\ e^2/\hbar c & \text{in Gaussian units} \end{Bmatrix} \approx 7.297\,352\,566 \times 10^{-3} \approx \frac{1}{137.035\,999\,14},$$

and is known with a much smaller rms uncertainty (currently, $\sim 3 \times 10^{-10}$) than those of the component constants.

References

[1] Gaiser C *et al* 2017 *Metrologia* **54** 280
[2] Newell D 2014 *Phys. Today* **67** 35–41

[2] Note that the selected value of μ_0 may be changed (a bit) in a few years—see, e.g. [2].

Bibliography

This section presents a partial list of the textbooks and monographs used in the work on the EAP series[1,2].

Part CM: Classical Mechanics

Fetter A L and Walecka J D 2003 *Theoretical Mechanics of Particles and Continua* (New York: Dover)

Goldstein H, Poole C and Safko J 2002 *Classical Mechanics* 3rd edn (Reading, MA: Addison Wesley)

Granger R A 1995 *Fluid Mechanics* (New York: Dover)

José J V and Saletan E J 1998 *Classical Dynamics* (Cambridge: Cambridge University Press)

Landau L D and Lifshitz E M 1976 *Mechanics* 3rd edn (Oxford: Butterworth-Heinemann)

Landau L D and Lifshitz E M 1986 *Theory of Elasticity* (Oxford: Butterworth-Heinemann)

Landau L D and Lifshitz E M 1987 *Fluid Mechanics* 2nd edn (Oxford: Butterworth-Heinemann)

Schuster H G 1995 *Deterministic Chaos* 3rd edn (New York: Wiley)

Sommerfeld A 1964 *Mechanics* (New York: Academic)

Sommerfeld A 1964 *Mechanics of Deformable Bodies* (New York: Academic)

Part EM: Classical Electrodynamics

Batygin V V and Toptygin I N 1978 *Problems in Electrodynamics* 2nd edn (New York: Academic)

Griffiths D J 2007 *Introduction to Electrodynamics* 3rd edn (Englewood Cliffs, NJ: Prentice-Hall)

Jackson J D 1999 *Classical Electrodynamics* 3rd edn (New York: Wiley)

Landau L D and Lifshitz E M 1984 *Electrodynamics of Continuous Media* 2nd edn (Auckland: Reed)

Landau L D and Lifshitz E M 1975 *The Classical Theory of Fields* 4th edn (Oxford: Pergamon)

Panofsky W K H and Phillips M 1990 *Classical Electricity and Magnetism* 2nd edn (New York: Dover)

[1] The list does not include the sources (mostly, recent original publications) cited in the ends of the chapters, and the mathematics textbooks and handbooks listed in appendix A, section A.16.

[2] Recently some high-quality teaching materials on advanced physics have become available online, including: R Fitzpatrick's text on classical electromagnetism (farside.ph.utexas.edu/teaching/jk1/Electromagnetism.pdf); B Simons' 'lecture shrunks' on advanced quantum mechanics (www.tcm.phy.cam.ac.uk/~bds10/aqp.html); and D Tong's lecture notes on several advanced topics (www.damtp.cam.ac.uk/user/tong/teaching.html).

doi:10.1088/978-0-7503-1398-8ch13

Stratton J A 2007 *Electromagnetic Theory* (New York: Wiley)
Tamm I E 1979 *Fundamentals of the Theory of Electricity* (Paris: Mir)
Zangwill A 2013 *Modern Electrodynamics* (Cambridge: Cambridge University Press)

Part QM: Quantum Mechanics

Abers E S 2004 *Quantum Mechanics* (London: Pearson)
Auletta G, Fortunato M and Parisi G 2009 *Quantum Mechanics* (Cambridge: Cambridge University Press)
Capri A Z 2002 *Nonrelativistic Quantum Mechanics* 3rd edn (Singapore: World Scientific)
Cohen-Tannoudji C, Diu B and Laloë F 2005 *Quantum Mechanics* (New York: Wiley)
Constantinescu F, Magyari E and Spiers J A 1971 *Problems in Quantum Mechanics* (Amsterdam: Elsevier)
Galitski V *et al* 2013 *Exploring Quantum Mechanics* (Oxford: Oxford University Press)
Gottfried K and Yan T-M 2004 *Quantum Mechanics: Fundamentals* 2nd edn (Berlin: Springer)
Griffith D 2005 *Quantum Mechanics* 2nd edn (Englewood Cliffs, NJ: Prentice Hall)
Landau L D and Lifshitz E M 1977 *Quantum Mechanics (Nonrelativistic Theory)* 3rd edn (Oxford: Pergamon)
Messiah A 1999 *Quantum Mechanics* (New York: Dover)
Merzbacher E 1998 *Quantum Mechanics* 3rd edn (New York: Wiley)
Miller D A B 2008 *Quantum Mechanics for Scientists and Engineers* (Cambridge: Cambridge University Press)
Sakurai J J 1994 *Modern Quantum Mechanics* (Reading, MA: Addison-Wesley)
Schiff L I 1968 *Quantum Mechanics* 3rd edn (New York: McGraw-Hill)
Shankar R 1980 *Principles of Quantum Mechanics* 2nd edn (Berlin: Springer)
Schwabl F 2002 *Quantum Mechanics* 3rd edn (Berlin: Springer)

Part SM: Statistical Mechanics

Feynman R P 1998 *Statistical Mechanics* 2nd edn (Boulder, CO: Westview)
Huang K 1987 *Statistical Mechanics* 2nd edn (New York: Wiley)
Kubo R 1965 *Statistical Mechanics* (Amsterdam: Elsevier)
Landau L D and Lifshitz E M 1980 *Statistical Physics, Part 1* 3rd edn (Oxford: Pergamon)
Lifshitz E M and Pitaevskii L P 1981 *Physical Kinetics* (Oxford: Pergamon)
Pathria R K and Beale P D 2011 *Statistical Mechanics* 3rd edn (Amsterdam: Elsevier)
Pierce J R 1980 *An Introduction to Information Theory* 2nd edn (New York: Dover)
Plishke M and Bergersen B 2006 *Equilibrium Statistical Physics* 3rd edn (Singapore: World Scientific)
Schwabl F 2000 *Statistical Mechanics* (Berlin: Springer)
Yeomans J M 1992 *Statistical Mechanics of Phase Transitions* (Oxford: Oxford University Press)

Multidisciplinary/specialty

Ashcroft W N and Mermin N D 1976 *Solid State Physics* (Philadelphia, PA: Saunders)
Blum K 1981 *Density Matrix and Applications* (New York: Plenum)
Breuer H-P and Petruccione E 2002 *The Theory of Open Quantum Systems* (Oxford: Oxford University Press)
Cahn S B and Nadgorny B E 1994 *A Guide to Physics Problems, Part 1* (New York: Plenum)

Cahn S B, Mahan G D and Nadgorny B E 1997 *A Guide to Physics Problems, Part 2* (New York: Plenum)

Cronin J A, Greenberg D F and Telegdi V L 1967 *University of Chicago Graduate Problems in Physics* (Reading, MA: Addison Wesley)

Hook J R and Hall H E 1991 *Solid State Physics* 2nd edn (New York: Wiley)

Joos G 1986 *Theoretical Physics* (New York: Dover)

Kompaneyets A S 2012 *Theoretical Physics* 2nd edn (New York: Dover)

Lax M 1968 *Fluctuations and Coherent Phenomena* (London: Gordon and Breach)

Lifshitz E M and Pitaevskii L P 1980 *Statistical Physics, Part 2* (Oxford: Pergamon)

Newbury N *et al* 1991 *Princeton Problems in Physics with Solutions* (Princeton, NJ: Princeton University Press)

Pauling L 1988 *General Chemistry* 3rd edn (New York: Dover)

Tinkham M 1996 *Introduction to Superconductivity* 2nd edn (New York: McGraw-Hill)

Walecka J D 2008 *Introduction to Modern Physics* (Singapore: World Scientific)

Ziman J M 1979 *Principles of the Theory of Solids* 2nd edn (Cambridge: Cambridge University Press)

www.ingramcontent.com/pod-product-compliance
Ingram Content Group UK Ltd.
Pitfield, Milton Keynes, MK11 3LW, UK
UKHW051902100926
8556IPUK00005B/734

9 780750 313995